I0057133

Glossary of Mathematical Terms and Concepts
(Part I)

Ram Bilas Misra

CWP

Central West Publishing

Glossary of Mathematical Terms and Concepts (Part I)

by

Prof. Dr. Ram Bilas Misra

Ex Vice Chancellor, Avadh University, Faizabad, U.P. (India);
Professor of Mathematics, Research & Strategic Studies Centre, Lebanese French University, Erbil, Kurdistan (Iraq).

Former: *Dean*, Faculty of Science, A.P. Singh University, Rewa, M.P. (**India**);
Prof., Dept. of Maths., Higher College of Edn., Aden Univ., Aden (**Yemen**);
Professor & Head, Dept. of Maths. & Stats., A.P.S. University, Rewa, M.P. (**India**);
Prof., Dept. of Maths., College of Science, Salahaddin University, Erbil (**Iraq**);
UGC Visiting Prof., Mahatma Gandhi Kashi *Vidyapith*, Varanasi, U.P. (**India**);
Professor, Dept. of Maths, Ahmadu Bello Univ., Zaria (**Nigeria**) – designate;
Prof. & Head, Dept. of Maths. & Comp. Sci., Univ. of Asmara, Asmara (**Eritrea**);
Director, Unique Inst. of Business & Technol., Modi Nagar, Ghaziabad, U.P. (**India**);
Prof. & Head, Dept. of Maths., Phys. & Stats., Univ. of Guyana, Georgetown (**Guyana**);
Prof. & Head, Dept. of Maths., Eritrea Inst. of Technology, Mai Nefhi (**Eritrea**);
Prof.& Head, Dept. of Maths., School of Engg., Amity Univ., Lucknow, U.P. (**India**);
Prof. & Head, Dept. of Maths. & Comp. Sci., PNG Univ. of Technology, Lae (**PNG**);
Prof. of Maths., Teerthankar Mahaveer University, Moradabad, U.P. (**India**);
Prof., Dept. of Maths, Oduduwa Univ., Ipetumodu, Osun State (**Nigeria**) – designate;
Prof., Dept. of Maths, Adama Science & Technology Univ., Adama (**Ethiopia**);
Prof. & Head, Dept. of Maths. & C.S., Bougainville Inst. of Bus. & Tech., Buka (**PNG**) – designate;
Prof. & Head, Dept. of Maths., J.J.T. University, Jhunjhunu, Rajasthan (**India**);
Dean, Faculty of Science, J.J.T. University, Jhunjhunu, Rajasthan (**India**);
Professor, Dept. of Maths, Wollo University, Dessie, Wollo (**Ethiopia**);
Professor, Dept. of Appld. Maths., State Univ. of New York, Incheon (**S. Korea**)
Prof., Dept. of Maths. & Computing Sci., Divine Word Univ., Madang (**PNG**);
Director, Maths., School of Sci. & Engg., Univ. of Kurdistan Hewler, Erbil (**Iraq**);
DAAD Fellow, University of Bonn, Bonn (**Germany**);
Visiting Professor, University of Turin, Turin (**Italy**);
Visiting Professor, University of Trieste, Trieste (**Italy**);
Visiting Professor, University of Padua, Padua (**Italy**);
Visiting Professor, International Centre for Theoretical Physics, Trieste (**Italy**);
Visiting Professor, University of Wroclaw, Wroclaw (**Poland**);
Visiting Professor, University of Sopron, Sopron (**Hungary**);
Reader, Dept. of Maths. & Stats., South Gujarat University, Surat, Gujarat (**India**);
Reader, Dept. of Maths. & Stats., University of Allahabad, Allahabad, U.P. (**India**);
Asst. Prof., Dept. of Maths., College of Sci., Mosul Univ., Mosul (**Iraq**) – designate;
Senior most *NCC Officer* (Naval Wing), Univ. of Allahabad, Allahabad, U.P. (**India**);
Lecturer, Dept. of Maths., KKV Degree College, Lucknow, U.P. (**India**).

2019

This edition has been published by Central West Publishing, Australia

© 2019 Central West Publishing

All rights reserved. No part of this volume may be reproduced, copied, stored, or transmitted, in any form or by any means, electronic, photo-copying, recording, or otherwise. Permission requests for reuse can be sent to editor@centralwestpublishing.com

For more information about the books published by Central West Publishing, please visit https://centralwestpublishing.com

Disclaimer
Every effort has been made by the publisher and author while preparing this book; however, no warranties are made regarding the accuracy and completeness of the content. The publisher and author disclaim without any limitation all warranties as well as any implied warranties about sales, along with fitness of the content for a particular purpose. Citation of any website and other information sources does not mean any en-dorsement from the publisher and author. For ascertaining the suitability of the contents contained herein for a particular lab or commercial use, consultation with the subject expert is needed. In addition, while using the information and methods contained herein, the practitioners and re-searchers need to be mindful for their own safety, along with the safety of others, including the professional parties and premises for which they have professional responsibility. To the fullest extent of law, the pub-lisher and author are not liable in all circumstances (special, incidental, and consequential) for any injury and/or damage to persons and proper-ty, along with any potential loss of profit and other commercial damages due to the use of any methods, products, guidelines, procedures con-tained in the material herein.

NATIONAL LIBRARY OF AUSTRALIA

A catalogue record for this book is available from the National Library of Australia

ISBN (print): 978-1-925823-68-4
ISBN (ebook): 978-1-925823-69-1

DEDICATED TO

MY TEACHERS AND ALL THOSE WHOSE BLESSINGS INSPIRED ME GLOBALLY

ESPECIALLY

Shri Chhote Lal Kumhar, Semrai
(who begged in Rsihi Bhardwaj Ashram, Prayagraj for my birth);

Shri Jeet Gossain, Semrai
(a simple village man advising me at 4 not to insist for unavailable things);

Shri Triveni Singh Rathore, Semrai
(my first teacher to have inculcated a mathematical vision);

Shrimati Gopa, Gola Gokarannath
(for saving my life in March 1956 from bees);

Shri Revati Nandan Pandey, Bhikhampur
(for my admission to Kanyakubja Degree College, Lucknow in 1958);

Shree Mohd. Wali Sher Khan, Bhallia Buzurg
(my most lovable Primary School Teacher);

Prof. Dr. Abhilash Kumar Agrawal, Gorakhpur
(for my admission to Ph.D.);

Shri Yogendra Nath Dixit, Prayagraj (Allahabad)
(for his god fatherly support at Allahabad and for his spiritual advices);

Pt. Shambhu Narain Misra, ex-M.P. & Advocate, Allahabad
(for his patronage and constant inspiration);

Prof. Dr. Udit Narain Singh, Delhi
*(Prof. of Maths., Delhi Univ. & ex VC, University of Allahabad -
for his honesty & integrity picking me up for DAAD-Fellowship
of Germany in 1971);*

Prof. Dr. Heera Lal Nigam, Allahabad
(ex VC, A.P.S. Univ., Rewa – for Founder Professorship of Maths.);

Prof. Dr. Franco Fava, Turin (Italy)
(for his true fraternal affection providing repeated opportunities at Turin).

CONTENTS

PREFACE   **xvii**

1   PRELIMINARIES   **1**

    1   Results referred in the text   1

2   ALGEBRA (CLASSICAL)   **5**

    1   Law of indices   5
    2   Factorization   5
    3   Equal fractions   5
    4   Progressions   5
        4.1 Arithmetic progression   5
        4.2 Geometric progression   6
        4.3 Harmonic progression   6
    5   First n natural numbers   7
    6   Factorials   7
    7   Permutations and combinations   8
    8   Binomial theorem   8
    9   Exponential function   9
   10 Logarithms   10
        10.1 Common logarithm   10
        10.2 Some rules for logarithms   10
        10.3 Change of base   10
        10.4 Expansion of logarithmic functions   11
        10.5 Log of some numbers   11

3   ALGEBRAIC STRUCTURES   **13**

    1   Group   13
        1.1 Monoid   13
        1.2 Groupoid (or semigroup)   13
        1.3 Semigroup with identity   13
        1.4 Commutative semigroup   13
        1.5 Group   13
        1.6 Commutative (or abelian) group   13
    2   Some properties of groups   14
    3   Some special groups   16
        3.1 Cyclic group   16
        3.2 Transformation group   16

4	Homomorphism and isomorphism of groups	17
5	Coset decomposition of a group and factor group	18
6	Rings, Fields and Integral domains	18
7	Some properties of rings	19
8	Some special rings	20
	8.1 Commutative ring	20
	8.2 Ring with identity	20
	8.3 Division ring	20
	8.4 Field	21
	8.5 Integral domain	21
	8.6 Subring	22
9	Ideals of a ring	22
10	*Coset* decomposition of a ring and quotient ring	23
11	Homomorphism and isomorphism	24

4   ARITHMETIC   **27**

1	Average	27
2	Percentage	27
3	Profit and loss	27
4	Ratio and proportion	27
5	Inequalities	27
6	Equal fractions	28
7	Simple interest	28
8	Compound interest	28
9	Work problems	28
10	Time and watch	28
11	Fundamental theorem of arithmetic	28

5   CALCULUS (DIFFERENTIAL)   **29**

1	Limit of a function	29
2	Properties of limits	30
3	Some important limits	31
	3.1 Properties of $f(x) = (1 + 1/x)^x$	32
4	Continuity of a function	32
5	Some properties of continuous functions	34
6	One-sided continuity	37
7	Differentiation of functions	40
8	Continuity of a differentiable function	42
9	Derivatives of some standard functions	44
	9.1 Algebraic functions	44

9.2	Exponential and logarithmic functions	45
9.3	Trigonometric functions	46
9.4	Hyperbolic functions	47
10	Chain rule (*methods of substitution*)	47
10.1	Logarithmic differentiation	47
11	Derivation of inverse trigonometric functions	48
12	Implicit functions and their derivation	49
13	Successive differentiation	49
14	Leibnitz theorem	51
15	Maxima/minima of a function on a closed interval	53
15.1	Extrema at the end points of interval	53
15.2	Local maxima / minima of a function	54
16	Derivative of $f(x)$ at local extremum points	55
17	Monotonic functions	61
18	Points of local extrema	62
19	Second derivative test for local extrema	63
20	Application of differentiation to calculate rate of change	66
21	Partial differentiation of scalar functions	66
22	Total derivative of a function	69
23	Euler's theorem on homogeneous functions	70
24	Envelope of family of curves	73
24.1	One-parameter family of curves	73
24.2	Envelope of family of curves involving two parameters	74
24.3	General method for more than one parameter	75
24.4	Envelope in polar coordinates	76
25	Involutes and evolutes	77
26	Series and expansion of functions	78
26.1	Power series	78
26.2	Interval of convergence of a power series	79
27	Maclaurin's series	80
28	Taylor's series	82

6	**CALCULUS (INTEGRAL)**	**87**
1	Introduction	87
2	Indefinite integrals of some standard functions	87
3	Different methods of integration	90
3.1	Substitution method	90
3.2	$\int \{ 1 / (ax^2 + bx + c)\}\, dx$	90
3.3	$\int \{(px + q) / (ax^2 + bx + c)\}\, dx$	91

	3.4	Some reduction formulae	92
4		Definite integrals	92
5		Some theorems on definite integrals	94
6		Area of a region by single integration	97
7		Displacement	100
8		Work done by a force	100
9		Volume of solids of revolution	102
10		Area of a region by double integration	103
	10.1	In rectangular Cartesian coordinates	103
	10.2	In polar coordinates	103
11		Change of order of integration	104
12		Changing order of integration when limits are defined by an inequality	107
13		Triple integration	109
14		Transformation of multiple integrals	111
15		Transformation into polar coordinates	114
	15.1	Double integrals	114
	15.2	Triple integrals	114
16		Dirichlet's and Liouville's integrals	115
	16.1	Particular cases	116

7   COMPLEX VARIABLE   119

1		Functions of a complex variable	119
2		Limit of $f(z)$	119
3		Derivative of $f(z)$ with respect to z	120
4		Analytic functions	120
5		Applications of C-R conditions to Laplace equation	122
6		Integral of a complex function along a curve	124
7		Cauchy's integral theorem	126
8		Cauchy's integral formula	126
9		Integration of a power series and Laurent's series	128
	9.1	Series of complex terms	128
10		Singular point of an analytic function. Residue	131
11		Evaluation of real definite integrals by contour integration	133
	11.1	Integration around a unit circle	133
	11.2	Integration around a small semi-circle	134
	11.3	Integration around rectangular contours	135
	11.4	Indenting the contours having poles on the real axis	137
12		Conformal transformation and geometrical	139

interpretation of $w = f(z)$

13	Some standard conformal transformations	141
	13.1 Translation	141
	13.2 Rotation	141
	13.3 Rotation	142
14	Bilinear transformation	144
	14.1 Invariant points of a bilinear transformation	144
15	Some special conformal transformations	146
	15.1 Transformation $w = z^2$	146
	15.2 The transformation $w = z^n$	147
	15.3 The transformation $w = e^z$	148
	15.4 Transformation $w = \cosh z$	148
	15.5 Joukowski's transformation $w = z + 1/z$	149
16	Complex potential and application of complex analysis to flow problems	151

8   DETERMINANTS

		157
		157
1	Determinant	158
2	Some properties of determinants	

9   DIFFERENTIAL EQUATIONS (ORDINARY)   161

1	Introduction	161
2	ODEs of 1st order and 1st degree (simple cases)	162
3	First order and first degree linear differential equations (general cases)	163
	3.1 Separable variables form	163
	3.2 Homogeneous form	163
	3.3 Reducible to homogeneous form	164
	3.4 A linear form	164
	3.5 Reducible to a linear form	165
	3.6 Exact form	166
	3.7 Change of variables	167
4	Differential equations of the 1st order but of any degree	167
	4.1 Equations solvable for p	167
	4.2 Equations solvable for y	168
	4.3 Equations solvable for x	168
	4.4 Clairaut's form	169
	4.5 A more general form	169
5	Linear differential equations of any order with	170

constant coefficients

    5.1   All real and distinct roots   170

    5.2   Some equal roots   170

    5.3   Complex roots   172

    5.4   Pair of roots $\alpha \pm \sqrt{\beta}$ 172

6   The particular integral   173

7   Homogeneous linear differential equations of any order   175

    7.1   Complementary function   176

    7.2   Particular integral   177

8   Differential equations reducible to homogeneous linear form   177

9   Simultaneous differential equations   178

    9.1   Simultaneous linear differential equations with constant coefficients   178

    9.2   Simultaneous differential eqs. in 3 variables   179

10   Solution of ODE by variation of parameters method   180

    10.1   Solution of a linear ODE of I order   180

    10.2   Solution of linear ODE of II order with constant coefficients   181

11   Normal form of an ODE   183

10   DIFFERENTIAL EQUATIONS (PARTIAL)   185

1   Introduction   185

    1.1   First order PDEs   185

    1.2   First order PDEs of general form   188

2   Classification of first order PDEs   188

    2.1   Solution of quasi-linear / linear equations   189

    2.2   Particular cases of linear ODEs   191

    2.3   (Quasi-) linear PDEs involving n independent variables   193

3   Non-linear PDEs of first order   194

4   Second order PDEs with variable coefficients   197

    4.1   First type   198

    4.2   Second type   199

    4.3   Third type   201

5   Second order linear PDEs in mechanics   204

6   Homogeneous linear PDE   207

    6.1   Separation of variables   207

7   One-dimensional wave equation   210

	7.1	Particular case	214
8	Second order PDEs in n independent variables	215
	8.1	Classification of linear PDEs of second order	216

11   DIFFUSION EQUATION   217

| 1 | The diffusion equation | 217 |
| 2 | Heat equation | 217 |

12   DYNAMICS   221

1	Motion of a particle in a straight line	221
2	Uniform acceleration	221
3	Vertical motion under gravity	224
	3.1	Downward motion	224
	3.2	Upward motion	225
4	Kinetics in two dimensions	226
5	Radial and transverse accelerations	227
6	Tangential and normal components of velocity and accelerations	228
7	Tangential and normal components of velocity and accelerations	229
	7.1	Above results in terms of vectors	229
	7.2	Circular motion with uniform angular velocity	230
8	Simple harmonic motion	234
	8.1	Distance in terms of time	235
	8.2	Motion in a vertical circle	236
9	Projectile on a horizontal plane	237
10	Projectile on an inclined plane	240
	10.1	Maximum range on inclined plane	241
	10.2	Particle projected down an inclined plane	242
11	Energy	244
	11.1	Kinetic energy	244
	11.2	Potential energy	245
12	Momentum of a particle	246
13	Conservation of linear momentum	247
14	Impact with a fixed surface	248
	14.1	Perpendicular impact	248
	14.2	Oblique impact	249
15	Direct impact of two spheres	251

16   Relative motion   252

13   FOURIER TRANSFORMS OF FUNCTIONS   255

1   Introduction   255
2   Properties of the Fourier transform   255
3   Fourier series   256
4   Evaluation of Fourier coefficients   257
5   Fourier series in any interval   260
    5.1   Interval $[-l, l\,]$ 260
    5.2   Interval $[0, \pi]$ 261
    5.3   A sine series in the interval $[0, \pi]$ 263
    5.4   Interval $[0, l\,]$ 263
    5.5   The interval $[0, 2\pi]$ 264
    5.6   The interval $[0, 1]$ 264
    5.7   The interval $[a, b]$ 265
6   Fourier series for piecewise functions   266

14   GEOMETRY (COORDINATE: 2-DIMENSIONAL)   269

1   Rectangular Cartesian coordinates   269
2   Distance between two points   269
3   Point dividing a line in a given ratio   270
4   Equation of a straight line   270
    4.1   Slope-intercept form   270
    4.2   Intercepts on axes   270
    4.3   Perpendicular form   271
    4.4   Line through a given point   272
    4.5   Line through two points   272
5   Angle between two lines   273
6   Perpendicular from a point to a line   274
7   Intersection of two lines   274
8   Circle   275
9   Parabola   275
10   Ellipse   276
11   Hyperbola   277
12   Polar coordinates   277
13   General equation of second degree in x and y 278

15   GEOMETRY (COORDINATE: 3-DIMENSIONAL)   285

1   Rectangular Cartesian coordinates (revisited)   285

	1.1	Two-dimensional	285
	1.2	Three-dimensional	285
2		Slope and direction cosines of a line	286
3		Line through a given point	286
4		Line through two points	287
5		The coordinate axes and the coordinate planes	288
6		A linear equation	289
	6.1	In 2 variables	289
	6.2	In 3 variables	290
7		Angle between two lines / planes	290
	7.1	Lines with slope m_1, m_2	290
	7.2	Lines with general equations	291
	7.3	Lines in 3-dimensional geometry	291
	7.4	Normal to a plane	291
	7.5	Angle between two planes	292
8		Perpendicular from a point	292
	8.1	To a line in E_2	292
	8.4	To a plane	293
9		Intersection of lines / planes	294
10		Shortest distance between two lines	294
11		Parametric equations of curves	294
12		Sphere	296
13		More general surfaces in E_3	296
	13.1	Ellipsoid and hyperboloid	297
	13.2	Central conicoid	297
	13.3	Paraboloid	297
	13.4	Tangent lines and normal to surface	298
14		General equation of second degree in three variables	298
15		Some standard cases	301
16		Intersection of a line and a conicoid	302
17		Tangent plane	303
18		Polar plane of a point with respect to a conicoid	306
19		Enveloping cone of a conicoid	308
20		Centre of a conicoid	309
21		Reduction of equation (13.1) into standard forms	310

BIBLIOGRAPHY 317

INDEX 321

PREFACE

The present book is the *first* issue of a series explaining various terms and concepts in Mathematics. It introduces the topics, definitions, main results and theorems (generally) avoiding proofs of the results. The topics are arranged in alphabetical order starting from Algebra (Classical) and cover up to Geometry (3-dimensional Coordinate) in the present volume. Further topics are included in the forthcoming volumes.

The subject matter is presented here in *fifteen* chapters of which the first one lists few results referred to in the later discussion. The next fourteen chapters cover the material on main topics of Algebra (Classical), Algebraic Structures, Arithmetic, Calculus (Differential), Calculus (Integral), Complex variable, Determinants, Differential Equations (Ordinary), Differential Equations (Partial), Diffusion Equation, Dynamics, Fourier Transforms of Functions, Geometry (Coordinate: 2-dimensional) and Geometry (Coordinate: 3-dimensional). The contents in these chapters are divided into Sections and the discussion within the Sections is presented in the form of Definitions, Theorems, Corollaries, Notes and Examples. The sub-titles within the Sections are numbered in decimal pattern. For instance, the equation number (*c.s.e.*) refers to the eth equation in the sth section of Chapter c. When c coincides with the chapter at hand, it is dropped. Adequate references to the results appeared earlier are made in the text avoiding their unnecessary repetition. Double slashes marked at the end of Theorems, Corollaries, Solutions of Exercises, etc., indicate their completion. For brevity, some set-theoretic notations and symbols are frequently used, e.g. the symbol $\Rightarrow$ means *implies*. The *logarithm* of a number to the exponential base e is denoted by *ln*. All the Latin mathematical symbols are normally *italicized*, while their Greek counterparts are in normal fonts. A selected bibliography of the subject is provided. The alphabetical index added at the end makes access to the contents easier.

My long teaching career of more than *five decades* at various universities round the globe and research expertise in different fields helped me for lucid presentation of the subject. In the preparation of the text, I have immensely benefited through books at Sr. nos. 3-28, 30, 32 and 33 in the bibliography. The book is dedicated to my teachers, mentors, colleagues and friends who have helped me for clarifications. Special thanks are given to my ex-colleague (Mr. Bipin Musle), Prof. Suraj Bhan Singh of G.B. Pant University of Agriculture & Technology, Pant

Nagar, Dist. Nainital (India), and to my ex-students: Prof. Dr. Raj Kumar Tiwari, Professor of Physics, Dr. R.M.L. Avadh University, Faizabad (India) and Dr. S.P. Khare, retd. Director, Joint Cipher Bureau, Govt. of India for their valuable assistance and suggestions. My thanks are also due to various Universities all over the world especially Univ. of Allahabad, Allahabad (India); University of Guyana, Georgetown (Guyana); P.N.G. University of Technology, Lae (Papua New Guinea); Adama Science & Technology University, Adama (Ethiopia); State University of New York, Incheon (South Korea); Divine Word University, Madang (P.N.G.) and the present employers (LFU) etc., where I gained a lot while exposing my expertise. It may be unfair if I do not record the sincere cooperation of my family especially the better-half (Mrs. Rekha Misra), whom I often exhausted for my academic passion. Sincere thanks are also due to the publisher for their valuable cooperation for bringing the book into limelight in a limited time.

Although proofs are read with utmost care and solutions to problems are verified repeatedly, yet an oversight or any discrepancy brought to the notice of the author by the inquisitive readers(s) shall be thankfully acknowledged. What a surprising coincidence of completing the first draft of the manuscript on the auspicious day of *Maha Shivratri*.

Lucknow (India): June 1, 2019   Ram Bilas Misra

CHAPTER 1

PRELIMINARIES

§ 1. Results referred in the text

Topic	Result	Equation	Eq. No.
Complex Nos.	$\log_e$ of complex number	$\ln(a+ib)$ $= (1/2)\ln(a^2+b^2) + i\tan^{-1}(b/a)$	(1.1)
Coordinate Geometry (2D)	Circle	$x^2+y^2 = a^2$	(1.2)
	Ellipse	$x^2/a^2 + y^2/b^2 = 1$	(1.3)
	Hyperbola	$x^2/a^2 - y^2/b^2 = 1$	(1.4)
	Hyperbola (rectangular)	$x^2-y^2 = 1, \ \text{or} \ xy = c^2$	(1.5)
Partial Differential Eqns.	Derivatives of z w.r.t. x and y	$p = \partial z/\partial x, \quad q = \partial z/\partial y$	(1.6)
		$r = \partial p/\partial x = \partial^2 z/\partial x^2,$ $s = \partial p/\partial y = \partial^2 z/\partial y\partial x = \partial q/\partial x,$ $t = \partial q/\partial y = \partial^2 z/\partial y^2$	(1.7)
Statics	Resultant of 2 forces P and Q inclined at angle α	$R^2 = P^2 + Q^2 + 2P\,Q\cos\alpha$	(1.8)
	Direction of resultant force R with P	$\tan\theta = Q\sin\alpha/(P+Q\cos\alpha)$	(1.9)
Vectors	Vectors in terms of their components along rectangular Cartesian Coordinate axes	$\mathbf{r} = x\,\hat{\mathbf{i}} + y\,\hat{\mathbf{j}} + z\,\hat{\mathbf{k}}$ $= x\,\hat{\mathbf{e}}_1 + y\,\hat{\mathbf{e}}_2 + z\,\hat{\mathbf{e}}_3,$ $\mathbf{a} = a_i\,\hat{\mathbf{e}}_i = a_1\,\hat{\mathbf{i}} + a_2\,\hat{\mathbf{j}} + a_3\,\hat{\mathbf{k}},$ $\mathbf{b} = b_i\,\hat{\mathbf{e}}_i = b_1\,\hat{\mathbf{i}} + b_2\,\hat{\mathbf{j}} + b_3\,\hat{\mathbf{k}}$	(1.10)

Dot product of unit vectors $\hat{\mathbf{e}}_i$ and $\hat{\mathbf{e}}_j$ acting along x and y axes	$\hat{\mathbf{e}}_i \cdot \hat{\mathbf{e}}_j = \delta_{ij} = \begin{cases} 1 \text{ when } i = j, \\ 0 \text{ when } i \neq j \text{ ;} \end{cases}$ $\hat{\mathbf{i}} \cdot \hat{\mathbf{i}} = \hat{\mathbf{j}} \cdot \hat{\mathbf{j}} = \hat{\mathbf{k}} \cdot \hat{\mathbf{k}} = 1,$ $\hat{\mathbf{i}} \cdot \hat{\mathbf{j}} = \hat{\mathbf{j}} \cdot \hat{\mathbf{i}} = \hat{\mathbf{j}} \cdot \hat{\mathbf{k}} = \hat{\mathbf{k}} \cdot \hat{\mathbf{j}}$ $= \hat{\mathbf{k}} \cdot \hat{\mathbf{i}} = \hat{\mathbf{i}} \cdot \hat{\mathbf{k}} = 0.$	(1.11)
Magnitude of vector $\mathbf{a} = a_i \hat{\mathbf{e}}_i$	$|\mathbf{a}| = \sqrt{(\mathbf{a} \cdot \mathbf{a})} = \sqrt{(a_i\, a_i)}$ $= a_1{}^2 + a_2{}^2 + a_3{}^2$	(1.12)
Dot product of vectors $\mathbf{a}$ and $\mathbf{b}$ inclined at an angle θ	$\mathbf{a} \cdot \mathbf{b} = |\mathbf{a}||\mathbf{b}| \cos\theta$ $= a_i\, b_j (\hat{\mathbf{e}}_i \cdot \hat{\mathbf{e}}_j) = a_i\, b_i$ $= a_1 b_1 + a_2 b_2 + a_3 b_3$	(1.13)
Angle between the vectors	$\cos\theta = a_i\, b_i / |\mathbf{a}||\mathbf{b}|$ $= (a_1 b_1 + a_2 b_2 + a_3 b_3) /$ $\sqrt{(a_1{}^2 + a_2{}^2 + a_3{}^2)} \sqrt{(b_1{}^2 + b_2{}^2 + b_3{}^2)}$	(1.14)
Cross product of vectors $\mathbf{a}$ and $\mathbf{b}$	$\mathbf{a} \times \mathbf{b} = |\mathbf{a}||\mathbf{b}|(\sin\theta)\hat{\mathbf{u}},$ $\hat{\mathbf{u}}$ being the unit vector orthogonal to both $\mathbf{a}$, $\mathbf{b}$ and $\mathbf{a}$, $\mathbf{b}$, $\hat{\mathbf{u}}$ form a right-handed system.	(1.15)
Cross product of vectors $\mathbf{a} = a_i \hat{\mathbf{e}}_i$ and $\mathbf{b} = b_i \hat{\mathbf{e}}_i$	$\begin{vmatrix} \hat{\mathbf{e}}_1 & \hat{\mathbf{e}}_2 & \hat{\mathbf{e}}_3 \\ a_1 & a_2 & a_3 \\ b_1 & b_2 & b_3 \end{vmatrix}$ $= (a_2 b_3 - a_3 b_2)\,\hat{\mathbf{i}} + (a_3 b_1 - a_1 b_3)\,\hat{\mathbf{j}}$ $+ (a_1 b_2 - a_2 b_1)\,\hat{\mathbf{k}}$	(1.16)
Cross product of vectors $\hat{\mathbf{e}}_i$ and $\hat{\mathbf{e}}_j$	$\hat{\mathbf{e}}_i \times \hat{\mathbf{e}}_j = \in_{ijk} \hat{\mathbf{e}}_k,$ where $\in_{ijk} \equiv \begin{cases} 1 \text{ when } ijk = 123, 231, 312; \\ -1 \text{ when } ijk = 213, 321, 132; \\ 0 \text{ otherwise; i.e.} \end{cases}$ $\hat{\mathbf{i}} \times \hat{\mathbf{j}} = \hat{\mathbf{k}}, \hat{\mathbf{j}} \times \hat{\mathbf{k}} = \hat{\mathbf{i}}, \hat{\mathbf{k}} \times \hat{\mathbf{i}} = \hat{\mathbf{j}}$	(1.17)

	Vectors $\mathbf{a} = a_i\,\hat{\mathbf{e}}_i$ and $\mathbf{b} = b_i\,\hat{\mathbf{e}}_i$ are:	orthogonal to each other iff $a_i\,b_i = a_1\,b_1 + a_2\,b_2 + a_3\,b_3 = 0$	(1.18)
		parallel to each other iff $a_1/b_1 = a_2/b_2 = a_3/b_3$	(1.19)
	Scalar product of 3 vectors $\mathbf{a} = a_i\,\hat{\mathbf{e}}_i$, $\mathbf{b} = b_i\,\hat{\mathbf{e}}_i$, $\mathbf{c} = c_i\,\hat{\mathbf{e}}_i$	$[\mathbf{a},\ \mathbf{b},\ \mathbf{c}] = [\mathbf{b}, \mathbf{c}, \mathbf{a}] = [\mathbf{c}, \mathbf{a}, \mathbf{b}]$ $= \begin{vmatrix} a_1 & a_2 & a_3 \\ b_1 & b_2 & b_3 \\ c_1 & c_2 & c_3 \end{vmatrix}$	(1.20) (1.21)
	Dot product of vectors $\hat{\mathbf{e}}_i, \hat{\mathbf{e}}_j, \hat{\mathbf{e}}_k$	$[\hat{\mathbf{e}}_i, \hat{\mathbf{e}}_j, \hat{\mathbf{e}}_k] = \in_{ijk}$, $[\hat{\mathbf{i}}, \hat{\mathbf{j}}, \hat{\mathbf{k}}] = 1$	(1.22)
	Cross product of 3 vectors $\mathbf{a}, \mathbf{b}, \mathbf{c}$	$(\mathbf{a} \times \mathbf{b}) \times \mathbf{c} = (\mathbf{a}\,.\,\mathbf{c})\,\mathbf{b} - (\mathbf{b}\,.\,\mathbf{c})\,\mathbf{a}$, $\mathbf{a} \times (\mathbf{b} \times \mathbf{c}) = (\mathbf{a}\,.\,\mathbf{c})\,\mathbf{b} - (\mathbf{a}\,.\,\mathbf{b})\,\mathbf{c}$	(1.23)
	Scalar product of 4 vectors $\mathbf{a}, \mathbf{b}, \mathbf{c}, \mathbf{d}$	$(\mathbf{a} \times \mathbf{b})\,.\,(\mathbf{c} \times \mathbf{d})$ $= (\mathbf{a}\,.\,\mathbf{c})\,(\mathbf{b}\,.\,\mathbf{d}) - (\mathbf{a}\,.\,\mathbf{d})\,(\mathbf{b}\,.\,\mathbf{c})$	(1.24)
Vector calculus	Gradient of a scalar function $f(x^i)$	$\mathrm{grad}\,f = \nabla f = (\partial f/\partial x^i)\,\hat{\mathbf{e}}_i$ $= (\partial f/\partial x^1,\ \partial f/\partial x^2, \partial f/\partial x^3)$ $= (\partial f/\partial x)\,\hat{\mathbf{i}} + (\partial f/\partial y)\,\hat{\mathbf{j}} + (\partial f/\partial z)\,\hat{\mathbf{k}}$	(1.25)

CHAPTER 2

ALGEBRA (CLASSICAL)

§ 1. Law of indices

Given a real number a and a natural number m the product of a with itself repeated m times is called the mth power of a. It is denoted by

$$a^m = a.\,a \ldots a. \quad (a \text{ repeated } m \text{ times}). \tag{1.1}$$

Particularly, when $m = 0$ but $a \neq 0$, $a^0 = 1$ (assumed). For negative integral value $-n$ of m:

$$a^m = a^{-n} = (1/a^n) = 1/a.\,a\ldots a. \quad (a \text{ repeated } n \text{ times}). \tag{1.2}$$

There hold the following laws of indices:

$$a^m.\,a^n = a^{m+n}, \quad a^m \div a^n = a^m/a^n = a^{m-n}, \quad (a^m)^n = a^{mn}; \tag{1.3}$$

and

$$a^{m/n} = (a^m)^{1/n} = \sqrt[n]{(a^m)}, \tag{1.4}$$

giving the nth root of a^m.

§ 2. Factorization

$$a^2 \pm 2ab + b^2 = (a \pm b)^2 \quad (2.1); \qquad a^2 - b^2 = (a+b).(a-b), \quad (2.2)$$

$$(a+b)^2 - 4ab = (a-b)^2 \quad (2.3); \quad a^3 + b^3 = (a+b).(a^2 - ab + b^2), \quad (2.4)$$

$$a^3 - b^3 = (a-b).(a^2 + ab + b^2), \tag{2.5}$$

$$a^4 - b^4 = (a^2 + b^2).(a^2 - b^2) = (a^2 + b^2).(a+b).(a-b). \tag{2.6}$$

§ 3. Equal fractions. When $b,\ d \neq 0$

$$a/b = c/d = (a \pm c)/(b \pm d) = \sqrt{(ac/bd)} = (ac/bd)^{1/2}. \tag{3.1}$$

§ 4. Progressions

4.1. Arithmetic progression: The numbers (usually integers but could be taken even real numbers)

$$a, \ a+d, \ a+2d,..., a+(n-1)\,d, \ ... \tag{4.1}$$

form a sequence called the *arithmetic progression* (A.P.) with the first term a and common difference (between any two successive terms) d. Sum to n terms of the series

$$a+(a+d)+(a+2d)+ \ ... \ + \{a+(n-1)\,d\}$$

is

$$S_n \equiv (n/2)\,\{2a+(n-1)\,d\} \ = \ (n/2)\,(T_1 + T_n), \tag{4.2}$$

where $T_1 \equiv a$ and $T_n \equiv a+(n-1)\,d$ are the first and the nth terms of the series. The arithmetic mean (or the average) of two numbers a, b is

$$\text{A.M.} \ = \ (a+b)/2. \tag{4.3}$$

4.2. Geometric progression: The numbers

$$a, \ ar, \ ar^2,...., ar^{n-1}, \ ... \tag{4.4}$$

form a *geometric progression* (G.P.) with first term a and common ratio (of any term to its preceding term) r. Sum to n terms of the series is

$$S_n \ = a+ar+ar^2+ \ ... \ + ar^{n-1} = \ a\,(r^n-1)/(r-1). \tag{4.5}$$

Note 4.1. An infinite geometric progression with first term a and common ratio $r < 1$ has a finite (definite) sum

$$S_\infty \ = \ a/(1-r), \tag{4.6}$$

as r^n tends to zero when $n \to \infty$.

The geometric mean of two numbers a, b is

$$\text{G.M.} \ = \ \sqrt{(a\,b)}. \tag{4.7}$$

4.3. Harmonic progression: The reciprocals of terms of an A.P. form a harmonic progression (H.P.):

$$1/a, \ 1/(a+d), \ 1/(a+2d), \ ... \tag{4.8}$$

The harmonic mean (H.M.) of two numbers a, b being the reciprocal of the Arithmetic Mean of $1/a$ and $1/b$ is

$$\text{H.M.} = 2ab / (a + b). \qquad (4.9)$$

Theorem 4.1. Let A, G, H be the arithmetic, geometric and harmonic means of two numbers, there hold the relations

$$A > G > H \qquad (4.10)$$

Note 4.2. Heron's mean of two numbers a, b is defined as

$$(a^2 + a\,b + b^2) / 3, \qquad (4.11)$$

which reduces to a^2 if $b = a$.

§ 5. First n natural numbers

Sum of first n natural numbers, by Eq. (4.2), is

$$\Sigma n \equiv 1 + 2 + 3 + \dots + n = n(n+1)/2. \qquad (5.1)$$

Theorem 5.1. Also, sum of squares of first n natural numbers is

$$\Sigma n^2 \equiv 1^2 + 2^2 + 3^2 + \dots + n^2 = n(n+1)(2n+1)/6. \qquad (5.2)$$

Theorem 5.2. Sum of cubes of first n natural numbers is

$$\Sigma n^3 = 1^3 + 2^3 + 3^3 + \dots + n^3 = n^2(n+1)^2/4. \qquad (5.3)$$

§ 6. Factorials

Given any natural number n the product of first n natural numbers, i.e. 1, 2, ... , n, is called the factorial of n. It is denoted by $n!$ or $\lfloor n$:

$$n! = 1.2.3.\dots n. \qquad (6.1)$$

Particularly, $1! = 1$, $2! = 2.1 = 2$, $3! = 3.2.1 = 6$ etc. Also,

$$n! = n(n-1)(n-2)\dots 2.1 = n(n-1)! \qquad (6.2)$$

and $0! = 1$ (accepted).

Theorems 6.1. Some of the numbers, which can be decomposed into sum of factorials of their digits, are

$$145 = 1! + 4! + 5! \quad \text{and} \quad 40585 = 4! + 0! + 5! + 8! + 5!.$$

Note 6.1. The single digit numbers 1 and 2 also possess above character.

§ 7. Permutations and combinations

Given n objects any r of them can be permuted in ^nP_r number of ways:

$$^nP_r = n! / (n-r)! = n(n-1)(n-2)....(n-r+1). \qquad (7.1)$$

Out of these having all distinct r objects are called *combinations* and are denoted by

$$^nC_r = n! / r! (n-r)! \qquad (7.2)$$

Note 7.1. ^nC_r are also called the *binomial coefficients*.

Theorem 7.1. These coefficients possess the properties

$$^nC_r = {}^nC_{n-r} \quad (7.3); \qquad ^nC_r + {}^nC_{r+1} = {}^{n+1}C_{r+1}. \qquad (7.4)$$

The following results are the direct consequences of above theorem:

Corollary 7.1. $^nC_0 = {}^nC_n = 1, \qquad ^nC_1 = {}^nC_{n-1} = n,$

$$^nC_2 = {}^nC_{n-2} = n(n-1)/2!,$$

$$^nC_3 = {}^nC_{n-3} = n(n-1)(n-2)/3!, \quad \text{etc.}$$

§ 8. Binomial theorem

Given any two real numbers a and b the nth (positive) integral power of their sum is given by

$$(a+b)^n = {}^nC_0\, a^n . b^0 + {}^nC_1\, a^{n-1} . b^1 + {}^nC_2\, a^{n-2} . b^2 + ...$$

$$+ {}^nC_r\, a^{n-r} . b^r + ... + {}^nC_n\, a^{n-n} . b^n$$

$$= a^n + {}^nC_1\, a^{n-1}.b + {}^nC_2\, a^{n-2}.b^2 + ... + {}^nC_r\, a^{n-r}.b^r + ... + b^n. \qquad (8.1)$$

In case n is negative integer or a fraction (of non-zero value) the expansion in right member of Eq. (8.1) becomes infinitely large (i.e.

it contains infinite number of terms).

Two special forms of Eq. (8.1) are:

$$(1 + x)^{-1} = 1 - x + x^2 - x^3 + \dots \infty, \qquad (8.2)$$

and

$$(1 - x)^{-1} = 1 + x + x^2 + x^3 + \dots \infty, \qquad (8.3)$$

where x is numerically less than 1 in either case in order to make the infinite series convergent (i.e. having finite and definite sums).

Note 8.1. The infinite series in Eqs. (8.2) and (8.3) are geometric progressions of common ratios $- x$ and x respectively; hence they are summable by Eq. (4.6), if $|x| < 1$.

If, however, we put $x = 1$ in Eq. (8.2) its left member becomes 1/2 while the right member $1 - 1 + 1 - 1 + \dots \infty$ oscillates between zero (for even number of terms) and 1 (for odd number of terms). Hence, the series in Eq. (8.2), for $x = 1$, does not converge to a finite (definite) sum. It is called an *oscillatory series*.

§ 9. Exponential function

The infinite series

$$e^x \equiv 1 + x / 1! + x^2 / 2! + x^3 / 3! + \dots + x^n / n! + \dots \infty \qquad (9.1)$$

is called the *exponential series* for some variable (number) x and is summable to the sum e^x (called the exponential function of x). Particularly,

$$e = e^1 = 1 + 1 / 1! + 1 / 2! + \dots \infty > 2, \qquad (9.2)$$

which is approximately equal to 2·7183. It may be noted that this number e is irrational.

Applying binomial expansion in Eq. (8.1) to $(1 + 1/x)^x$ and finally making x to tend to $\pm \infty$ so that $1/x$ tends to zero one gets the following important limit:

$$\lim_{x \to \pm \infty} (1 + 1/x)^x = e. \qquad (9.3)$$

Following conclusions can also be derived from above result:

$$\lim_{y \to 0} (1 + y)^{1/y} = e \quad (9.4); \qquad \lim_{x \to \pm\infty} (1 + a / x)^x = e^a, \quad (9.5)$$

$$\lim_{x \to \pm\infty} (1 - 1 / x)^x = e^{-1} = 1/ e, \tag{9.6}$$

and

$$\lim_{y \to 0} (1 + y)^{-1/y} = \lim_{y \to 0} \{(1 + y)^{1/y}\}^{-1} = e^{-1} = 1/ e. \tag{9.7}$$

§ 10. Logarithms

Putting $e^x = y$, where $y \neq 0$, we define a new function called the *logarithm* of y to the base e. If is denoted by $\ln y$:

$$e^x = y \qquad \Rightarrow \qquad x = \ln y. \tag{10.1}$$

e being the natural base (number) $\ln y$ is a *natural logarithm* of y and is also denoted by $\ln y$.

Note 10.1. Since the second relation in Eq. (10.1) also implies the former one the exponential and logarithm functions are inverses to each other.

10.1. Common logarithm: Putting $(10)^x = y$ we also define the common logarithm of y to the base 10:

$$(10)^x = y \qquad \Rightarrow \qquad x = \log_{10} y. \tag{10.2}$$

Note 10.2. For any number $a \neq 0, \infty$, $a^0 = 1$ and $a^1 = a$ causing

$$0 = \log_a 1 \qquad (10.3); \qquad 1 = \log_a a. \tag{10.4}$$

10.2. Some rules for logarithms: Given any two non-zero numbers m, n there hold the relations

$$\log (m\, n) = \log m + \log n, \tag{10.5}$$

$$\log (m/n) = \log m - \log n; \tag{10.6}$$

$$\log (m)^n = n \log m, \tag{10.7}$$

for any base $a \neq 0, \infty$.

10.3. Change of base: Putting $a^x = y$ and taking its logarithm to the bases e and a we derive

$$x \ln a = \ln y \quad \Rightarrow \quad x = (\ln y) / (\ln a), \quad (10.8)$$

and

$$x \log_a a = \log_a y \quad \Rightarrow \quad x = \log_a y, \quad \text{by Eq. (10.4)}.$$

The last two relations yield

$$\log_a y = (\ln y) / (\ln a) \quad \Rightarrow \quad (\log_a y)(\ln a) = \ln y. \quad (10.9)$$

Thus, for any two bases $a, b \neq 0, \infty$, there also hold

$$(\log_b a)(\log_a b) = \log_b b = 1. \quad (10.10)$$

10.4. Expansion of logarithmic functions: Natural logarithmic functions of $(1 \pm x)$ can be expanded in infinite series as follows:

$$\ln(1 + x) = x - x^2/2 + x^3/3 - x^4/4 + \dots \infty, \quad (10.11)$$

and

$$\ln(1 - x) = -(x + x^2/2 + x^3/3 + x^4/4 + \dots \infty). \quad (10.12)$$

Their difference also yields

$$(1/2) \ln \{(1 + x)/(1 - x)\} = x + x^3/3 + x^5/5 + \dots \infty. \quad (10.13)$$

10.5. Log of some numbers:

$$\ln 2 = 0{\cdot}6931, \qquad \log_{10} 2 = 0{\cdot}3010,$$

$$\ln 3 = 1{\cdot}0986, \qquad \log_{10} 3 = 0{\cdot}4771.$$

CHAPTER 3

ALGEBRAIC STRUCTURES

§ 1. Group

A non-empty set G with an operation $\circ$ satisfying some of the algebraic laws forms an algebraic structure. Following are some of the simple algebraic structures:

1.1. Monoid: The structure $(G, \circ)$ is called *monoid*, if $\circ$ is binary in G.

1.2. Groupoid (or semigroup): If $\circ$ is both binary and associative in G, then $(G, \circ)$ is called a *groupoid* or *semigroup*.

1.3. Semigroup with identity: If there also holds the identity law with respect to a binary and associative operation $\circ$ in a non-empty set G then $(G, \circ)$ is called a *semi-group with identity*.

1.4. Commutative semigroup: If there hold binary, associative and commutative laws w.r.t. the operation $\circ$ in G, then $(G, \circ)$ is called a *commutative semigroup*.

1.5. Group: If $\circ$ satisfies binary, associative, identity and inverse laws in a non-empty set G, then $(G, \circ)$ is called an *algebraic group*.

1.6. Commutative (or abelian) group: If o satisfies the first five algebraic laws given above in a non-empty set G, then $(G, \circ)$ is called a *commutative (or abelian) group*.

Example 1.1. (E, +), (Z, +), (Q, +), (R, +) and (C, +) are all algebraic groups.

Example 1.2. The singleton sets $\{0\}$ and $\{1\}$ form groups w.r.t. addition and multiplication respectively.

Example 1.3. The sets of non-zero rational numbers, non-zero real numbers and non-zero complex numbers also form groups w.r.t. multiplication.

Example 1.4. The set
$$G = \{1, \omega, \omega^2\} \tag{1.1}$$

of cube roots of unity forms a group w.r.t. multiplication.

Note 1.1. All above groups are also *abelian* groups.

Example 1.5. The set of 2×2 non-zero matrices defined over the set of real (or complex) numbers also forms a group w.r.t. matrix multiplication. However, it is *not commutative* in general.

§ 2. Some properties of groups

Definition 2.1. An element of an algebraic structure G is called *unit* if its inverse (w.r.t. binary operation $\circ$ in G) also exists in G.

Example 2.1. The element 1 is unit in the semigroup $(N, \cdot)$.

Theorem 2.1. The units of a semi-group $(G, \circ)$ with *identity* form a group w.r.t. the same binary operation $\circ$.

Theorem 2.2. A semigroup $(G, \circ)$ becomes a group if the equations

$$a \circ x = b \quad \text{and} \quad y \circ a = b \qquad (2.1)$$

are uniquely soluble in $G \; \forall \; a, b \; \varepsilon \; G$.

Note 2.1. Both the identity element and the inverse element of a given element in a group are unique.

Definition 2.2. A subset H of a group $(G, \circ)$ forms a *subgroup* of $(G, \circ)$ if $(H, \circ)$ itself is a group.

Example 2.2. The following groups form a chain of subgroups of respective groups:

$$(\{0\}, +) \subseteq (E, +) \subseteq (Z, +) \subseteq (Q, +) \subseteq (R, +) \subseteq (C, +).$$

Example 2.3. The following groups form a chain of subgroups:

$$(\{1\}, \cdot) \subseteq (Q^*, \cdot) \subseteq (R^*, \cdot) \subseteq (C^*, \cdot).$$

Example 2.4. The square roots $1, -1$ of unity form a subgroup of a group $G = \{1, i, -i, -1\}$ of the fourth roots of unity w.r.t. multiplication.

Theorem 2.3. A subset H of a group $(G, \circ)$ forms a subgroup of $(G, \circ)$ if $\forall\, a, b \,\varepsilon\, H \,\exists\; a \circ b^{-1} \,\varepsilon\, H$.

Corollary 2.1. A subset H of a group $(G, +)$ forms a subgroup of $(G, +)$ if there holds

$$\forall\, a, b \,\varepsilon\, H \;\exists\; a - b \,\varepsilon\, H.$$

Corollary 2.2. A subset H of a group $(G, \cdot)$ forms a subgroup of $(G, \cdot)$ if there holds

$$\forall\, a, b \,\varepsilon\, H \;\exists\; ab^{-1} \;\varepsilon\; H. \tag{2.2}$$

Definition 2.3. (*Centre of a group*) Let $(G, \cdot)$ be a group and a subset C of G be defined by

$$C = \{x : xa = ax \,\forall\, a \,\varepsilon\, G\}, \tag{2.3}$$

then C is called the *centre* of the group $(G, \cdot)$.

Theorem 2.4. The centre of a group forms a subgroup of the group.

Example 2.5. (*Reversal rule*) Given a group $(G, \cdot)$, there hold the rules:

$$\forall\, x, y \,\varepsilon\, G \;\Rightarrow\; (xy)^{-1} = y^{-1}x^{-1}, \tag{2.4}$$

and

$$(x^{-1})^{-1} = x. \tag{2.5}$$

Example 2.6. A group $(G, \cdot)$ with property $a^2 = e \,\forall\, a \,\varepsilon\, G$ is necessarily commutative.

Example 2.7. For any positive integer m the set of residue classes w.r.t. modulo m:

$$I_m = \{\overline{0}, \overline{1}, \overline{2}, ..., \overline{m-1}\} \tag{2.6}$$

forms a group w.r.t. addition defined by

$$\overline{p} + \overline{q} = \overline{r}, \tag{2.7}$$

where r is the remainder for division of the sum $p + q$ by m and all p, q, r vary from 0 to $m - 1$.

Example 2.8. The set of non-zero elements in Eq. (2.6):

$$I_m{}^* = \{\overline{1}, \overline{2}, ..., \overline{m-1}\} \tag{2.8}$$

also forms a group w.r.t. the multiplication defined by

$$\bar{p}.\bar{q} = \bar{r}, \tag{2.9}$$

where r is the remainder in the division of p . q by m and m is a prime number.

§ 3. Some special groups

3.1. Cyclic group: Let $(G, \cdot)$ be a group and a be its any element. The integral powers of a (i.e. repeated multiples of a with itself) form a subset

$$\mathcal{C} = \{ a^p : p \varepsilon \; Z\} \tag{3.1}$$

of G. The set $\mathcal{C}$ also forms a group w.r.t. multiplication, called the *cyclic group*. The identity element of $(\mathcal{C}, \cdot)$ is the identity element of $(G, \cdot)$ namely $1 = a^\circ$ (for $p = 0$) and the inverses of a^p are

$$a^{-p} = (a^{-1})^p = a^{-1}. a^{-1}... a^{-1} \text{ (repeated } p \text{ times).}$$

If for some finite value of p ($\neq 0$), a^p coincides with the identity element 1, the group is called a *finite cyclic group* of *order p*, otherwise the order of the group is infinite. Also, a is called a *generator* of the group.

Example 3.1. The cube roots of unity form a cyclic group of order 3 of either of ω and ω^2 as its generators.

Example 3.2. The set of residue classes with respect to modulo 5 forms a cyclic group of order 5 w.r.t. addition of residue classes having every non-zero element as its generator.

3.2. Transformation group

Given a non-empty set G we consider one-to-one mappings of G onto itself. The set of these mappings, denoted by $\mathcal{T}(G)$, forms a group w.r.t. product of mappings. Such a group is called a *transformation group* of G. In case G is finite the transformation group $\mathcal{T}(G)$ is often called a *permutation group*.

Example 3.3. The permutations of the elements of the set $S = \{1, 2, 3\}$ form the permutation group S_3 of degree 3 and order 3!.

Note 3.1. The permutation group in above example is called a *symmetric group of order* 3!.

Similarly, the permutation group S_n of the set $\{1, 2, ..., n\}$ of n elements will be a symmetric group of order $n!$.

§ 4. Homomorphism and isomorphism of groups

Let $(G, \circ)$ and $(G', \circ')$ be two groups. A mapping f of G into G', written as $f: G \rightarrow G'$, will be called a *homomorphism* if there holds

$$\forall\, x, y \,\varepsilon\, G, \qquad f\,(x \circ y) = f\,(x) \circ' f(y). \qquad (4.1)$$

Further, if f is also one-to-one and onto so that there hold the relations

$$f\,(x) = f(y) \quad \Rightarrow \quad x = y, \qquad (4.2)$$

and

$$\forall\, x' \,\varepsilon\, G' \,\exists\, \text{some } x \,\varepsilon\, G \;\; \text{s.t.} \;\; f(x) = x', \qquad (4.3)$$

it becomes an *isomorphism*. The two groups are then called *isomorphic* to each other and we write $G \cong G'$.

Definition 4.1. The set of elements of G mapping onto the identity element e' of G' is called the *kernel* of the homomorphism f:

$$\mathcal{K} = \{\, x : x \,\varepsilon\, G, \;\; f\,(x) = e' \,\varepsilon\, G' \,\}. \qquad (4.4)$$

Theorem 4.1. A homomorphism preserves the identity and inverses of the elements, i.e. it maps the identity (respectively inverse) element of G onto the identity (resp. inverse) element of G':

$$f\,(e) = e', \qquad e \,\varepsilon\, G, \;\; e' \,\varepsilon\, G'; \qquad (4.5)$$

$$f\,(a^{-1}) = \{f\,(a)\}^{-1}, \quad a \,\varepsilon\, G, \;\; f\,(a) \,\varepsilon\, G'. \qquad (4.6)$$

Theorem 4.2. The image set $f\,(G)$ is a subgroup of the group $(G', \circ')$ and is called the *homomorphic group* of $(G, \circ)$.

Theorem 4.3. The kernel set $\mathcal{K}$ of the homomorphism forms a group w.r.t. $\circ$.

§ 5. Coset decomposition of a group and factor group

Let H be any non-empty subset of a group $(G, \cdot)$. We consider the set of multiples of H by every element of G:

$$S = \{aH, bH, cH, \dots, H, \dots, a^{-1}H, \ b^{-1}H, c^{-1}H, \dots\}.$$

These multiples of H are called the *cosets* of G and their union spans back G itself. We defining multiplication process in S by

$$(aH)(bH) = (a\,b)\,H \ \forall \ a, b \ \varepsilon \ G. \tag{5.1}$$

It may be seen that H acts as the identity element w.r.t. above multiplication in S:

$$(H)(aH) = aH = (aH)(H).$$

Also, the inverse of aH is $a^{-1}H$, for a^{-1} being the inverse of a in G. The structure $(S, \cdot)$ forms a group called the *factor group* of G w.r.t. its subset H and is denoted by G / H.

Theorem 5.1. (*Fundamental theorem on homomorphism*) Any group is isomorphic to its factor group.

§ 6. Rings, fields and integral domains

Some algebraic structures consisting of a non-empty set and an operation satisfying certain algebraic laws are discussed in previous sections. Now, we consider a non-empty set with two binary operations. Such an algebraic structure may form a *ring*, an *integral domain*, or a *field* subject to certain postulates satisfied by the operations.

Definition 6.1. (*Ring*) A non-empty set $R = \{a, b, c, \dots\}$ of some elements forms a ring with respect to addition and multiplication if there hold:

(i) $(R, +)$ is an abelian group, i.e. there hold *five* postulates of a commutative group w.r.t. $(+)$, namely binary, associative, identity, inverse and commutative laws;

(ii) $(R, \cdot)$ is a semigroup, i.e. the multiplication is binary and associative in R;

(iii) The multiplication is distributive over addition: both by left and by right.

Example 6.1. The following sets forms a ring w.r.t. (+) and (·):

(i) singleton set $\{0\}$; **(ii)** set of integers Z;

(iii) set of even integers E; **(iv)** set of rational numbers Q;

(v) set of real numbers R; **(vi)** set of complex numbers C;

(vii) set of Gaussian integers $a + i\, b$, where both a, b are integers;

(viii) set $\mathbb{I}_m$ of residue classes w.r.t. modulo number m.

§ 7. Some properties of rings

Theorem 7.1. For any element x of a ring R there holds the relation

$$x.\,0 = 0, \tag{7.1}$$

where 0 is the additive identity in R.

In a ring the product of two zero-valued elements is zero. However, in some of the rings the converse of this fact need not hold, i.e. the product of two non-zero elements is not always zero. For example, the ring $(\mathbb{I}_m, +, \cdot)$ in Example 6.1 above, for a composite modulo number m (having proper factors other than 1 and m) so that $m = m_1 \cdot m_2$, has two non-zero elements $\overline{m_1}, \overline{m_2}$ but their product is zero element of the ring:

$$\overline{m_1}.\overline{m_2} = \overline{m} = \overline{0}.$$

Definition 7.1. An element z in a ring R is called a *divisor of zero* if there exists some non-zero elements x and y in R satisfying

$$z \cdot x = 0, \qquad \text{or} \qquad y \cdot z = 0. \tag{7.2}$$

In any ring with non-zero elements the element 0 (zero) itself is a divisor of zero.

Theorem 7.2. For any elements x, y, z in a ring R there also hold the relations

$$- (-x) = x \quad (7.3); \qquad x(-y) = (-x)y = -(xy); \quad (7.4)$$

$$(-x)(-y) = xy \quad (7.5); \qquad x(y-z) = xy - xz; \quad (7.6)$$

$$(x-y)z = xz - yz. \quad (7.7)$$

§ 8. Some special rings

8.1. A ring $(R, +, \cdot)$ is called *commutative* if the multiplication is commutative in R, i.e. $\forall\, x, y \,\varepsilon\, R \,\exists\, xy = yx$.

Indeed, all the rings in Example 6.1 are commutative. However, the set of 2×2 matrices defined over the set of real numbers (cf. Example 1.5) forms a ring w.r.t. matrix addition and matrix multiplication which is not commutative.

8.2. A ring is called a *ring with identity* if it contains the multiplicative identity 1 with the property $x \cdot 1 = 1 \cdot x = x \,\forall\, x \,\varepsilon\, R$.

The rings in Examples 6.1 (i) and (iii) do not have the multiplicative identity. Rest all the rings therein have multiplicative identity, i.e.1. It is very much probable that in a ring with identity some elements may possess their multiplicative inverses as well satisfying $xy = yx = 1$ for some x and y in R. Such elements possessing their multiplicative inverses (i.e. units in R w.r.t. $(\cdot)$) are called *regular* or *invertible* or *non-singular*. On contrary, the elements which are not regular (i.e. which do not possess their multiplicative inverses in the ring) are called *singular*. It may be noted that the additive identity (i.e. zero) is always a *singular* element while the multiplicative identity 1 is always *regular* in a ring with identity. 1 and -1 are the only regular elements in the ring of integers (cf. Example 6.1 (ii)). In Examples 6.1 (iv) to (vi) all the non-zero elements are regular in their respective rings. In Example 6.1 (vii) only regular elements are $1, -1, i$ and $-i$. In Example 6.1 (viii) every non-zero element is regular if m is prime (cf. Example 2.8).

8.3. Division ring: A ring with identity becomes a division ring if all its non-zero elements are *regular*. In other words, there hold the following postulates for a division ring:

(i) $(R, +)$ is an abelian group, **(ii)** $(R^*, \cdot)$ is a group, **(iii)** $(\cdot)$ is distributive over $(+)$.

The ring of 2 × 2 matrices (cf. Subsection 8.1), for non-commutative property, apparently may appear to be a division ring but it is not so as the product of two non-null matrices may be zero in this ring:

$$\begin{bmatrix} 1 & 0 \\ 0 & 0 \end{bmatrix} \cdot \begin{bmatrix} 0 & 0 \\ 0 & 1 \end{bmatrix} = \begin{bmatrix} 0 & 0 \\ 0 & 0 \end{bmatrix}.$$

Hence such matrix (elements) are not regular in the ring.

Theorem 8.1. A ring with identity is a division ring iff its non-zero elements form a group w.r.t. multiplication.

8.4. Field

A commutative division ring becomes an algebraic *field*. Thus, there hold the following postulates in a field $(R, +, \cdot)$:

(i) $(R, +)$ is an abelian group, **(ii)** $(R^*, \cdot)$ is also an abelian group,

(iii) $(\cdot)$ is distributive over $(+)$.

Example 8.1. The rings in Examples 6.1 (iv) to (vi) form fields.

Example 8.2. The ring in Example 6.1 (viii) also forms a field when m is prime..

Example 8.3. The ring of 2 × 2 matrices does not form a field.

Note 8.1. The fields formed by the sets of real (respectively complex) numbers are called *real* (resp. *complex*) *fields*.

8.5. Integral domain

A ring (with identity) which is free from proper zero divisors is called an *integral domain*. Thus, there also holds the additional postulate: "For any non-zero elements $a, b \in R$ their product $a\,b$ is also non-zero", in a ring with identity to become an integral domain.

Example 8.4. The rings in Examples 6.1 (ii) to (vii) are integral domains.

Example 8.5. The ring in Example 6.1 (viii) also forms an integral domain when m is prime. But, for a composite $m = m_1 m_2$ the ring $(\mathbb{Z}_m$ $, +, \cdot)$, in spite of having the multiplicative identity 1, does not become an integral domain, for

$$\overline{m_1} \neq \overline{0}, \quad \overline{m_2} \neq \overline{0} \quad \text{but} \quad \overline{m_1} . \overline{m_2} = \overline{m} = \overline{0}.$$

The following theorem describes an interesting link between an integral domain and a field.

Theorem 8.2. A finite integral domain becomes a field.

8.6. Subring

A non-empty subset S of a ring $(R, +, \cdot)$ can form a *subring* of R w.r.t. the same binary operations $(+)$ and $(\cdot)$ of R if $(S, +, \cdot)$ is a ring by itself.

Example 8.6. In view of Example 2.2, various sets in Example 6.1 form subrings of the corresponding rings.

Example 8.7. The ring in Example 6.1 (vii) is a subring of that in Example 6.1 (vi).

Similarly, a subfield can be defined. A non-empty subset S of a field $(F, +, \cdot)$ forms a *subfield* of F w.r.t. the same binary operations of F if $(S, +, \cdot)$ itself is a field.

Example 8.8. The fields in Example 8.1 are subfields of the corresponding ones: $(Q, +, \cdot) \subset (R, +, \cdot) \subset (C, +, \cdot)$.

§ 9. Ideals of a ring

Let $(I, +, \cdot)$ be a subring of a ring $(R, +, \cdot)$. The subring I becomes an ideal in R if there holds an additional postulate:

$$\forall x \varepsilon I \quad \text{and} \quad \forall a \varepsilon R \implies \quad a x \quad \text{and} \quad x a \varepsilon I. \tag{9.1}$$

When I is a proper subset of R forming an ideal in R then it is called a *proper ideal* in R.

Both the singleton set $\{0\}$ and the whole set R form subrings of the ring $(R, +, \cdot)$. Further, the postulate in Eq. (9.1) holds in both of these subrings so both the sets {0} and R form trivial ideals in the ring $(R, +, \cdot)$. Thus, every ring with some non-zero elements has at least two distinct ideals.

Example 9.1. The ring of even integers $(E, +, \cdot)$ forms an ideal of ring of integers $(Z, +, \cdot)$.

Example 9.2. Given any integer m the set of its integral multiples:

$$\overline{m} = \{ km : k \varepsilon E \} = \{0, \pm 2m, \pm 3m, \pm 4m, ...\}$$

forms an ideal in the ring of integers.

Note 9.1. Some rings may have many non-trivial ideals while others have none at all.

Theorem 9.1. A commutative ring with identity becomes a field iff it has no non-trivial ideals.

§ 10. *Coset* decomposition of a ring and quotient ring

Let $(I, +, \cdot)$ be an ideal in a ring $(R, +, \cdot)$. Using this ideal we introduce an equivalence relation in R:

Two elements x and y in R are said to be *congruent* (w.r.t. modulo I) if $x - y \varepsilon I$. This relation can easily be seen as an equivalence relation decomposing the set R into disjoint classes. These classes are called the *cosets* of R. Their union describes the whole set R again. As per definition, the coset containing (or determined by) an element x of R with respect to above ideal I is the set of those elements y of R which are congruent to x. Denoting it by $[x]$, we thus have

$$[x] = \{y : y \equiv x \,(\text{mod } I), \text{ i.e. } y - x = p \ \varepsilon I \text{ for some } p \varepsilon I\}. \quad (10.1)$$

In the set R / I of such cosets

$$R / I = \{x + I, \ y + I, ...\}, \quad (10.2)$$

where $x, y,... \varepsilon R$, the operations of addition and multiplication are defined:

$$(x + I) + (y + I) = (x + y) + I, \quad (10.3)$$

and

$$(x + I) \cdot (y + I) = x y + I. \tag{10.4}$$

The set R / I forms a ring w.r.t. above operations. The additive identity in the ring is $0 + I = I$ and the negative of $x + I$ is $(-x) + I$. Such a ring is called the *quotient ring* of R w.r.t. its ideal I. Further, if R is commutative so is R / I; and if R has the identity 1 and I is a proper ideal then R / I has the identity $1 + I$.

§ 11. Homomorphism and isomorphism

Let R and R' be two rings. A mapping f of R into R' is called a *homomorphism* of R into R' if there hold the postulates:

$$f (x + y) = f (x) + f (y), \tag{11.1}$$

and

$$f (x y) = f (x) \, f (y) \tag{11.2}$$

$\forall \, x, y \, \varepsilon \, R$. A homomorphism of a ring into another ring preserves the ring operations.

Theorem 11.1. A homomorphism maps:

(i) the additive identity of R onto the additive identity of R':

$$f (0) = 0; \,^{1)} \tag{11.3}$$

(ii) the negative of some element $x \, \varepsilon \, R$ onto the negative of $f(x)$ in R' :

$$f (-x) = -f (x). \tag{11.4}$$

Note 11.1. The image set $f(R)$ can be seen as a subring of R' and it is called a *homomorphic image* of R.

Definition 11.1. The inverse image in R of the zero ideal in R' is called the *kernel* of the homomorphism f of R into R':

$$\mathcal{K} = \{x : x \, \varepsilon \, R \text{ and } f (x) = 0\}. \tag{11.5}$$

---

[1] For simplicity the additive identities in both the rings R and R' are denoted by the same symbol 0.

Theorem 11.2. The kernel set of a homomorphism forms a subring of R and an ideal in R.

As a consequence of above theorem there results the quotient ring $R / \mathcal{K}$. In the following we observe that $R / \mathcal{K}$ is a homomorphic image of R under a homomorphism.

Theorem 11.3. There exists the natural homomorphism g of R into $R / \mathcal{K}$ defined by

$$\forall x \, \varepsilon \, R, \quad g\,(x) = x + \mathcal{K}. \tag{11.6}$$

Definition 11.2. A homomorphism f of R into R' becomes an *isomorphism* if it is one-to-one and onto.

Thus, for an isomorphism f of R onto R' the kernel set $\mathcal{K}$ becomes a singleton $\{0\}$ containing only the additive identity of R. Also, the image set $f(R)$ then covers the entire set R' and the two rings R and R' are said to be *isomorphic* to each other.

Definition 11.3. An ideal in a ring is called a *maximal ideal* if it is a proper ideal and is not contained in any other proper ideal.

An interesting property of a maximal ideal is exhibited by the

Theorem 11.4. Given a commutative ring with identity R an ideal I in R is maximal iff R / I is a field.

CHAPTER 4

ARITHMETIC

§ 1. Average

The average of n numbers $a_1, a_2, \ldots, a_n$ is $(a_1 + a_2 + \ldots + a_n) / n$.

§ 2. Percentage

The ratio $x / 100$ is called x percent and is denoted by $x\%$ for instance $x\%$ of some number y is $x y / 100$.

§ 3. Profit and loss

The purchase value of a certain item is called its *cost price* (say x). If the item is sold for some price say y then:

$y - x$ (if positive) is called the *profit* earned on the cost price x.

On the other hand, if it is negative it is called the *loss*.

Note 3.1. The percentage of profit (or loss) is calculated on the cost price of an item.

§ 4. Ratio and proportion

Given two numbers a and b ($\neq 0$) their ratio a / b is denoted by $a : b$. Two ratio a / b and c / d, (where $b, d \neq 0$), are equal (or in the same proportion) if

$$a : b = c : d, \quad \text{i.e} \quad a/b = c/d \quad \Rightarrow \quad a d = b c. \qquad (4.1)$$

§ 5. Inequalities

Given any two numbers a and b:

$$a > b \quad \Rightarrow \quad a + x > b + x, \qquad (5.1)$$

and

$$ax > bx \quad (\text{provided } x > 0). \qquad (5.2)$$

Also

$$a/b = a x / b x, \quad \text{if } x \neq 0, \infty. \qquad (5.3)$$

§ 6. Equal fractions

Given two *equal* fractions a/b and c/d (where $b, d \neq 0$) each of them is also equal to $(a \pm c)/(b \pm d)$ as well as to $(ac/bd)^{1/2}$:

$$a/b = c/d = (a \pm c)/(b \pm d) = (ac/bd)^{1/2}. \qquad (6.1)$$

§ 7. Simple interest

The simple interest accrued on a principal amount P at the rate of $r\%$ per annum after n years is $nPr/100$. Therefore, the total amount payable after n years is

$$P(1 + nr/100). \qquad (7.1)$$

§ 8. Compound interest

The compound interest on the principal amount P at the rate $r\%$ per annum after n years (interest compounded annually) is

$$P\{1 + r/100\}^n - P. \qquad (8.1)$$

If interest compounded quarterly, the above sum becomes

$$P\{1 + r/400\}^{4n} - P. \qquad (8.2)$$

§ 9. Work problems

If x number of persons complete a job in n days then a single person requires $x\,n$ days to complete the same job (if working daily for same number of hours).

§ 10. Time and watch

In the duration of 12 hours the hour hand and minute hand of a watch overlap over each other at 12:00 hours and once between each of the pairs of hours 1, 2; 2, 3; ... ; 10, 11. Thus, they overlap over each other 11 times in 12 hours duration and 22 times in 24 hours duration.

§ 11. Fundamental theorem of arithmetic

Every integer is factorizable into prime factors.

CHAPTER 5

CALCULUS (DIFFERENTIAL)

§ 1. Limit of a function

Let $f(x)$ be a real valued function of a real variable defined in some domain $a \le x \le b$. Let the graph of the function be represented by a curve C and P is a point with abscissa c on the graph. If P_l $(x = c - \varepsilon)$ is a point on C in the neighbourhood of P (lying to the left of P) where the function possesses value $f(c - \varepsilon)$. As $c - \varepsilon$ tends to c, i.e. ε tends to 0 (along x-axis), the point P_l approaches P (along C), this value of the function (if it is some definite real value) is said to be the *limit* of $f(x)$ at P *from left*. In notation it is written as:

$$\lim_{x \to c^-} f(x) = \lim_{\varepsilon \to 0} f(c - \varepsilon) = L_l \text{ (say)}, \tag{1.1}$$

where L_l is some definite real number. In case it is infinite or indeterminate or not real we say that the function does not possess its limit from left at P.

Similarly, if P_r $(x = c + \varepsilon)$ is a point on C in the neighbourhood of P (lying to the right of P) where the function possesses value $f(c + \varepsilon)$. As $c + \varepsilon$ tends to c, i.e. ε tends to 0 (along x-axis), the point P_r approaches P (along C), this value of the function (if some definite real value) is said to be the *limit* of $f(x)$ at P *from right*. We write it as:

$$\lim_{x \to c^+} f(x) = \lim_{\varepsilon \to 0} f(c + \varepsilon) = L_r \text{ (say)}, \tag{1.2}$$

where L_r is some definite real number. In case it is infinite or indeterminate or not real we say that the function does not possess its limit from right at P.

When both L_l and L_r exist and are equal to each other the function is said to possess limit at P. Denoting the common value of the *limit* by L we then write:

$$\lim_{x \to c} f(x) = L \text{ (say)}. \tag{1.3}$$

Example 1.1. The function $f(x) = x + 1$ (if $x \ne 2$), 6 (if $x = 2$) possesses limit 3 at $x = 2$.

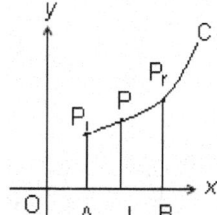
Fig 1.1

Example 1.2. The function $f(x) = x^2$ (if $x \geq 2$), $x/2 - 3$ (if $x < 2$) does not possess limit at $x = 2$.

Example 1.3. The function $f(x) = |x| (x^2 + 1)/x$ does not possess limit at $x = 0$.

Example 1.4. The function $f(x) = (\sqrt{x} - 1) / (x - 1)$ possesses limit at $x = 1$.

Example 1.5. The function $f(x) = 1/x$, $x \neq 0$ does not possess limit at $x = 0$ but it possesses limit as $x \to \pm \infty$.

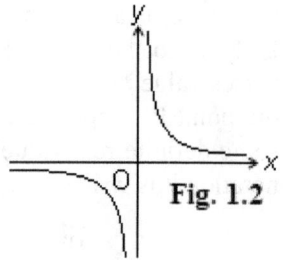

Example 1.6. The function $f(x) = 3^{1/x}$, $x \neq$ 0 does not possess limit at $x = 0$ but it possesses limit as $x \to \pm \infty$.

§ 2. Properties of limits

Fig. 1.2

Theorem 2.1. If $\lim_{x \to c} f(x) = L$,

$\lim_{x \to c} g(x) = M$ and k is any real number then

$$\lim_{x \to c} \{k f(x)\} = k L, \quad \lim_{x \to c} \{f(x) \pm g(x)\} = L \pm M, \quad (2.1)$$

$$\lim_{x \to c} \{f(x).g(x)\} = L.M, \lim_{x \to c} \{f(x) /g(x)\} = L /M, M \neq 0, (2.2)$$

and
$$\lim_{x \to c} \{f(x)\}^a = L^a, \text{ if } L \neq 0, \; a, L^a \; \varepsilon \; P. \quad (2.3)$$

Theorem 2.2. Let $\lim_{x \to c} f(x) = L$, $\lim_{x \to c} g(x) = M$ and $f(x) \leq g(x)$ then $L \leq M$.

Further, if there exists a function $h(x)$ between $f(x)$ and $g(x)$:

$f(x) \leq h(x) \leq g(x)$, and if $L = M$, then $\lim_{x \to c} h(x) = L$ or M.

Definition 2.1. (i) If $\lim_{x \to c^-} f(x) = \lim_{x \to c^+} f(x) = \pm \infty$, the line $x = c$ is called a *vertical asymptote* of the graph of the function f.

(ii) $\lim_{x \to \pm \infty} f(x) = L$ (a real number) then the line $y = L$ is called a *horizontal asymptote* of the graph of f.

(iii) If $\lim_{x \to \pm \infty} \{ f(x) - (ax + b) \} = 0$ for some a, b ε P then the line $y = ax + b$ is called an *oblique asymptote* of the graph of the function f.

§ 3. Some important limits

Theorem 3.1. $\lim_{x \to 0} (\sin x) / x = 1.$ (3.1)

Corollary 3.1. The quotient function $(\sin x) / x$ behaves as follows:

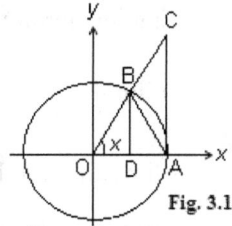

Fig. 3.1

(i) $\lim_{x \to \pi/2} (\sin x)/x = 2/\pi < 1,$ **(ii)** $\lim_{x \to \pi} (\sin x) / x = 0,$

(iii) $\lim_{x \to 3\pi/2} (\sin x)/x = (2/3\pi). \{ - \sin (\pi/2) \} = - 2/3\pi,$

(iv) $\lim_{x \to 2\pi} (\sin x) / x = 0,$ **(v)** $\lim_{x \to -\pi/2} (\sin x) /x = 2/\pi < 1,$

(vi) $\lim_{x \to -\pi} (\sin x) / x = 0,$ **(vii)** $\lim_{x \to -2\pi} (\sin x)/x = 0,$

(viii) $\lim_{x \to -3\pi/2} (\sin x) / x = - (2/3\pi) \sin (\pi/2) = - 2/3\pi.$

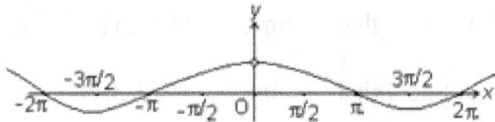

Fig. 3.2

Theorem 3.2. $\lim_{x \to \pm \infty} (1 + 1/x)^x = e.$ (3.2)

Proof. Expanding $(1 + 1/x)^x$ by binomial theorem:

$$(1 + y)^n = 1 + ny/1! + n (n - 1)y^2/2! + n(n - 1)(n - 2)y^3/3! + \ldots \quad (3.3)$$

we get

$$(1 + 1/x)^x = 1 + 1/1! + (1 - 1/x) /2! + (1 - 1/x) (1 - 2/x) /3! + \ldots$$

Finally, taking limit as $x \to \pm \infty$ above expansion yields

$$\lim_{x \to \pm \infty} (1 + 1/x)^x = 1 + 1/1! + 1/2! + 1/3! + \ldots \quad \infty = e. //$$

Corollary 3.2. $\lim_{y \to 0} (1 + y)^{1/y} = e.$ (3.4)

Proof. Replacing x by $1/y$ so that $1/x = y$ in Eq. (3. 2), we derive the result. //

3.1. Properties of $f(x) = (1 + 1/x)^x$

(i) We note that $f(0) = (1+1/0)^0 = \infty^0$, being indeterminate, does not exist.

(ii) $\lim_{x \to -1^-} f(x)$

$y = e$

$x = -1$ **Fig. 3.3**

$= \lim_{\varepsilon \to 0} \{1 + 1/(-1-\varepsilon)\}^{(-1-\varepsilon)}$

$= \lim_{\varepsilon \to 0} \{1 - 1/(1+\varepsilon)\}^{-(1+\varepsilon)} = \lim_{\varepsilon \to 0} \{(1+\varepsilon)/\varepsilon\}^{1+\varepsilon} = \infty.$

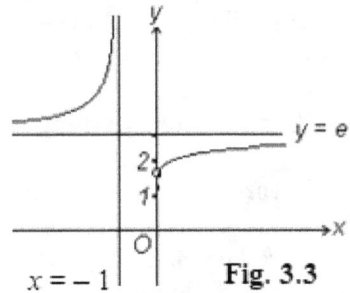

Thus, $x = -1$ is a vertical asymptote of the graph of $y = f(x)$.

(iii) $\lim_{x \to -1^+} f(x) = \lim_{\varepsilon \to 0} \{1 + 1/(-1+\varepsilon)\}^{(-1+\varepsilon)}$

$= \lim_{\varepsilon \to 0} \{\varepsilon/(-1+\varepsilon)\}^{-(1-\varepsilon)} = \lim_{\varepsilon \to 0} \{(-1+\varepsilon)/\varepsilon\}^{1-\varepsilon} = (-\infty)^{<1}$

does not exist.

Example 3.1. Show that $\quad \lim_{x \to \infty} (1 + 1/x)^{x+3} = e.$ (3.5)

Example 3.2. Show that $\quad \lim_{x \to \infty} (1 + 3/x)^x = e^3.$ (3.6)

Example 3.3. Show that $\quad \lim_{x \to \infty} \{x/(1+x)\}^x = 1/e.$ (3.7)

§ 4. Continuity of a function

Let $f(x)$ be a real valued function of a real variable x defined in some domain $a \le x \le b$ and c is any point in the domain. Let the limit of the function when $x \to c$ be denoted by some real number l. There arise the following cases:

(i) The limit exists and equals the value of the function $f(c)$ at the point $x = c$. In such a case, the function is said to be *continuous* at $x = c$.

(ii) The limit exists but does not equal to $f(c)$ then the function is said to be discontinuous at $x = c$.

(iii) The limit exists but the function is not defined at the point $x = c$.

(iv) The limit does not exist but the value $f(c)$ is defined at $x = c$.

(v) Neither the limit exists nor the value of the function is defined at $x = c$.

Example 4.1. The function $f(x) = 2x$ is continuous at $x = 3$.

Solution. The value of the function at $x = 3$ is 6 which equals the limit:

$$\lim_{x \to 3} f(x) = \lim_{x \to 3} (2x) = 6,$$

proving that the function is continuous at $x = 3$. //

Example 4.2. The function $f(x) = x$ (if $x \neq 2$), 1 (if $x = 2$) is not continuous at $x = 2$.

Solution. $\lim_{x \to 2} f(x) = \lim_{x \to 2} f(x) = 2$, while $f(2) = 1$. Therefore, the function is discontinuous. //

Example 4.3. Check the continuity of the function

$$f(x) = (x^2 - 1) / (x - 1) \qquad \text{at} \qquad x = 1.$$

Solution. $(x^2 - 1) / (x - 1) = (x - 1)(x + 1) / (x - 1) = (x + 1)$, if $x \neq 1$.

Therefore,

$$\lim_{x \to 1^-} f(x) = \lim_{x \to 1^+} f(x) = 1/2,$$

but $f(1)$ is not defined as it becomes indeterminate. Hence, the function is not continuous at $x = 1$. //

Example 4.4. Check the continuity of the function $f(x) = 2$ (if $x < 2$), 1 (if $x \geq 2$) at $x = 2$.

Solution. $\lim_{x \to 2^-} f(x) = 2$, while $\lim_{x \to 2^+} f(x) = 1$. These limits being unequal, we conclude that $\lim_{x \to 2} f(x)$ does not exist so its comparison with $f(2) = 1$ cannot be made. Eventually, the function cannot be continuous. //

Example 4.5. Check the continuity of the function $f(x) = |x| / x$ at $x = 0$.

Solution. $\lim_{x \to 0^-} f(x) = \lim_{\varepsilon \to 0} |-\varepsilon| / (-\varepsilon)$

$$= \lim_{\varepsilon \to 0} \varepsilon / (-\varepsilon) = -1;$$

while

$$\lim_{x \to 0^+} f(x) = \lim_{\varepsilon \to 0} |\varepsilon| / \varepsilon = \lim_{\varepsilon \to 0} \varepsilon / \varepsilon = 1.$$

These limits being unequal, we conclude that $\lim_{x \to 0} f(x)$ does not exist. Also, $f(0) = |0| / 0$, being indeterminate is not defined. Hence, the function is not continuous. //

Note 4.1. A continuous function is the one whose graph is without any break, gap, hole or jump.

Example 4.6. The function $f(x) = (x^2 - x - 2) / (x - 2)$ is not continuous at $x = 2$. On the other hand, the following function is continuous at $x = 2$:

$$f(x) = (x^2 - x - 2) / (x - 2), \text{ if } x \neq 2; \qquad \text{but } f(2) = 3,$$

Solution. (i) $\lim_{x \to 2} f(x) = \lim_{x \to 2} (x - 2)(x + 1) / (x - 2)$

$$= \lim_{x \to 2} (x + 1) = 3;$$

but $f(2)$ being indeterminate is not defined. So, the function is not continuous.

(ii) In the second case, the value of the function is already given as 3 which is same as the limit of the function. This makes the function continuous. //

Example 4.7. Show that the exponential function e^x is continuous everywhere in the set of real numbers; while the logarithmic function ln x (also denoted by ln x) is continuous at any point in the set of positive real numbers only.

Solution. (i) $\lim_{x \to a} f(x) = e^a = f(a)$ for any real number a. This makes the function e^x continuous at any real point $x = a$.

(ii) On the other hand, $\lim_{x \to a} (\ln x) = \ln a$, which is real when a is positive real number. So, the function becomes continuous at any point in the set of positive real numbers. //

§ 5. Some properties of continuous functions

Theorem 5.1. Let $f(x)$ and $g(x)$ be two functions defined in a common domain and both be continuous at some point $x = c$ of the domain, then

    (i) kf (for any real number k),   **(ii)** $f \pm g$,   **(iii)** fg

are also continuous at $x = c$; but,

    (iv) f/g is continuous at $x = c$ only when $g(c) \neq 0$.

Theorem 5.2. Every polynomial $P(x)$ is continuous in the interval $(-\infty, \infty)$.

Theorem 5.3. Every rational function $R(x) = P(x) / Q(x)$ is continuous in the interval $(-\infty, \infty)$ except at the points where $Q(x)$ is zero.

Definition 5.1. (*Substitution rule*) Let $f(x)$ be continuous at a point $x = l$ and $g(x)$ be a function such that $\lim_{x \to c} g(x) = l$, then

$$\lim_{x \to c} f\{g(x)\} = f\{\lim_{x \to c} g(x)\} = f(l);$$

i.e. if $g(x)$ is continuous at $x = c$ and f is also continuous at $x = g(c)$ then $f\{g(x)\}$ is continuous at $x = c$.

Theorem 5.4. (*Intermediate value theorem*) If $f(x)$ is continuous on a closed interval $[a, b]$ then $\forall y_0$ within $(a, b) \exists$ a point x_0 in (a, b) such that $f(x_0) = y_0$.

Example 5.1. Determine α so that the function

$$f(x) = \alpha x^2 + 2 \ \text{(if } x \geq 3) \text{ and } 2\alpha x + 11 \ \text{(if } x < 3)$$

is continuous at $x = 3$.

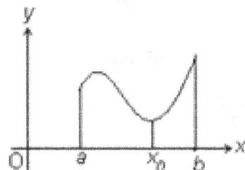
Fig. 5.1

Solution. $\lim_{x \to 3^+} f(x) = \lim_{x \to 3^+} (\alpha x^2 + 2) = 9\alpha + 2,$
while
$$\lim_{x \to 3^-} f(x) = \lim_{x \to 3^-} (2\alpha x + 11) = 6\alpha + 11.$$

As the function is continuous, both these limits must be equal: $9\alpha + 2 = 6\alpha + 11$ determining $\alpha = 3$. Thus,

$$\lim_{x \to 3} f(x) = \lim_{x \to 3} (\alpha x^2 + 2) = 9\alpha + 2 = 9 \times 3 + 2 = 29,$$

together with

$$f(3) = \alpha\,(3)^2 + 2 = 3\,(3^2) + 2 = 29. \,//$$

Example 5.2. Show that the function $R\,(x) = (3x^5 + 2x^2 - 8x - 1) \,/\, (x^2 + 1)$ is continuous in $(-\infty, \infty)$.

Solution. Since $Q\,(x) \equiv x^2 + 1 \neq 0$ at any point on the real line $(-\infty, \infty)$, either of the polynomials

$$P\,(x) = 3x^5 + 2x^2 - 8x - 1 \quad \text{and} \quad Q\,(x) = (x^2 + 1)$$

are continuous on the real line. This makes $R\,(x)$ continuous on the real line. //

Example 5.3. Show that $\cos x$ is a continuous function.

Solution. For $\cos x = \sin\,(\pi/2 - x)$ for every real number x, and $\sin x$ being continuous by substitution rule:

$$\lim\nolimits_{x \to c}\,\cos x = \lim\nolimits_{x \to c}\,\sin\,(\pi/2 - x) = \sin\,\{\lim\nolimits_{x \to c}\,(\pi/2 - x)\}$$

$$= \sin\,(\pi/2 - c) = \cos c \quad \Rightarrow \quad \cos x \text{ is continuous at } x = c.$$

But, c being an arbitrary real number we have the statement. //

Example 5.4. Find k so that the following function is continuous at $x = 2$:

$$f(x) = x + 1 \text{ (when } x < 2), \quad x^2 + k\,x + 1 \text{ (when } x \geq 2)$$

Solution. For continuity, limits both from left and right should be equal:

$$\lim\nolimits_{x \to 2^-} f(x) = \lim\nolimits_{x \to 2^-} f(x+1) = 3;$$

$$\lim\nolimits_{x \to 2^+} f(x) = \lim\nolimits_{x \to 2^+} (x^2 + k\,x + 1) = 2k + 5;$$

i.e.

$$3 = 2k + 5 \quad \Rightarrow \quad k = -1.$$

Further, both limits should also be equal to the value of the function at

$$x = 2: \qquad f(2) = 2k + 5 = 2\,(-1) + 5 = 3. \,//$$

§ 6. One-sided continuity

A function $f(x)$ has been called *continuous* at a point $x = c$ when its limits from both left and right exist and are equal to each other as well as equal to the value of the function $f(c)$, that also exists at the point. However, in the following, we observe that certain functions possess limits either from left or right only but in either case they are equal to the value of the function. This gives rise to *one-sided continuity* of the functions.

Definition 6.1. A function is said to be continuous at $x = c$:

(i) *from left*, if $\qquad \lim_{x \to c^-} f(x) = f(c)$;

and

(ii) *from right*, if $\qquad \lim_{x \to c^+} f(x) = f(c)$.

Note 6.1. When a function is continuous at a point from both left and right it is simply called continuous at the point. Conversely, a continuous function at a point is continuous from both left and right.

Definition 6.2. If a function is continuous on an open interval (a, b), continuous from right at $x = a$, and continuous from left at $x = b$, then it is continuous on the closed interval $[a, b]$.

Example 6.1. The function $f(x) = \sqrt{(x-1)}$ is continuous at $x = 1$ from right only.

Solution. $\lim_{x \to 1^+} f(x) = \lim_{\varepsilon \to 0} f(1 + \varepsilon) = \lim_{\varepsilon \to 0} \sqrt{(1 + \varepsilon - 1)}$

$$= \lim_{\varepsilon \to 0} \sqrt{\varepsilon} = 0 = f(1).$$

So, the function is continuous from right at $x = 1$. On the other hand,

$$\lim_{x \to 1^-} f(x) = \lim_{\varepsilon \to 0} f(1 - \varepsilon) = \lim_{\varepsilon \to 0} \sqrt{(1 - \varepsilon - 1)} = \lim_{\varepsilon \to 0} \sqrt{(-\varepsilon)},$$

does not exist (as it is not real). So, the function is not continuous at $x = 1$ from left. //

Example 6.2. The function $f(x) = -1$ (when $x < 0$), 1 (when $x \geq 0$) is continuous at $x = 0$ from right only.

Solution. $\lim_{x \to 0^-} f(x) = -1 \neq f(0)$. But, $\lim_{x \to 0^+} f(x) = 1 = f(0)$.

Hence, we have the statement. //

Example 6.3. The function $f(x) = \sqrt{(4 - x^2)}$ is continuous:

(i) from left at $x = 2$, and **(ii)** from right at $x = -2$.

Solution. (i) $\lim_{x \to 2^-} f(x) = \lim_{\varepsilon \to 0} \sqrt{\{4 - (2 - \varepsilon)^2\}}$

$$= \lim_{\varepsilon \to 0} \sqrt{\{4\varepsilon - \varepsilon^2\}} = 0,$$

as ε being very small makes $4\varepsilon - \varepsilon^2$ positive. Also, same is the value of $f(2)$. Contrary to that

$$\lim_{x \to 2^+} f(x) = \lim_{\varepsilon \to 0} \sqrt{\{4 - (2 + \varepsilon)^2\}} = \lim_{\varepsilon \to 0} \sqrt{\{-4\varepsilon - \varepsilon^2\}}$$

does not exist (as it is not real). Hence, we have the statement.

(ii) $\lim_{x \to -2^-} f(x) = \lim_{\varepsilon \to 0} \sqrt{\{4 - (-2 - \varepsilon)^2\}} = \lim_{\varepsilon \to 0} \sqrt{\{-4\varepsilon - \varepsilon^2\}}$,

does not exist (as it is not real). However,

$$\lim_{x \to 2^+} f(x) = \lim_{\varepsilon \to 0} \sqrt{\{4 - (-2 + \varepsilon)^2\}} = \lim_{\varepsilon \to 0} \sqrt{\{4\varepsilon - \varepsilon^2\}} = 0,$$

does exist for the reasons stated above and equals the value of the function $f(-2)$. Hence, we have the statement. //

Note 6.2. Above example (of a semi-circular function) is continuous on the closed interval $[-2, 2]$ as per Definition 3.2.

Example 6.4. Check the continuity of the function $\tan x$ on the interval $[0, 2\pi]$.

Solution. (i) $\lim_{x \to 0^-} f(x) = \lim_{\varepsilon \to 0} \tan(-\varepsilon) = -\tan 0 = 0 = f(0)$.
Also,

$$\lim_{x \to 0^+} f(x) = \lim_{\varepsilon \to 0} \tan \varepsilon = \tan 0 = 0 = f(0).$$

Therefore, the function is continuous at $x = 0$.

(ii) $\lim_{x \to (2\pi)^-} f(x) = \lim_{\varepsilon \to 0} \tan(2\pi - \varepsilon) = \lim_{\varepsilon \to 0}(-\tan \varepsilon)$

$$= -\tan 0 = 0 = f(2\pi) = \tan 2\pi.$$

Also,

$$\lim_{x \to 2\pi^+} f(x) = \lim_{\varepsilon \to 0} \tan(2\pi + \varepsilon) = \lim_{\varepsilon \to 0} \tan \varepsilon$$

$$= \tan 0 = 0 = f(2\pi) = \tan 2\pi.$$

So, the function is continuous at $x = 2\pi$ also.

Note 6.3. We note in the following that the function has one-sided continuity at $x = \pi/2$ and $3\pi/2$.

(iii) $\lim_{x \to (\pi/2)^-} f(x) = \lim_{\varepsilon \to 0} \tan(\pi/2 - \varepsilon) = \lim_{\varepsilon \to 0} \cot \varepsilon$

$$= \cot 0 = \infty = f(\pi/2),$$

$$\lim_{x \to (\pi/2)^+} f(x) = \lim_{\varepsilon \to 0} \tan(\pi/2 + \varepsilon) = \lim_{\varepsilon \to 0}(-\cot \varepsilon)$$

$$= -\cot 0 = -\infty \neq f(\pi/2).$$

Hence, $\tan x$ is continuous at $x = \pi/2$ from left only.

(iv) $\lim_{x \to (3\pi/2)^-} f(x) = \lim_{\varepsilon \to 0} \tan(3\pi/2 - \varepsilon) = \lim_{\varepsilon \to 0} \cot \varepsilon = \infty,$

$$\lim_{x \to (3\pi/2)^+} f(x) = \lim_{\varepsilon \to 0} \tan(3\pi/2 + \varepsilon) = \lim_{\varepsilon \to 0}(-\cot \varepsilon) = -\infty,$$

and

$$\tan(3\pi/2) = \tan(\pi + \pi/2) = \tan \pi/2 = \infty.$$

Therefore, $\tan x$ is continuous at $x = 3\pi/2$ from left only. //

Example 6.5. The function $f(x) = \sqrt{x}$ is continuous on the closed interval $[0, 2]$.

Solution. (i) For any point c in the open interval $(0, 2)$, the limits both from left and right (of c) exist and are equal. They are also equal to the value of the function at $x = c$:

$$\lim_{x \to c^-} f(x) = \lim_{\varepsilon \to 0} \sqrt{(c - \varepsilon)} = \sqrt{c} = f(c),$$

and

$$\lim_{x \to c^+} f(x) = \lim_{\varepsilon \to 0} \sqrt{(c + \varepsilon)} = \sqrt{c} = f(c),$$

as $c > 0$. Hence, the function is continuous at $x = c$. The point c being arbitrarily chosen in (0, 2), the function is continuous throughout the open interval.

(ii) Secondly, $\lim_{x \to 0^+} f(x) = \lim_{\varepsilon \to 0} \sqrt{(\varepsilon)} = 0 = f(0)$ exists and equals the value of the function; but, $\lim_{x \to 0^-} f(x) = \lim_{\varepsilon \to 0} \sqrt{(-\varepsilon)}$ does not exist. So, the function is continuous at $x = 0$ from right only.

(iii) Finally, both

$$\lim_{x \to 2^-} f(x) = \lim_{\varepsilon \to 0} \sqrt{(2 - \varepsilon)} = \sqrt{2} = f(2)$$

and

$$\lim_{x \to 2^+} f(x) = \lim_{\varepsilon \to 0} \sqrt{(2 + \varepsilon)} = \sqrt{2} = f(2)$$

exist and are equal to each other as well as to the value of the function at $x = 2$. Hence, the function is continuous (from both left and right) at $x = 2$.

Thus, all the requirements of Definition 6.2 are fulfilled, making the function continuous on [0, 2]. //

Example 6.6. Show that the following function is continuous on the clopen interval [0, ∞):

$$f(x) = 5 - x \text{ (when } x \geq 0), \quad x + 1 \text{ (when } x < 0).$$

§ 7. Differentiation of functions

Let $y = f(x)$ be a real valued continuous function of a real variable x. In general, it represents a curve C in the xOy-plane. Let P (x, y) and Q $(x + \delta x, y + \delta y)$ be two neighbouring points on the curve. As P moves to Q along C its abscissa x experiences an infinitesimal change (indeed, increase as per the adjacent figure) PL = AB = OB − OA = δx. Correspondingly the ordinate of P changes by LQ = BQ − BL = BQ − AP = δy. The limit of the ratio $\delta y/\delta x$ as $\delta x \to 0$ defines the *derivative* (or *differential coefficient*) of y with respect to x. If is denoted by

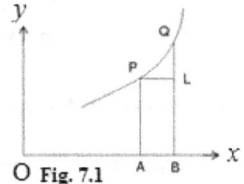

$$dy / dx \equiv \lim_{\delta x \to 0} (\delta y / \delta x) = df / dx \equiv f'(x). \tag{7.1}$$

Example 7.1. Velocity of a particle is the derivative of the displacement of the particle with regard to time.

Example 7.2. Acceleration of a particle is the derivative of its velocity with respect to time.

Definition 7.1. Let $f(x)$ be a function of a variable x, continuous at some point x_0. It is said to be *differentiable* at x_0 if the limit

$$\lim_{x \to x_0} \{f(x) - f(x_0)\} / (x - x_0)$$

$$\equiv \lim_{h \to 0} \{f(x_0 + h) - f(x_0)\} / h \equiv l \text{ (say)} \qquad (7.2)$$

exists. As in Eq. (7.1), it is then denoted by $\{df / dx\}_x$ or $f'(x_0)$ or $Df(x_0)$.

This definition of derivative of a function is also called the *differentiation by first principle*.

Example 7.3. The identity function $f(x) = x$ is differentiable in $(-\infty, \infty)$.
Solution. $\lim_{x \to x_0} (x - x_0)/(x - x_0) = 1$ exists. Hence, the statement. //

Example 7.4. The *sine* function $\sin x$ is differentiable.

Solution. The limit $\lim_{h \to 0} \{\sin(x_0 + h) - \sin(x_0)\} / h$

$$= \lim_{h \to 0} 2.\cos(x_0 + h/2).\{\sin(h/2)\} / h = \cos x_0,$$

exists, by § 3. Hence, the statement. //

Example 7.5. The function $f(x) = |x - 1|$ is not differentiable at $x = 1$.
Solution. The mod function, being always positive, satisfies

$$f(x) = x - 1 \text{ (if } x \geq 1); \text{ or, } 1 - x \text{ (if } x < 1), \text{ and } f(1) = 0. \text{ (7.3)}$$

Hence,

$$\lim_{x \to 1^-} \{f(x) - f(1)\}/(x - 1) = \lim_{x \to 1^-} \{1 - x - 0\}/(x - 1) = -1,$$

while

$$\lim\nolimits_{x\to1^+} \{f(x) - f(1)\}/(x-1) = \lim\nolimits_{x\to1^+} \{x - 1 - 0\}/(x-1) = 1.$$

The two limits being different imply that $\lim_{x\to1}\{f(x) - f(1)\}/(x-1)$ does not exist proving the statement. //

Example 7.6. Check if the following function is differentiable at $x = 1$:

$$f(x) = \sqrt{x} \text{ (if } 0 \le x \le 1); \text{ or, } (x+1)/2 \text{ (if } x > 1).$$

Solution. For $f(1) = \sqrt{1} = 1$,

$$\lim\nolimits_{x\to1^-} \{f(x) - f(1)\}/(x-1) = \lim\nolimits_{x\to1^-} (\sqrt{x} - 1)/(x-1)$$

$$= \lim\nolimits_{x\to1^-} 1/(\sqrt{x} + 1) = 1/2,$$

and

$$\lim\nolimits_{x\to1^+} \{f(x) - f(1)\}/(x-1) = \lim\nolimits_{x\to1^+} \{(x+1)/2 - 1\}/(x-1)$$

$$= \lim\nolimits_{x\to1^+} \lim\nolimits_{x\to1^+} (x-1)/2(x-1) = 1/2.$$

Two limits being equal imply that $\lim_{x\to1} \{f(x) - f(1)\}/(x-1)$ exists making the function differentiable at $x = 1$. //

Example 7.7. Show that the following function is differentiable at $x = 0$:

$$f(x) = x \text{ (if } x \le 0); \text{ or, } \sin x \text{ (if } x > 0).$$

Solution. For $f(0) = 0$ and Theorem 1.3.1,

$$\lim\nolimits_{x\to0^-} \{f(x) - f(0)\}/(x-0) = \lim\nolimits_{x\to0} (x-0)/x = 1,$$

and

$$\lim\nolimits_{x\to0^+} \{f(x) - f(0)\}/(x-0) = \lim\nolimits_{x\to0} (\sin x - 0)/x = 1.$$

Two limits being equal imply that $\lim_{x\to0} \{f(x) - f(0)\}/(x-1)$ exists making the function differentiable at $x = 0$. //

§ 8. Continuity of a differentiable function

Theorem 8.1. If a function $f(x)$ is differentiable at a point $x = a$ then it is also continuous at the point.

Note 8.1. The converse of above theorem need not be true.

The following example deals with a function which is *continuous but not differentiable*.

Example 8.1. The function $f(x) = |x|$ is continuous at $x = 0$ but not differentiable at the point.

Solution. The absolute value, being always positive, we have

$$f(x) = x \ (\text{if } x \geq 0); \ \text{or,} \ -x \ (\text{if } x < 0), \ \text{and} \ f(0) = 0.$$

The value of the function and its limits: both from left and right at $x = 0$:

$$\lim_{x \to 0^-} f(x) = \lim_{x \to 0} (-x) = 0, \ \lim_{x \to 0^+} f(x) = \lim_{x \to 0} (x) = 0,$$

being equal establish the continuity of the function. On the other hand,

$$\lim_{x \to 0^-} \{f(x) - f(0)\} / (x - 0) = \lim_{x \to 0^-} (-x - 0) / (x - 0) = -1,$$

and

$$\lim_{x \to 0^+} \{f(x) - f(0)\} / (x - 0) = \lim_{x \to 0^+} (x - 0) / (x - 0) = 1,$$

being different imply that $\lim_{x \to 0} \{f(x) - f(0)\} / (x - 0)$ does not exist hence the function is not differentiable at $x = 0$. //

Example 8.2. Find a function which is continuous at $x = 3$ but not differentiable at the point.

Solution. Let us consider the function

$$f(x) = |x - 3| = x - 3 \ (\text{if } x \geq 3); \ \text{or,} \ -x + 3 \ (\text{if } x < 3), \ \text{and} \ f(3) = 0.$$

Its limits as $x \to 3$ from left and right:

$$\lim_{x \to 3^-} f(x) = \lim_{x \to 3^-} (-x + 3) = 0,$$

$$\lim_{x \to 3^+} f(x) = \lim_{x \to 3^+} (x - 3) = 0,$$

are equal to each other as well as to the value of the function. Therefore, the function is continuous at the point $x = 3$. On the other hand, the limits:

$$\lim_{x\to 3^-} \{f(x) - f(3)\}/(x - 3) = \lim_{x\to 3^-} (-x + 3 - 0)/(x - 3) = -1,$$

and

$$\lim_{x\to 3^+} \{f(x) - f(3)\}/(x - 3) = \lim_{x\to 3^+} (x - 3 - 0)/(x - 3) = 1,$$

being different imply that $\lim_{x\to 3} \{f(x) - f(3)\}/(x - 3)$ does not exist hence the function is not differentiable at $x = 3$. //

Theorem 8.2. (*Algebraic properties of derivatives*) Let $f(x)$ and $g(x)$ be two functions both derivable at the point x then there hold the results:

(i) Additive property: $(f + g)'(x) = f'(x) + g'(x)$,   (8.1)

(ii) Multiplicative property:

$$(f.g)'(x) = f'(x).g(x) + f(x).g'(x),   (8.2)$$

(iii) Division property:

$$(f/g)'(x) = \{f'(x). g(x) - f(x). g'(x)\} / \{g(x)\}^2,   (8.3)$$

if $g(x) \neq 0$.

Corollary 8.1. For a constant (i.e. a fixed real number) c and a function $f(x)$ there holds

$$(cf)'(x) = cf'(x).   (8.4)$$

§ 9. Derivatives of some standard functions

9.1. Algebraic functions: Presently, we consider the derivatives of the functions expressible as polynomials of certain degree in a real variable x.

Theorem 9.1. $dx^n / dx = n. x^{n-1}$ (9.1)

Corollary 9.1. The derivatives of an identity function $f(x) = x$, $\sqrt{x}$ and of a constant c are given by

(i) $(x)' = 1$, **(ii)** $(\sqrt{x})' = 1/2\sqrt{x}$,   **(iii)** $(c)' = 0$.   (9.2)

Corollary 9.2. The derivative of a polynomial

$$P(x) = c_n x^n + c_{n-1} x^{n-1} + \ldots + c_1 x + c_0 \qquad (9.3)$$

with respect to x is

$$P'(x) = n\,c_n x^{n-1} + (n-1)\,c_{n-1} x^{n-2} + \ldots + 2c_2 x + c_1. \qquad (9.4)$$

Example 9.1. Find the derivative of $(ax + b)^n$.

Solution. Putting $ax + b = t$ and differentiating it with respect to t we get $a = dt/dx$. Hence, by (3.1),

$$d(ax+b)^n / dx = d\,t^n / dx = (d\,t^n / dt)(dt/dx) = (n\,t^{n-1})\,a$$

$$= n\,a\,(ax+b)^{n-1}.\ // \qquad (9.5)$$

Note 9.1. Above derivation is due to *method of substitution*.

9.2. Exponential and logarithmic functions: We consider the derivatives of the functions e^x, $\ln x$, $\log_a x$ and a^x in this Sub-section.

Theorem 9.2. We have

$$(e^x)' = e^x \qquad (9.6); \qquad (\ln x)' = 1/x \qquad (9.7);$$

$$(\log_a x)' = 1/x . \log_e a. \qquad (9.8)$$

Proof. (i) By first principle,

$$(e^x)' = \lim_{h \to 0} (e^{x+h} - e^x)/h = e^x. \lim_{h \to 0} (e^h - 1)/h.$$

Expanding e^h in powers of h:

$$e^h = 1 + h/1! + h^2/2! + \ldots \infty, \qquad (9.9)$$

above limit simplifies to as in Eq. (3.6).

(ii) By first principle,

$$(\ln x)' = \lim_{h \to 0} \{\ln(x+h) - \ln x\}/h$$

$$= \lim_{h \to 0} \{\ln(1 + h/x)\}/h. \qquad (9.10)$$

Expanding the log function

$$\ln(1 + h/x) = h/x - (h/x)^2/2 + (h/x)^3/3 + \dots \infty \qquad (9.11)$$

and simplifying the limit, one gets the result.

Alternately, putting $h/x = t$, we get $h = t\,x$. Therefore, $h \to 0 \Rightarrow t \to 0$, the limit in Eq. (9.10) reduces to

$$\lim_{t\to 0} (1/x)\{\ln(1 + t)^{1/t}\} = (1/x).\ln\{\lim_{t\to 0}(1 + t)^{1/t}\}$$

$$= (1/x).\ln(e) = 1/x, \quad \text{by Eq. (3.4).}$$

(iii) Proceeding as above and putting $h/x = t$, or $h = t\,x$, $h \to 0 \Rightarrow t \to 0$, we get

$$(\log_a x)' = \lim_{h\to 0}\{\log_a(x + h) - \log_a x\}/h$$

$$= \lim_{h\to 0}\{\log_a(1 + h/x)\}/h = (1/x).\log_a\{\lim_{t\to 0}(1 + t)^{1/t}\}$$

$$= (1/x).\log_a(e) = 1/x.\ln(a).$$

Alternately, changing the base from a to e:

$$\log_a x = (\ln x)(\log_a e) = (\ln x)/(\ln a), \qquad (9.12)$$

differentiating it with respect to x, and using Eq. (9.7), we obtain Eq. (9.8). //

Corollary 9.3. $d\,a^x/dx = a^x(\ln a). \qquad\qquad\qquad\qquad (9.13)$

Corollary 9.4. $d\,e^{ax}/dx = a.\,e^{ax}. \qquad\qquad\qquad\qquad (9.14)$

9.3. Trigonometric functions: Presently we discuss derivation of (circular) trigonometric functions: $\sin x$, $\cos x$, $\tan x$, $\sec x$ and $\operatorname{cosec} x$ with respect to x.

Theorem 9.3. The basic trigonometric functions have following derivatives:

$(\sin x)' = \cos x \qquad (9.15);\qquad (\cos x)' = -\sin x, \qquad\qquad (9.16)$

$(\tan x)' = \sec^2 x \qquad (9.17);\qquad (\cot x)' = -\operatorname{cosec}^2 x; \qquad (9.18)$

$(\sec x)' = \sec x.\tan x \quad (9.19);\quad (\operatorname{cosec} x)' = -\operatorname{cosec} x.\cot x. \qquad (9.20)$

9.4. Hyperbolic functions: The hyperbolic *sine* and *cosine* functions are defined by

$$\sinh x = (e^x - e^{-x}) / 2, \tag{9.21}$$

$$\cosh x = (e^x + e^{-x}) / 2. \tag{9.22}$$

Other hyperbolic functions: tanh x, coth x, sech x and cosech x are defined similarly as their (circular) counterparts. These functions satisfy the following identities:

$$\cosh^2 x - \sinh^2 x = 1; \tag{9.23}$$

$$\cosh^2 x + \sinh^2 x = \cosh 2x; \tag{9.24}$$

$$1 - \tanh^2 x = \mathrm{sech}^2 x \quad (9.25); \quad \coth^2 x - 1 = \mathrm{cosech}^2 x; \quad (9.26)$$

Theorem 9.4. The derivatives of hyperbolic functions are given by:

$$(\sinh x)' = \cosh x \qquad (9.27); \qquad (\cosh x)' = \sinh x, \qquad (9.28)$$

$$(\tanh x)' = \mathrm{sech}^2 x \qquad (9.29); \qquad (\coth x)' = -\mathrm{cosech}^2 x; \quad (9.30)$$

$$(\mathrm{sech}\, x)' = -\,\mathrm{sech}\, x.\tanh x; \tag{9.31}$$

$$(\mathrm{cosech}\, x)' = -\,\mathrm{cosech}\, x.\coth x. \tag{9.32}$$

§ 10. Chain rule (*methods of substitution*)

Theorem 10.1. Let f and g be two continuous functions such that g (x) is differentiable w.r.t. x and $f\{g(x)\}$ is differentiable w.r.t. $g(x)$, then

$$[f\{g(x)\}]' = f'\{g(x)\}.\, g'(x), \tag{10.1}$$

where prime (') over f denotes its derivative w.r.t. $g(x)$.

Note 10.1. The rule given by Eq. (10.1) can also be generalized as

$$Df[g\{h(x)\}] = Df\{g(h)\}.D\{g(h).\,Dh(x). \tag{10.2}$$

10.1. Logarithmic differentiation

Let $f(x)$ and $g(x)$ be two functions of a variable x: both differentiable with respect to x. They define another function

$$F(x) = \{f(x)\}^{g(x)}. \tag{10.3}$$

To obtain its derivative, first we evaluate its natural logarithm:

$$\ln\{F(x)\} = g(x).\ln\{f(x)\}, \tag{10.4}$$

and differentiate it with respect to x:

$$\{F'(x)\}/F(x) = g'(x).\ln\{f(x)\} + g(x).\{f'(x)\}/f(x)$$

$$\Rightarrow$$

$$F'(x)\} = F(x).[g'(x).\ln\{f(x)\} + g(x).\{f'(x)\}/f(x)]. \tag{10.5}$$

Example 10.1. If $f(x) = \cos x$, $g(x) = x^3$, find $[f\{g(x)\}]'$.

Solution. $[f\{g(x)\}]' = \{\cos x^3\}' = (d\cos t)/dx = \{(d\cos t)/dt\}(dt/dx)$,

where $t = x^3$ so that $dt/dx = 3x^2$. Putting from Eq. (9.16), we get the result

$$(-\sin t).3x^2 = -3x^2.\sin x^3. //$$

Example 10.2. If $f(x) = \operatorname{cosec} x$, $g(x) = 1 + x^2$, find $[f\{g(x)\}]'$.

Solution. $[f\{g(x)\}]' = \{\operatorname{cosec}(1+x^2)\}' = (d\operatorname{cosec} t)/dx$

$$= \{(d\operatorname{cosec} t)/dt\}(dt/dx),$$

where $t = 1 + x^2$ so that $dt/dx = 2x$. Putting from (3.20), we get the result:

$$(-\operatorname{cosec} t.\cot t).2x = -2x.\operatorname{cosec}(1+x^2).\cot(1+x^2). //$$

§ 11. Derivation of inverse trigonometric functions

The inverse (circular) trigonometric functions are:

$$\sin^{-1}(x),\ \cos^{-1}(x),\ \tan^{-1}(x),\ \cot^{-1}(x),\ \sec^{-1}(x) \text{ and } \operatorname{cosec}^{-1}(x).$$

Theorem 11.1. We have

$$(\sin^{-1}x)' = 1/\sqrt{(1-x^2)} \quad (11.1); \quad (\cos^{-1}x)' = -1/\sqrt{(1-x^2)}; \quad (11.2)$$

$(\tan^{-1}x)' = 1/(1+x^2)$ (11.3); $(\cot^{-1}x)' = -1/(1+x^2)$; (11.4)

$(\sec^{-1}x)' = 1/x\sqrt{(x^2-1)}$ (11.5); $(\csc^{-1}x)' = -1/x\sqrt{(x^2-1)}$. (11.6)

Theorem 11.2. The derivatives of inverse hyperbolic functions are given by:

$(\sinh^{-1}x)' = 1/\sqrt{(1+x^2)}$ (11.7); $(\cosh^{-1}x)' = 1/\sqrt{(x^2-1)}$; (11.8)

$(\tanh^{-1}x)' = 1/(1-x^2)$ (11.9); $(\coth^{-1}x)' = 1/(1-x^2)$; (11.10)

$$(\operatorname{sech}^{-1}x)' = -1/x\sqrt{(1-x^2)};$$ (11.11)

$$(\operatorname{cosech}^{-1}x)' = -1/x\sqrt{(1+x^2)}.$$ (11.12)

§ 12. Implicit functions and their derivation

So far we have considered functions $f(x)$ expressed in terms of a variable x. As such, they are called *explicit functions* of the variable. On the other hand, when an equation involving two variables, say x and y, is given which is not soluble for y in terms of x:

$$f(x, y) \equiv ax^2 + by^2 + 2hx\,y + 2gx + 2fy + c = 0.$$ (12.1)

Then y is called an *implicit function* of the variable x.

Differentiating above equation with respect to x, we obtain

$$2ax + 2byy' + 2h(y + xy') + 2g + 2fy' = 0,$$

or

$$(hx + by + f)\,y' + (ax + hy + g) = 0$$

$\Rightarrow$

$$y' = -(ax + hy + g)/(hx + by + f).$$

Note 12.1. When both the variables x and y are expressed as functions of a single parameter, say t, the value of dy/dx is found by the formula

$$dy/dx = (dy/dt)/(dx/dt).$$ (12.2)

§ 13. Successive differentiation

Let $f(x)$ be a continuous function of a variable x which is differentiable with respect to x at a number of times. These higher order derivatives of the function are called *successive derivatives*. Given any positive integer n we consider the nth order differentiation of some functions especially algebraic, exponential, logarithmic, trigonometric and their combinations in this chapter.

Theorem 13.1. For any real number m, the nth derivative of $(ax + b)^m$ with respect to x is given by

$$D^n (ax + b)^m = m (m-1) (m-2) \ldots (m-n+1).a^n (ax + b)^{m-n}. \quad (13.1)$$

Corollary 13.1. $D^n (ax + b)^{-1} = (-1)^n .n!.a^n / (ax + b)^{n+1}. \qquad (13.2)$

Corollary 13.2. $D^n x^n = n!. \qquad\qquad\qquad (13.3)$

Corollary 13.3. $D^n \ln x = (-1)^{n-1}.(n-1)! / x^n. \qquad\qquad (13.4)$

Proof. The first order derivative of $\ln x$ is given by Eq. (9.7). Differentiating the same further by $n - 1$ times and applying Eq. (13.2) we get the desired result. //

Theorem 13.2. $D^n (e^{ax+b}) = a^n. e^{ax+b}. \qquad\qquad (13.5)$

Theorem 13.3. $D^n (a^{mx}) = m^n. (\ln a)^n. a^{mx}. \qquad\qquad (13.6)$

Theorem 13.4. $D^n \sin (bx + c) = b^n \sin (bx + c + n\pi/2), \qquad (13.7)$
and
$$D^n \cos (bx + c) = b^n \cos (bx + c + n\pi/2). \qquad (13.8)$$

Proof. (i) Putting $bx + c = y$ so that $b = dy / dx$, the first order derivative of the function can be obtained by the chain rule:

$$D \sin (bx + c) = b. \cos (bx + c) = b. \sin (bx + c + \pi/2).$$

We notice that the effect of the differential operator D on the *sine* function increases its angle by $\pi/2$. Differentiating it once more and following the similar method we derive

$$D^2 \sin (bx + c) = b^2 \cos (bx + c + \pi/2) = b^2 \sin (bx + c + \pi/2 + \pi/2).$$

Continuing the process of differentiation further by $n - 2$ times we derive Eq. (13.7).

(ii) Following the same substitution the first order derivative of cos $(bx + c)$ is

$$D\cos (bx + c) = - b.\sin (bx + c) = b.\cos (bx + c + \pi/2).$$

Thus, the effect of the differential operator D on the *cosine* function also increases its angle by $\pi/2$. Differentiating it further by $n - 1$ times more and following the similar method we derive (1.8). //

§ 14. Leibnitz theorem

Theorem 14.1. The nth order derivative of the product of two functions is given by

$$D^n (uv) = (D^n u)\, v + {}^n C_1 (D^{n-1} u)\,(Dv) + {}^n C_2 (D^{n-2} u)\,(D^2 v) + \dots$$

$$+ {}^n C_r (D^{n-r} u)\,(D^r v) + \dots + u\,(D^n v). \tag{14.1}$$

Proof. (i) For $n = 1$, the theorem is already seen true vide Eq. (8.2), i.e.

$$D\,(u\,v) = (D u)\, v + u.\, Dv. \tag{14.2}$$

Operating it again by D, we get

$$D^2 (u\,v) = (D^2 u)\, v + 2\,(D u)\,(Dv) + u.\, D^2 v, \tag{14.3}$$

which is same as Eq. (14.1) for $n = 2$. Thus, the theorem holds for $n = 2$.

(ii) Next, let the theorem hold for some lesser value of n, say m, i.e.

$$D^m (u\,v) = (D^m u)\, v + {}^m C_1 (D^{m-1} u)\,(Dv) + \dots$$

$$+ {}^m C_r (D^{m-r} u)\,(D^r v) + \dots + u\,(D^m v). \tag{14.4}$$

Differentiating it further w.r.t. x and applying Eq. (14.2), we derive

$$D^{m+1} (u\,v) = (D^{m+1} u)\, v + (1 + {}^m C_1)\,(D^m u)\,(D v) + \dots$$

$$+ ({}^m C_{r-1} + {}^m C_r)\,(D^{m-r+1} u)\,(D^r v) + \dots$$

$$+ ({}^m C_{m-1} + 1)\,(D u)\,(D^m v) + u\,(D^{m+1} v).$$

Using the properties of binomial coefficients (cf. Chapter 1), above equation reduces to

$$D^{m+1}(u\,v) = (D^{m+1}u)\,v + {}^{m+1}C_1\,(D^m\,u)\,(D\,v) + \ldots$$

$$+ {}^{m+1}C_r\,(D^{m+1-r}\,u)\,(D^r v) + \ldots + {}^{m+1}C_m\,(D\,u)\,(D^m v) + u\,(D^{m+1}v),$$

which is in agreement with Eq. (14.1) for $n = m + 1$. Thus, the theorem is also true for $n = m + 1$. Combining the conclusions expressed in above two parts we, therefore, conclude that the theorem holds for any positive integral value of n. //

Theorem 14.2. $D^n\,\{e^{ax}.\sin(bx+c)\} = r^n.e^{ax}.\sin(bx+c+n\varphi)$,   (14.5)
and

$$D^n\,\{e^{ax}.\cos(bx+c)\} = r^n.\,e^{ax}.\,\cos(bx+c+n\varphi); \qquad (14.6)$$

where

$$r = \surd(a^2+b^2) \qquad (14.7); \qquad \varphi = \tan^{-1}(b/a). \qquad (14.8)$$

Proof. (i) The first order derivative of the product $e^{ax}.\sin(bx+c)$, for Eqs. (9.14), (9.15) and (14.2), is

$$D\,\{e^{ax}.\sin(bx+c)\} = e^{ax}\,\{a.\sin(bx+c) + b.\cos(bx+c)\}.$$

Putting

$$a = r\cos\varphi \qquad \text{and} \qquad b = r\sin\varphi, \qquad (14.9)$$

so that there hold Eqs. (14.7) and (14.8), above equation reduces to

$$D\,\{e^{ax}.\sin(bx+c)\} = r.\,e^{ax}.\{\sin(bx+c).\cos\varphi + \cos(bx+c).\sin\varphi\}$$

$$= r.\,e^{ax}.\{\sin(bx+c+\varphi).$$

Its further order derivation also yields

$$D^2\{e^{ax}.\sin(bx+c)\} = r.\,e^{ax}\{a.\sin(bx+c+\varphi) + b.\cos(bx+c+\varphi)\}$$

$$= r^2.\,e^{ax}.\sin(bx+c+2\varphi). \qquad (14.10)$$

Next, operating it by D^{n-2} we may similarly derive Eq. (14.6).

(ii) Proceeding similarly, the first order derivative of $e^{ax}.\cos(bx+c)$ is

$$D\,\{e^{ax}.\cos(bx+c)\} = e^{ax}.\{a.\cos(bx+c) - b.\sin(bx+c)\},$$

which, for Eq. (14.9), reduces to

$$D\{e^{ax}.\cos(bx+c)\} = r.e^{ax}\{\cos(bx+c).\cos\varphi - \sin(bx+c).\sin\varphi\}$$

$$= r.e^{ax}.\{\cos(bx+c+\varphi).$$

Operating it by D and proceeding as above we derive

$$D^2\{e^{ax}.\cos(bx+c)\} = r.e^{ax}\{a.\cos(bx+c+\varphi) - b.\sin(bx+c+\varphi)\}$$

$$= r^2.e^{ax}.\cos(bx+c+2\varphi).$$

Operating it further by D^{n-2} we may similarly derive Eq. (14.6). //

Theorem 14.3. The nth order derivative of $\tan^{-1}(x/a)$ is given by

$$D^n \tan^{-1}(x/a) = (-1)^{n-1}.(n-1)!.r^{-n}.\sin(n\alpha), \qquad (14.11)$$

where
$$r = \sqrt{(a^2+x^2)}, \qquad \alpha = \tan^{-1}(a/x). \qquad (14.12)$$

§ 15. Maxima / minima of a function on a closed interval

15.1. Extrema at the end points of interval

Let $y = f(x)$ be a continuous function defined on a closed interval $[a, b]$
of a real line P (along Ox-axis). The
points A (where $x_1 = a$) and B (where
$x_2 = b$) are the points of maxima and
minima respectively.

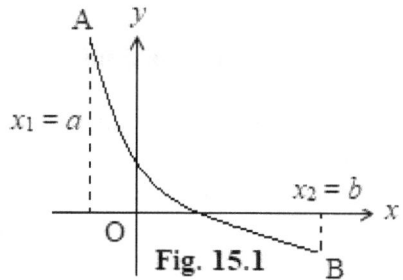

Fig. 15.1

Theorem 15.1. (*Extremum value theorem*) Let $f(x)$ be a continuous function on a closed interval $[a, b]$ then there exist points x_1 and x_2 in $[a, b]$ such that

$$f(x_2) \le f(x) \le f(x_1) \qquad \forall x \varepsilon [a, b], \qquad (15.1)$$

where $f(x_1)$ is maximum and $f(x_2)$ minimum on $[a, b]$.

Note 15.1. For above theorem the function $f(x)$ must be :

(i) defined on $[a, b]$, and **(ii)** continuous on $[a, b]$.

In absence of any of these conditions, above theorem may not hold good. A function continuous on an open interval possess neither maxima nor minima is observed in the following example.

Example 15.1. The function $f(x) = x \forall x \varepsilon (0, 1)$ is continuous on the open interval $(0, 1)$. It possesses neither maxima nor minima on $(0, 1)$. As such, it is not defined on the closed interval $[0, 1]$. //

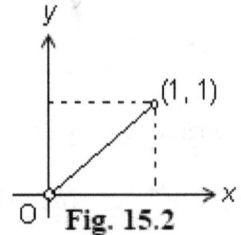

Fig. 15.2

Example 15.2. Let $f(x) = 1/x$ (if $x \neq 0$), 0 (if $x = 0$) be a function from $[-1, 1]$ to P. As such, it is defined on $[-1, 1]$. We see in the following that it is not continuous at $x = 0$:

$$\lim_{x \to 0^-} f(x) = \lim_{x \to 0^-} (1/x) = -\infty,$$

$$\lim_{x \to 0^+} f(x) = \lim_{x \to 0^+} (1/x) = \infty,$$

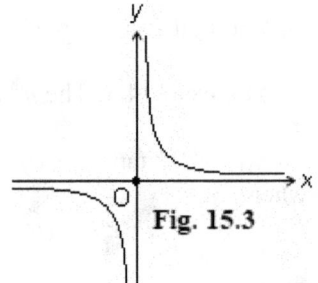

Fig. 15.3

while $f(0) = 0$ (by definition). The two limits being different (and also not equal to $f(0)$) make the function discontinuous. Thus, the second condition of above Note is not satisfied. So, there exist neither maxima nor minima. (cf. Fig. 15.3.).

15.2. Local maxima / minima of a function

In this case, the function does not attain extremum values at the end points of a closed interval (where it is defined) instead at some points in the interval.

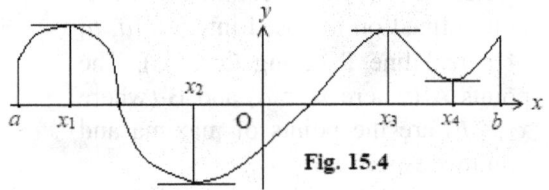

Fig. 15.4

Definition 15.1. Let f be a continuous function on some interval I of the real line P.

(i) It is said to possess a *local maximum* at a point x_1 if there exists an open interval (c, d) containing x_1 and contained in I so that $f(x)$ attains a maximum at x_1 in (c, d):

$$f(x) \leq f(x_1) \ \forall x \varepsilon (c, d).$$

(ii) $f(x)$ is said to possess a *local minimum* at a point x_2 if there exists an open interval (c, d) containing x_2 and contained in I so that $f(x)$ attains a minimum at x_2 in (c, d):

$$f(x_2) \leq f(x) \qquad \forall\, x \,\varepsilon\, (c, d).$$

Definition 15.2. A point where the function attains a local maxima or a local minima is called a point of *local extrema*.

In Fig. 15.4, the function $f(x)$ has local maxima at the points x_1 and x_3; local minima at the points x_2 and x_4; and f attains its minimum value at x_2 and maximum value at x_1 on $[a, b]$.

Note 15.2. Sometimes, maximum / minimum values of a function are also called *absolute maximum* and *absolute minimum* respectively.

§ 16. Derivative of $f(x)$ at local extremum points

We notice in Fig. 15.4, that the tangents to the curve at local extrema, viz. x_1, x_2, x_3, x_4 are clearly horizontal lines. So, the values of $f'(x)$ at these points are zero.

Definition 16.1. Let I be any interval: finite or infinite. The set of all points of I excluding its end points is called the *interior* of I and it is denoted by I_0.

Theorem 16.1. (*Fundamental theorem of local extrema*) Let $f(x)$ be a continuous function on the interval I and differentiable in the interior of I. If x_0 is a local extremum point of I_0 on I, then

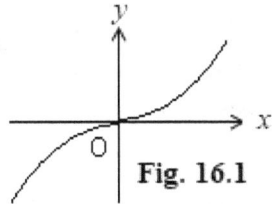

Fig. 16.1

$$f'(x_0) = 0. \tag{16.1}$$

Note 16.1. The converse of above theorem need not be true in general, i.e. Eq. (16.1) does not necessarily make x_0 a point of local maxima or minima.

Example 16.1. Although the function $f(x) = x^3$ possesses zero derivative at origin yet there exists neither maxima nor minima.

Solution. $\lim_{x \to 0^-} \{f(x) - f(0)\}/(x - 0) = \lim_{x \to 0^-} \{x^3 - 0\}/(x - 0) = 0$, and

$$\lim_{x \to 0^+} \{f(x) - f(0)\} / (x - 0) = \lim_{x \to 0^+} (x^3 - 0) / x = 0,$$

being equal imply differentiability of the function at $x = 0$ and $f'(0) = 0$. But, $x = 0$ is neither a local maxima nor a local minima. //

Theorem 16.2. (*Rolle's theorem*) Let f be a function:

(i) continuous on a closed interval $[a, b]$,

(ii) differentiable on open interval (a, b), and

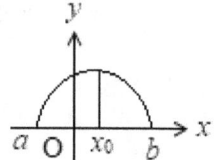

Fig. 16.2

(iii) $f(a) = f(b) = 0$; then there exists at least one point $x_0 \, \varepsilon \, (a, b)$ satisfying Eq. (16.1).

Proof. As per hypothesis (i), by Theorem 16.1, there exist points x_1, $x_2 \, \varepsilon \, (a, b)$ satisfying Eq. (16.1). In case, both x_1, x_2 are the end–points of $[a, b]$, then $f(x)$ is a constant function (along x-axis). This implies Eq. (16.1) $\forall \, x \, \varepsilon \, (a, b)$ in view of hypothesis (ii). If both x_1, x_2 are not end–points of (a, b) then at least one of these (say denoted by x_0) is in (a, b). Hence, by Theorem 16.1, there follows Eq. (16.1). //

Note 16.2. The condition (ii) in above theorem is crucial. In absence of the same, the theorem may not hold as observed in the following Example.

Example 16.2. Let a function $f(x) = -|x| + 1$ be defined on the interval $[-1, 1]$. There does not exist any point x_0 in $[-1, 1]$ satisfying Eq. (16.1).

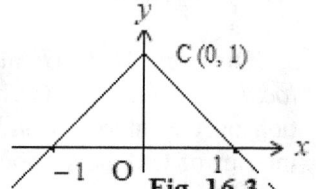

Fig. 16.3

Solution. The function so defined is

$$f(x) = -x + 1 \ (\text{if } x \geq 0), \ x + 1 \ (\text{if } x < 0).$$

Hence,

$$\lim_{x \to 0^-} f(x) = \lim_{x \to 0^-} (x + 1) = 1,$$

$$\lim_{x \to 0^+} f(x) = \lim_{x \to 0^+} (-x + 1) = 1,$$

and $f(0) = 1$ being equal establish continuity of the function at $x = 0$. Also, there hold

$\lim_{x \to c} f(x) = c + 1 = f(c) \; \forall c \, \varepsilon \, [-1, 0), \; \lim_{x \to c} f(x) = -c + 1 = f(c)$

$\forall c \, \varepsilon \, (0, 1]$ proving the continuity of $f(x)$ at $x = c$. So, the function is continuous on the entire interval $[-1, 1]$. Also, it satisfies the third condition of the Theorem 16.2: $f(-1) = f(1) = 0$. However, the following limits are not equal:

$$\lim_{x \to 0^-} \{f(x) - f(0)\}/(x - 0) = \lim_{x \to 0^-} \{x + 1 - 1\}/(x - 0) = 1,$$

$$\lim_{x \to 0^+} \{f(x) - f(0)\}/(x - 0) = \lim_{x \to 0^+} \{-x + 1 - 1\}/(x - 0) = -1,$$

implying that the function is not differentiable at $x = 0$. Also, for any point $c \, \varepsilon \, (-1, 1)$, the limits

$$\lim_{x \to c} \{f(x) - f(c)\}/(x - c) = \lim_{x \to c} \{-x + 1 - (-c + 1)\}/(x - c) = -1,$$

and
$$\lim_{x \to c} \{f(x) - f(c)\}/(x - c) = \lim_{x \to c} \{x + 1 - (c + 1)\}/(x - c) = 1,$$

are different. So, there does not exist any point in the interval $(-1, 1)$ satisfying Eq. (16.1). Thus, Rolle's theorem is not satisfied by this function. //

Example 16.3. Find x so that the function $f(x) = x^2 - 5x - 6$ satisfies the Rolle's theorem and both $f(x)$ and $f'(x)$ vanish there.

Solution. $f(x) = (x - 6)(x + 1)$ vanishes at $x = 6, -1$. The function $f(x)$ being a polynomial is continuous on the interval $[-1, 6]$ and is differentiable on $(-1, 6)$; indeed, it is differentiable on the entire interval $[-1, 6]$. Thus, it satisfies all the three conditions of the Rolle's theorem.

Hence, there exists some point x_0 in $(-1, 6)$ where

$$f'(x) \equiv 2x - 5 = 0 \quad \Rightarrow \quad x = 5/2. \; //$$

Example 16.4. Does the function $f(x) = 2 - |x|$ defined on the interval $[-2, 2]$ contradict Rolle's theorem?

Solution. As such, the function defined by

$$f(x) = 2 - x, \text{ (when } x \geq 0\text{), } 2 + x \text{ (when } x < 0\text{)}.$$

so that,

$$f(-2) = 2 + (-2) = 0, \quad f(2) = 2 - 2 = 0.$$

satisfies the third condition of Rolle's theorem. Also, the limits

$$\lim_{x \to (-2)^-} f(x) = \lim_{x \to (-2)^-} (2 + x) = 0,$$

$$\lim_{x \to 2^+} f(x) = \lim_{x \to 2^+} (2 + x) = 0,$$

and the value of the function at $x = -2$ are all equal to each other. Hence, the function is continuous at $x = -2$. Similarly, its continuity can also be verified at $x = 0$ and 2. So, the function is continuous on $[-2, 2]$. Thus, the function also satisfies the first condition of the theorem. However,

$$\lim_{x \to 0^-} \{f(x) - f(0)\} / (x - 0) = \lim_{x \to 0^-} \{2 + x - 2\} / (x - 0) = 1,$$

and

$$\lim_{x \to 0^+} \{f(x) - f(0)\} / (x - 0) = \lim_{x \to 0^+} \{2 - x - 2\} / (x - 0) = -1,$$

being different imply that the function is *not* differentiable at $x = 0$.

Also, at any point $c \, \varepsilon \, (-2, 0)$, $f'(c) = (2 + x)'_{x = c} = 1 \neq 0$. Similarly, at any point $c \, \varepsilon \, (0, 2)$, $f'(c) = (2 - x)'_{x = c} = -1$, which is also not zero. Thus, there does not exist any point x_0 in the interval $(-2, 2)$ where the function may satisfy Eq. (16.1). Hence, the function does not contradict the Rolle's theorem. //

Theorem 16.3. (*Mean Value Theorem*) If a function $f(x)$ is

(i) continuous on $[a, b]$, and

(ii) differentiable on (a, b);

then there exists a point $x_0 \, \varepsilon \, (a, b)$ where

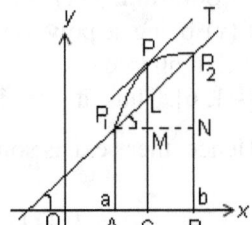
Fig. 16.4

$$f'(x_0) = \{f(b) - f(a)\} / (b - a). \tag{16.2}$$

Proof. Let $P_1 \{a, f(a)\}$ and $P_2 \{b, f(b)\}$ be two points on the graph of the function $f(x)$. The equation to line P_1, P_2 is

$$y - f(a) = [\{f(b) - f(a)\} / (b - a)]. (x - a). \tag{16.3}$$

Let P $\{x,\ f(x)\}$ be any point on the graph of $f(x)$. We define a function $g(x) = LP = CP - CL$, where

$$CL = CM + ML = AP_1 + (P_1M)\tan\psi = f(a) + (x - a).\tan\psi,$$

and

$$\tan\psi = NP_2/P_1N = (BP_2 - AP_1)/AB = \{f(b) - f(a)\}/(b-a).$$

Hence,

$$g(x) = f(x) - f(a) - (x-a).\{f(b) - f(a)\}/(b-a)\}. \qquad (16.4)$$

As per hypothesis, $g(x)$ is also continuous on $[a, b]$ and differentiable on (a, b).

Moreover, $g(a) = g(b) = 0$, by Eq. (16.4). Thus, all the three conditions of Rolle's theorem are satisfied by $g(x) \Rightarrow$ that there exists a point x_0 in (a, b) where

$$g'(x_0) \equiv f'(x_0) - \{f(b) - f(a)\}/(b-a) = 0 \Rightarrow \text{Eq. (16.2). //}$$

Corollary 16.1. If $f(x)$ is continuous on an interval I, differentiable on the interior I_o of I and $f'(x) = 0\ \forall\ x\ \varepsilon\ I$, then $f(x)$ is constant.

Example 16.5. Check if the *Mean Value Theorem* holds for the function $f(x) = (x-1)/x$ on the interval $[1, 3]$?

Solution. Both limits $\lim_{x\to1^-} f(x) = \lim_{\varepsilon\to0} (1-\varepsilon-1)/(1-\varepsilon) = 0,$

$$\lim_{x\to1^+} f(x) = \lim_{\varepsilon\to0} (1+\varepsilon-1)/(1+\varepsilon) = 0,$$

and the value of the function $f(1) = 0$ are all equal. Hence, the function is continuous at $x = 1$. Similarly, its continuity can also be verified at $x = 3$ and at any point $c\ \varepsilon\ (1,3)$:

$$\lim_{x\to c^-} f(x) = (c-1)/c = f(c),$$

and

$$\lim_{x\to c^+} f(x) = (c-1)/c = f(c).$$

So, the function is continuous on $[1, 3]$. Next,

$$\lim_{x\to c^-} \{f(x) - f(c)\}/(x-c) = \lim_{x\to c^-} \{(x-1)/x - (c-1)/c\}/(x-c)$$

$$= \lim_{x\to c^-} (1/cx) = 1/c^2, \quad \text{and} \quad \lim_{x\to c^+} \{f(x) - f(c)\}/(x-c)$$

$= \lim_{x \to c^+} \{(x-1)/x - (c-1)/c\}/(x-c) = \lim_{x \to c^+} (1/cx) = 1/c^2,$

being equal imply differentiability of the function at $x = c \, \varepsilon \, (1, 3)$. Thus, both the conditions of Theorem 16.3 are satisfied. So, there exists a point $x \, \varepsilon \, (1, 3)$ where

$$f'(x) = 1/x^2 = \{f(3) - f(1)\}/(3-1)$$

$$= (2/3 - 0)/2 = 1/3 \implies x = \sqrt{3} \, \varepsilon \, (1, 3).$$

The alternate square-root $-\sqrt{3}$ is neglected as it $\notin (1, 3)$. //

Example 16.6. Check if the *Mean Value Theorem* holds for the function

$$f(x) = |x - 3| \text{ on the interval } [1, 4] ?$$

Solution. (i) The function so defined is

$$f(x) = x - 3 \text{ (when } x \geq 3), -x + 3 \text{ (when } x < 3).$$

Hence, $f(3) = 0$,

$$\lim_{x \to 3^-} f(x) = \lim_{\varepsilon \to 0} \{-(3 - \varepsilon) + 3\} = 0,$$

and

$$\lim_{x \to 3^+} f(x) = \lim_{\varepsilon \to 0} \{3 + \varepsilon - 3\} = 0$$

make the function continuous at $x = 3$.

(ii) Further, at any point c in the *clopen* interval $[1, 3)$, $f(c) = 3 - c$,

$$\lim_{x \to c^-} f(x) = \lim_{\varepsilon \to 0} \{-(c - \varepsilon) + 3\} = 3 - c,$$

and

$$\lim_{x \to c^+} f(x) = \lim_{\varepsilon \to 0} \{-(c + \varepsilon) + 3\} = 3 - c,$$

are equal to each other. Hence, the function is continuous at $x = c$ in $[1, 3)$.

(iii) Similarly, at any point c in the *open-close* interval $(3, 4]$, $f(c) = c - 3$,

$$\lim_{x \to c^-} f(x) = \lim_{\varepsilon \to 0} \{(c - \varepsilon) - 3\} = c - 3,$$

and

$$\lim_{x \to c^+} f(x) = \lim_{\varepsilon \to 0} \{(c + \varepsilon) - 3\} = c - 3,$$

being equal also establish continuity of the function at $x = c$ in (3, 4]. Hence, the function is continuous on the entire interval [1, 4].

(iv) Finally, the limits

$$\lim_{x \to 3^-} \{f(x) - f(3)\}/(x - 3) = \lim_{x \to 3^-} (-x + 3 - 0)/(x - 3) = -1,$$

and

$$\lim_{x \to 3^+} \{f(x) - f(3)\}/(x - 3) = \lim_{x \to 3^+} (x - 3 - 0)/(x - 3) = 1,$$

being different cause non-differentiability of the function at $x = 3$. So, the function is not differentiable throughout the interval (1, 4). As such, the second condition of the Theorem 16.3 is not satisfied. //

§ 17. Monotonic functions

Definition 17.1. A function $f(x)$ is said to be

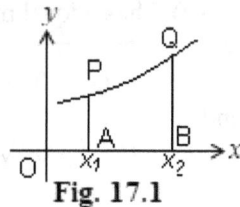

Fig. 17.1

(i) *increasing* if $f(x_1) \leq f(x_2)$,

(ii) *decreasing* if $f(x_1) \geq f(x_2)$,

(iii) *strictly increasing* if $f(x_1) < f(x_2)$,

(iv) *strictly decreasing* if $f(x_1) > f(x_2)$,

$\forall\, x_1 < x_2$ on an interval I of the real line P.

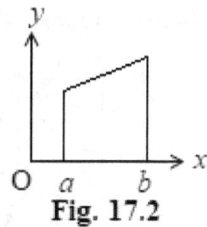

Fig. 17.2

Definition 17.2. An increasing or decreasing function is called *monotonic* on I. Similarly, a strictly increasing or strictly decreasing function is called *strictly monotonic* on I.

Example 17.1. A constant function is monotonic but not strictly monotonic.

Example 17.2. The Figures 17.2–17.5 depict graphs of *strictly increasing, strictly decreasing, increasing* and *decreasing* functions respectively on the interval [a, b].

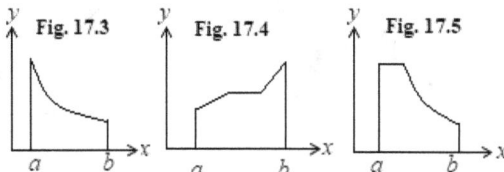

Fig. 17.3   Fig. 17.4   Fig. 17.5

§ 18. Points of local extrema

Theorem 18.1. (*First derivative test for local extrema*) Let $f(x)$ be a continuous function on an interval I, and $x_0 \, \varepsilon \, I_0$. Consider the open interval $(x_0 - \delta, x_0 + \delta)$ for some $\delta > 0$. Then

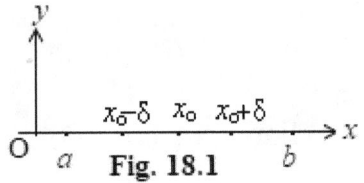

(i) f has a local maximum value at x_0, if

$$f'(x) > 0 \quad \forall \, x \, \varepsilon \, (x_0 - \delta, x_0)$$

and

$$f'(x) < 0 \quad \forall \, x \, \varepsilon \, (x_0, x_0 + \delta);$$

Fig. 18.1

(ii) f has a local minimum value at x_0, if

$$f'(x) < 0 \quad \forall \, x \, \varepsilon \, (x_0 - \delta, x_0),$$

and

$$f'(x) > 0 \quad \forall \, x \, \varepsilon \, (x_0, x_0 + \delta).$$

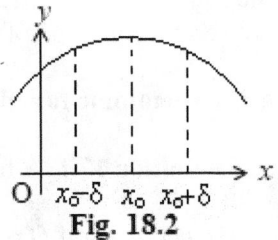

Fig. 18.2

Example 18.1. Figures 18.2 – 18.3 depict the points of local maxima whereas Figures 18.4 – 18.5 indicate points of local minima.

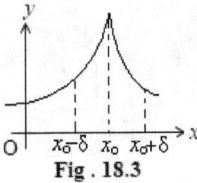
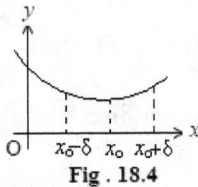
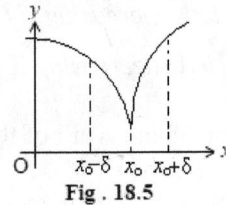

Fig . 18.3   Fig . 18.4   Fig . 18.5

Example 18.2. Find the local extremum points for the function $f(x) = |x|$.

Solution. We note that $f(x)$ is continuous at $x = 0$ and so at P too.

(i) $\forall \, x > 0, \qquad f(x) = |x| = x \Rightarrow f'(x) = 1 \neq 0.$
So, by Theorem 16.1, $x > 0$ is not a point of local extrema for f.

(ii) $\forall \, x < 0, \; f(x) = |x| = -x$

$\Rightarrow$

$$f'(x) = -1 \neq 0.$$

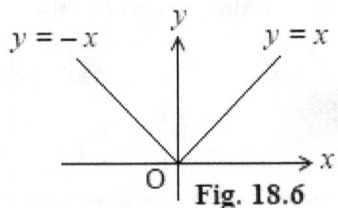

Fig. 18.6

So, again by the same theorem, $x < 0$ is also not a point of local extrema for f.

(iii) $\lim_{x \to 0^-} \{f(x) - f(0)\}/(x - 0) = \lim_{x \to 0^-} (-x - 0)/x = -1,$

$\lim_{x \to 0^+} \{f(x) - f(0)\}/(x - 0) = \lim_{x \to 0^+} (x - 0)/x = 1,$

being different, imply that $f(x)$ is not differentiable at $x = 0$. However,

$f'(x) = -1 < 0$ for $x < 0$, and $f'(x) = 1 > 0$ for $x > 0$,

(as seen in the two parts above). Hence, by Theorem 18.1, $x = 0$ becomes a point of local minima for f. //

Example 18.3. Find the local extremum points of the function $f(x) = 2 - |x|$.

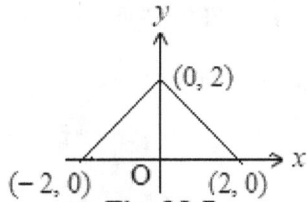
Fig. 18.7

Solution. (i) $\forall\, x > 0, f(x) = 2 - x$ $\Rightarrow$ $f'(x) = -1 \neq 0$.

So, by Theorem 16.1, $x > 0$ is not a point of local extrema for f.

(ii) $\forall\, x < 0,\ f(x) = 2 + x \Rightarrow f'(x) = 1 \neq 0$. So, again by the same theorem, $x < 0$ is also not a point of local extrema for f.

(iii) $\lim_{x \to 0^-} \{f(x) - f(0)\}/(x - 0) = \lim_{x \to 0^-} (2 + x - 2)/x = 1,$

$\lim_{x \to 0^+} \{f(x) - f(0)\}/(x - 0) = \lim_{x \to 0^+} (2 - x - 2)/x = -1,$

being different, imply that $f(x)$ is not differentiable at $x = 0$. However,

$f'(x) = 1 > 0$ for $x < 0$,   and $f'(x) = -1 < 0$ for $x > 0$,

(as seen in the two parts above). Hence, by Theorem 18.1, $x = 0$ becomes a point of local maxima for f. //

§ 19. Second derivative test for local extrema

Theorem 19.1. Let $f(x)$ be a continuous function on an interval I, $f'(x_0) = 0$ and $f''(x_0)$ exists at some $x_0\ \varepsilon\ I_0$. Then

(i) f has a local maxima at x_0, if $f''(x_0) < 0$, and

(ii) f has a local minima at x_0, if $f''(x_0) > 0$.

Proof. (i) By definition,

$$f''(x_0) = \lim_{x \to x_0} \{f'(x) - f'(x_0)\} / (x - x_0)$$

$$= \lim_{x \to x_0} \{f'(x)\} / (x - x_0), \qquad (19.1)$$

as per hypothesis. There arise two cases:

(*a*) For $x \, \varepsilon \, (x_0 - \delta, x_0)$, $x - x_0 < 0$, above limit is negative if $f'(x) > 0$.

(*b*) For $x \, \varepsilon \, (x_0, x_0 + \delta)$, $x - x_0 > 0$, above limit is negative if $f'(x) < 0$.

Hence, by Theorem 18.1, $f(x)$ has a local maxima at x_0.

(ii) This part can also be proved similarly. //

Note 19.1. In case, the limit in Eq. (19.1) vanishes, we obtain higher order derivatives of f. For the points of extrema for f, its *third order derivative* must be zero and we test the conditions (i) and (ii) of above theorem for the fourth order derivative. Further, if fourth order derivative also vanishes proceed for the next order derivatives. In brief, the odd order derivatives of f must vanish at the points of extrema of f and the next (even) order derivative should determine the points of maxima or minima as per the conditions of above theorem.

Example 19.1. Find the points of local extrema for the function

$$f(x) = x^3/3 + x^2/2 - 2x + 1.$$

Solution. $f'(x) = x^2 + x - 2 = (x + 2)(x - 1) = 0 \Rightarrow x = -2, 1.$

$f''(x) = 2x + 1,$ so that $f''(-2) = -3$ and $f''(1) = 3.$

Hence, by Theorem 19.1, f has a local maximum at $x = -2$ and a local minimum at $x = 1$. //

Theorem 19.2. (*Fundamental theorem of maxima or minima*) Let f be a continuous function on a closed interval $[a, b]$. Then:

(i) *f* attains a *maxima* at x_0 ⇒ either x_0 coincides with the end points *a* or *b*, or *f* attains a local maximum at x_0;

(ii) *f* attains a *minima* at x_0 ⇒ either x_0 coincides with the end points *a* or *b*, or *f* attains a local minimum at x_0.

Example 19.2. Find the points of extrema and the extremum values of the function $f(x) = |x|$ for $x \varepsilon [-3, 4]$.

Solution. As seen in Example 18.2, the only local extremum (indeed minimum) point of the function is $x = 0$. The minimum value of the function there is $f(0) = 0$. Further, for $x \varepsilon [-3, 4]$, $f(x)$ possesses the following values:

$$x = -3, -2, -1, 0, 1, 2, 3, 4; \quad f(x) = 3, 2, 1, 0, 1, 2, 3, 4.$$

Thus, by Theorem 19.2, $x = 4$ gives the maximum value of the function: $f(4) = 4$, and $x = 0$ gives the minimum value: $f(0) = 0$. //

Example 19.3. Find the absolute extrema of the function $f(x) = (x^2 - 21)/(x - 5)$, $x \varepsilon [-2, 5)$.

Solution. It may be seen that f is continuous on the interval. Also,

$$f'(x) = \{2x(x - 5) - (x^2 - 21)\}/(x - 5)^2$$

$$= 0 \Rightarrow (x^2 - 10x + 21) \equiv (x - 3)(x - 7)$$

Fig. 19.1

$= 0 \Rightarrow x = 3, 7$; out of which only the first critical number ($x = 3$) lies in the given interval. The second order derivative is

$$f''(x) = (2x - 10)/(x - 5)^2 - 2(x^2 - 10x + 21)/(x - 5)^3,$$

so that

$$f''(3) = -4/(-2)^2 = -1 \Rightarrow f(3) = (-12)/(-2) = 6 \text{ is maximum.}$$

Note 19.2. $f'(x)$ exists $\forall x \varepsilon [-2, 5)$. It is positive for $x \varepsilon [-2, 3)$ but negative for $x \varepsilon (3, 5)$. So, it is strictly increasing on $[-2, 3]$ and strictly decreasing on $[3, 5)$. Hence, $f(x)$ possesses absolute maxima 6 at $x = 3$. //

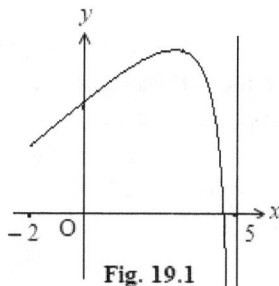

§ 20. Application of differentiation to calculate rate of change

Let a particle move along a straight line from point O and describes the distance OP $= s$ in time t seconds given by $s = f(t)$. Then the average speed of the particle after t secs. is

$$\{s(t + h) - s(t)\} / (t + h - t),$$

where h is very small positive real number. To get exact speed at time t we have to make h sufficiently small. Thus, speed at time t is

$$\lim_{h \to 0} \{s(t + h) - s(t)\} / h,$$

which is just the derivative of the distance function with respect to t.

Thus, the rate of change of any quantity q (for instance, distance, area, volume, weight, population, etc.) which is a function of time parameter is obtained by differentiating q with respect to t.

§ 21. Partial differentiation of scalar functions

Let $f(x, y)$ be a scalar function of two independent variables x and y. Unless these variables are connected by some relation they are treated as independent to each other. Like a vector function the partial derivation of scalar function $f(x, y)$ can also be defined similarly. For instance,

$$\lim_{\delta x \to 0} \{f(x + \partial x, y) - f(x, y)\} / \delta x \qquad (21.1)$$

shall be called the *partial derivative* of f with respect to x (treating y constant) and is denoted by $\partial f / \partial x$. This derivative is also denoted by putting a lower suffix x to f: f_x or more explicitly $(\partial f / \partial x)_{y = \text{const.}}$

Similarly, the partial derivatives of f with respect to its other variable y (treating x constant) can be defined:

$$f_y \equiv (\partial f / \partial y)_{x = \text{const.}} = \lim_{\delta y \to 0} \{f(x, y + \partial y) - f(x, y)\} / \delta y.$$

Further, the second order partial derivatives of f with respect to x and y are denoted by

$$f_{xx} \equiv (\partial / \partial x) f_x = (\partial / \partial x)(\partial f / \partial x) = \partial^2 f / \partial x^2 = (\partial^2 f / \partial x^2)_{y = \text{const.}}; \qquad (21.2)$$

$$f_{xy} \equiv (\partial / \partial y) f_x = (\partial / \partial y) (\partial f / \partial x) = \partial^2 f / \partial y \, \partial x \, ; \qquad (21.3a)$$

$$f_{yx} \equiv (\partial / \partial x) f_y = (\partial / \partial x) (\partial f / \partial y) = \partial^2 f / \partial x \, \partial y \, ; \qquad (21.3b)$$

$$f_{yy} \equiv (\partial / \partial y) f_y = (\partial / \partial y) (\partial f / \partial y) = \partial^2 f / \partial y^2 = (\partial^2 f / \partial y^2)_{x \, = \, \text{const}}. \qquad (21.4)$$

Note 21.1. In general, the second order derivatives f_{xy} and f_{yx} are not equal but in our discussion we shall consider only those functions for which they are equal.

Above discussion can also be extended to the functions of more than two independent variables. Also, the derivatives of order higher than two can be defined similarly.

Example 21.1. Compute the partial derivatives of the function

$$f(x, y, z) = x^4 + x^2 y^2 + y^4,$$

and show that the derivatives f_{xy} and f_{yx} are equal for this function.

Solution. Differentiating f partially with respect to x and y independently, we get

$$\partial f / \partial x = 4 x^3 + 2x \, y^2, \qquad \partial f / \partial y = 2x^2 y + 4 y^3,$$

$$\partial^2 f / \partial x^2 = 12 x^2 + 2 y^2, \qquad \partial^2 f / \partial y \, \partial x = 4x \, y = \partial^2 f / \partial x \, \partial y,$$

$$\partial^2 f / \partial y^2 = 2x^2 + 12 y^2.$$

Partial derivatives of higher order can also be obtained similarly. Equivalence of the derivatives f_{xy} and f_{yx} may also be noted. //

Example 21.2. Find the first order partial derivatives of the functions:

(i) $\tan^{-1} \{(x^2 + y^2) / (x + y)\}$, **(ii)** x^y, **(iii)** $\ln (x^2 + y^2)$.

Solution. (i)

$$f_x = \frac{1}{1 + (x^2 + y^2)^2 / (x + y)^2} \cdot \frac{2x(x + y) - (x^2 + y^2)}{(x + y)^2}$$

$$= \frac{x^2 + 2xy - y^2}{(x+y)^2 + (x^2+y^2)^2},$$

$$f_y = \frac{1}{1+(x^2+y^2)^2/(x+y)^2} \cdot \frac{2y(x+y)-(x^2+y^2)}{(x+y)^2}$$

$$= \frac{y^2 + 2xy - x^2}{(x+y)^2 + (x^2+y^2)^2}.$$

Note 21.2. The function being symmetric in the variables, the derivative f_y can be directly obtained by interchanging x and y in the value of f_x.

(ii) Writing $f = x^y$, taking its natural logarithm: $\ln f = y.\ln x$ and differentiating it partially with respect to x, y we obtain

$$(1/f)(\partial f/\partial x) = y/x \implies \partial f/\partial x = f.(y/x) = yx^{y-1},$$

and

$$(1/f)(\partial f/\partial y) = \ln x \implies \partial f/\partial y = f.(\ln x) = x^y.\ln x.$$

(iii) $\quad f_x = 2x/(x^2+y^2), \quad f_y = 2y/(x^2+y^2).$ //

Example 21.3. If $f = e^{xyz}$, show that

$$\partial^3 f/\partial x\,\partial y\,\partial z = (1 + 3xyz + x^2y^2z^2)\,e^{xyz}.$$

Solution. Differentiating the function partially with respect to z, we get

$$\partial f/\partial z = xy.\,e^{xyz}.$$

Next, differentiating it partially w.r.t. y, we have

$$\partial^2 f/\partial y\,\partial z = (x + x^2yz)\,e^{xyz};$$

which, on further differentiation w.r.t. x, yields

$$\partial^3 f/\partial x\,\partial y\,\partial z = \{1 + 2xyz + (x + x^2yz)yz\}\,e^{xyz},$$

proving the desired result. //

Example 21.4. If $x^x y^y z^z = c$, show that at the point $x = y = z$,

$$\partial^2 z / \partial x \, \partial y = - 1 / \{x.\ln \ (e.x)\}.$$

Solution. Since the variables x, y, z are connected by a relation all of them cannot be independent to each other. Treating z as a function of the remaining variables x and y, we derive the partial derivatives of z w.r.t. x and y. Taking the natural log of the relation we get

$$x. \ln x + y. \ln y + z. \ln \ z = \ln c.$$

Differentiating it partially w.r.t. x, y, we get

$$1 + \ln x + (1 + \ln z) \, (\partial z / \partial x) \ = \ 1 + \ln y + (1 + \ln z) \, (\partial z / \partial y) = 0. \ (21.5)$$

Differentiating the second of above relations further partially w.r.t. x, we get
$$(1/z) \, (\partial z / \partial x) \, (\partial z / \partial y) + (1 + \ln z) \, (\partial^2 z / \partial x \, \partial y) = 0,$$

or, by Eq. (1.5)

$$(1 + \ln x) \, (1 + \ln y) / z \, (1 + \ln z)^2 + (1 + \ln z) \, (\partial^2 z / \partial x \, \partial y) = 0,$$

$\Rightarrow$

$$\partial^2 z / \partial x \, \partial y \ = \ - \ (1 + \ln x) \, (1 + \ln y) / \ z \, (1 + \ln z)^3.$$

Evaluating it at the point $x = y = z$, we derive

$$\partial^2 z / \partial x \, \partial y = - 1 / x. \, (1 + \ln x) = - 1 / x. \ln (ex). \ //$$

§ 22. Total derivative of a function

Let $f(x, y)$ be a function of two variables x and y. Let these variables be also expressible as functions of some common parameter t (say):

$$x = g \, (t), \qquad y = h \, (t).$$

Substitutions for x and y finally reduce f as a function of single parameter t. As such, its differential coefficient w.r.t. t may be called the *total derivative* in order to distinguish the same with partial derivatives $\partial f / \partial x$ and $\partial f / \partial y$. An useful formula for the total derivative df / dt is derived by means of Taylor's theorem (applied to the first order of approximation):

$$df \,/\, dt = (\partial f \,/\, \partial x)\,(dx \,/\, dt) + (\partial f \,/\, \partial y)\,(dy \,/\, dt). \qquad (22.1)$$

Note 22.1. Above result can also be generalized for a function $f(x_1, x_2, \ldots, x_n)$ of more than two variables:

$$df \,/\, dt \;=\; (\partial f \,/\, \partial x_1)\,(dx_1 \,/\, dt) + (\partial f \,/\, \partial x_2)\,(dx_2 \,/\, dt) + \ldots$$

$$+\, (\partial f \,/\, \partial x_n)\,(dx_n \,/\, dt). \qquad (22.2)$$

As a special case, let the parameter t be replaced by x itself so that y becomes a function of x. As such, u becomes a function of x alone. However, its total derivative w.r.t. x can also be calculated by the formula (22.1):

$$df \,/\, dx \;=\; \partial f \,/\, \partial x + (\partial f \,/\, \partial y)\,(dy \,/\, dx). \qquad (22.3)$$

Theorem 22.1. Given a functional relation $f(x, y) = c$, the derivative of y w.r.t. x can be computed by

$$dy \,/\, dx \;=\; -\,(\partial f \,/\, \partial x) \,/\, (\partial f \,/\, \partial y). \qquad (22.4)$$

Example 22.1. Given the functional relations $f(x, y) = 0$ and $\varphi(y, z) = 0$, show that

$$(\partial f \,/\, \partial y)\,(\partial \varphi \,/\, \partial z)\,(dz \,/\, dx) = (\partial f \,/\, \partial x)\,(\partial \varphi \,/\, \partial y). \qquad (22.5)$$

Solution. Eq. (22.4) determines

$$\partial f \,/\, \partial x \;=\; -\,(\partial f \,/\, \partial y)\,(dy \,/\, dx). \qquad (22.6)$$

Similarly, the total derivative of φ w.r.t. y is given by

$$0 = \partial \varphi \,/\, \partial y + (\partial \varphi \,/\, \partial z)\,(dz \,/\, dy)$$

$$\Rightarrow$$

$$\partial \varphi \,/\, \partial y = -\,(\partial \varphi \,/\, \partial z)\,(dz \,/\, dy). \qquad (22.7)$$

Multiplying Eqs. (22.6) and (22.7), we get (22.5). //

§ 23. Euler's theorem on homogeneous functions

Let $f(x, y, z)$ be an algebraic function involving certain terms in the variables x, y, z. If each term in the function has same (resultant) degree in the variables, the function is said to be homogeneous. Thus,

$$f(x, y, z) \equiv ax^2 + by^2 + cz^2 + 2\,(l\,yz + m\,zx + n\,xy)$$

is a homogeneous function of degree 2 in the variables x, y, z. Also,

$$f(x, y) \equiv a_0\,x^n + a_1\,x^{n-1}.y + a_2\,x^{n-2}.y^2 + \ldots + a_{n-1}.x.y^{n-1} + a_n.y^n, \quad (23.1a)$$

where a_i's are constants, is a homogeneous function of degree n in variables x and y. Above function may be rewritten as

$$f(x, y) \equiv x^n \{a_0 + a_1\,(y/x) + a_2\,(y/x)^2 + \ldots + a_{n-1}\,(y/x)^{n-1} + a_n\,(y/x)^n\}$$

$$= x^n\,F\,(y/x), \quad (23.1b)$$

$F\,(y/x)$ being a polynomial of degree n in y/x.

Theorem 23.1. (*Euler's theorem*) For a homogeneous function $f\,(x, y)$ of degree n in variables x and y, there holds the relation

$$x\,(\partial f / \partial x) + y\,(\partial f / \partial y) = n\,f(x, y). \quad (23.2)$$

Note 23.1. Above theorem can also be generalized for a homogeneous function of any degree in any number of variables. Thus, for a function $f\,(x_1, x_2, \ldots, x_m)$ of degree n, there holds the result

$$x_1\,(\partial f / \partial x_1) + x_2\,(\partial f / \partial x_2) + \ldots + x_m\,(\partial f / \partial x_m) = n\,f. \quad (23.3)$$

Theorem 23.2. For a homogeneous function $f\,(x, y)$ of degree n, there also hold the relations

$$x\,(\partial^2 f / \partial x^2) + y\,(\partial^2 f / \partial x\,\partial y) = (n-1)\,(\partial f / \partial x), \quad (23.4)$$

and

$$x\,(\partial^2 f / \partial x\,\partial y) + y\,(\partial^2 f / \partial y^2) = (n-1)\,(\partial f / \partial y). \quad (23.5)$$

Proof. Differentiating Eq. (23.2) partially w.r.t. x and y, we get

$$\partial f / \partial x + x\,(\partial^2 f / \partial x^2) + y\,(\partial^2 f / \partial x\,\partial y) = n\,(\partial f / \partial x) \Rightarrow \text{Eq. (23.4)};$$

and

$$x\,(\partial^2 f / \partial y\,\partial x) + \partial f / \partial y + y\,(\partial^2 f / \partial y^2) = n\,(\partial f / \partial y) \Rightarrow \text{Eq. (23.5)}.$$

Alternately, derivation process diminishes the degree of a function by 1. Thus, the derivatives $\partial f / \partial x$ and $\partial f / \partial y$ become homogeneous functions of degree $n-1$ in their variables x, y. Applying Euler's theorem to these functions, we therefore, have

$$x\,(\partial/\partial x)\,(\partial f/\partial x) + y\,(\partial/\partial y)\,(\partial f/\partial x) = (n-1)\,(\partial f/\partial x) \Rightarrow \text{Eq. (23.4)}$$

and

$$x\,(\partial/\partial x)\,(\partial f/\partial y) + y\,(\partial/\partial y)\,(\partial f/\partial y) = (n-1)\,(\partial f/\partial y) \Rightarrow \text{Eq. (23.5). //}$$

Theorem 23.3. For a homogeneous function $f\,(x,\ y)$ of degree n, there holds the relation

$$x^2\,(\partial^2 f/\partial x^2) + 2x\,y\,(\partial^2 f/\partial x\,\partial y) + y^2\,(\partial^2 f/\partial y^2) = n\,(n-1)\,f. \qquad (23.6)$$

Example 23.1. If $f = \sin^{-1}(x/y) + \tan^{-1}(y/x)$, prove that

$$x\,(\partial f/\partial x) + y\,(\partial f/\partial y) = 0.$$

Solution. Putting

$$f_1 = \sin^{-1}(x/y)\ \text{and}\ u = x/y \Rightarrow \partial u/\partial x = 1/y,\ \partial u/\partial y = -x/y^2,$$

the partial derivatives of $f_1 = \sin^{-1} u$ are

$$\partial f_1/\partial x = \{1/\sqrt{(1-u^2)}\}\,(\partial u/\partial x) = 1/\sqrt{(y^2-x^2)}$$

$$\Rightarrow$$

$$x\,\partial f_1/\partial x = x/\sqrt{(y^2-x^2)};$$

and

$$\partial f_1/\partial y = \{1/\sqrt{(1-u^2)}\}\,(\partial u/\partial y) = -x/y\sqrt{(y^2-x^2)}$$

$$\Rightarrow$$

$$y\,\partial f_1/\partial y = -x/\sqrt{(y^2-x^2)}.$$

Adding the last two relations we get

$$x\,(\partial f_1/\partial x) + y\,(\partial f_1/\partial y) = 0. \qquad (23.7)$$

Similarly for $f_2 = \tan^{-1}(y/x)$ and $v = y/x$, we have

$$\partial f_2/\partial x = \{1/(1+v^2)\}\,(\partial v/\partial x) = -y/(x^2+y^2)$$

$$\Rightarrow$$

$$x\,\partial f_2/\partial x = -x\,y/(x^2+y^2).$$

Also,

$$\partial f_2/\partial y = \{1/(1+v^2)\}\,(\partial v/\partial y) = x/(x^2+y^2)$$

$$\Rightarrow$$

$$y\,\partial f_2/\partial y = x\,y/(x^2+y^2),$$

Adding the last two relations we get

$$x \, \partial f_2 / \partial x + y \, \partial f_2 / \partial y = 0. \tag{23.8}$$

Lastly, addition of Eqs. (23.7) – (23.8) establishes the desired result. //

§ 24. Envelope of family of curves

24.1. One-parameter family of curves: Let the equation

$$f(t) \equiv f(x, y, t) = 0, \tag{24.1}$$

represent a family of curves, say C_t, on xOy plane, where t is a parameter (i.e. a real number) having double character: *fixed* on a particular member of the family but *varying* from one member to the other. On contrary, x, y represent the rectangular Cartesian coordinates of a point describing any member of the family and, thus, are always variables.

Definition 24.1. A curve Γ possessing two properties:

(i) It touches every member of the given family (of curves); and

(ii) At its each point it is touched by some member of the family,

is called the *envelope* (or enveloping curve) of the family.

Theorem 24.1. The equation of the envelope is obtained by eliminating the parameter t from the equations and

$$f(t) = 0 = \partial f(t) / \partial t. \tag{24.2}$$

Theorem 24.2. Both the conditions in Definition 24.1 are fulfilled by the enveloping curve obtained in previous theorem.

Proof. The slope of the tangent line to any specific member curve C_a (where t has a has a fixed value a for this member) of the family is given by Eq. (22.3):

$$\partial f / \partial x + (\partial f / \partial y)(dy / dx) = 0. \tag{24.3}$$

On the other hand, the slope of the tangent to the enveloping curve Γ (satisfying both Eqs. (24.2)) at the same point Q is given by

$$\partial f / \partial x + (\partial f / \partial y)(dy / dx) + (\partial f / \partial t)(dt / dx) = 0.$$

But, there also holds $\partial f(t)/\partial t = 0$ at every point of Γ causing reduction of above equation to Eq. (24.3). Thus, both the curves C_a (a member of the family) and Γ (the envelope of the family) have same tangent line at Q. In other words, they touch each other. //

Corollary 24.1. A family of curves whose equation is expressible as a quadratic relation in the parameter t :

$$f(t) \equiv F t^2 + G t + H = 0, \qquad (24.4)$$

has its envelope given by

$$G^2 = 4 F.H, \qquad (24.5)$$

where F, G, H are functions of variables x, y.

Example 24.1. Envelope of the family of straight lines $y = t x + a / t$ is the parabola

$$y^2 = 4 ax. \qquad (24.6)$$

Example 24.2. Find the envelope of the family of straight lines

$$y = m x + \sqrt{(a^2 m^2 + b^2)}, \qquad (24.7)$$

where m is a parameter and a, b are constant.

24.2. Envelope of family of curves involving two parameters

When the family of curves involves more than one parameter which are related to each other by requisite number of conditions the method to find the envelope of the family can be tackled by elimination of additional parameters by means of such conditions, if feasible, and reducing the equation of family to one parameter only so that the method discussed earlier can be applied. The following Examples illustrate this process.

Example 24.3. Find the envelope of family of ellipses centered at origin when the sum of squares of their semi-axes remains constant.

Solution. Taking the family of ellipses represented by Eq. (1.1.3), where both the semi-axes a, b are parameters connected by $a^2 + b^2 = c^2$ (constant) and eliminating one of the parameters, say b, above equation reduces to

$$x^2 / a^2 + y^2 / (c^2 - a^2) = 1; \quad \text{or,} \quad a^4 + (y^2 - x^2 - c^2) a^2 + c^2 x^2 = 0,$$

which is quadratic in (single) parameter a^2. Hence, the envelope is found by Eq. (24.5):

$$(y^2 - x^2 - c^2)^2 = 4c^2 x^2.$$

Taking its square-root, we have

$$y^2 - x^2 - c^2 = \pm 2c\, x \,; \text{ or,} \qquad y^2 = x^2 \pm 2c\, x + c^2 = (x \pm c)^2$$

$$\Rightarrow \qquad\qquad y = \pm (x \pm c)\, x,$$

which represent 4 straight lines $x + y = \pm c$ and $x - y = \pm c$. //

24.3. General method for more than one parameter

There is a more general method to find the envelope of a family of curves, say

$$f(x, y, a, b) = 0, \tag{24.8}$$

where the parameters a, b are connected by additional relation

$$\phi\,(a, b) = 0. \tag{24.9}$$

The relation (24.9) implies the mutual dependence of two parameters. Regarding both as functions of some single parameter, say t, we can now apply Eq. (22.1) to compute the derivatives of f and ϕ w.r.t. t:

$$df/dt = (\partial f/\partial a)\,(da/dt) + (\partial f/\partial b)\,(db/dt) = 0$$

$$\Rightarrow$$

$$(db/da) \equiv (db/dt) / (da/dt) = -(\partial f/\partial a) / (\partial f/\partial b),$$

and

$$d\phi/dt = (\partial\phi/\partial a)\,(da/dt) + (\partial\phi/\partial b)\,(db/dt) = 0$$

$$\Rightarrow$$

$$(db/da) \equiv (db/dt) / (da/dt) = -(\partial\phi/\partial a) / (\partial\phi/\partial b).$$

Thus, equating values of db/da obtained from above two relations, we get

$$(\partial f/\partial a) / (\partial f/\partial b) = (\partial\phi/\partial a) / (\partial\phi/\partial b). \tag{24.10}$$

Finally, elimination of a, b from above equations yields the envelope.

24.4. Envelope in polar coordinates

Example 24.4. Derive the equation of envelope of the family of circles described on the radii vectors (as diameters) of the following curves

$$r = a\,(1 + \cos\theta) \qquad (24.11); \quad \text{and} \quad r^n = a^n.\cos n\theta \qquad (24.12)$$

Solution. (i) Let A (l, α) be a point on the cardioid whose (polar) coordinates (l, α) satisfy Eq. (24.11):

$$l = a\,(1 + \cos\alpha). \qquad (24.13)$$

A circle C is described on OA as the diameter. Let P (r, θ) be a variable point on C. By a property of the circle, $\angle$ OPA is a right angle. So, length OP $= r$ (a radii vector of the circle C) is determined:

$$r = l \cos(\theta - \alpha) = a\,(1 + \cos\alpha).\cos(\theta - \alpha). \qquad (24.14)$$

For different choices of vectorial angle α the Eq. (24.14) represents a family of such circles. To find their envelope we follow the method given in Sub-section 24.1. Differentiating Eq. (24.14) partially w.r.t. parameter α :

$$-\sin\alpha .\cos(\theta - \alpha) + (1 + \cos\alpha).\sin(\theta - \alpha) = 0,$$

i.e.

$$\sin(\theta - 2\alpha) + \sin(\theta - \alpha) = 0 \quad \Rightarrow \quad \theta - 2\alpha = \alpha - \theta$$

$$\Rightarrow$$

$$\alpha = 2\theta/3 \qquad \text{and} \qquad \theta - \alpha = \theta/3.$$

Eliminating α from Eqs. (24.13) and (24.14), we get the envelope

$$r = a\,\{1 + \cos(2\theta/3)\}.\cos(\theta/3) = 2\,a.\cos^3(\theta/3).$$

(ii) Proceeding similarly, the equation of the family of circles is

$$r = l \cos(\theta - \alpha) = a.(\cos n\alpha)^{1/n}.\cos(\theta - \alpha). \qquad (24.15)$$

Differentiating it partially w.r.t. α, we get

$$-(\cos n\alpha)^{1/n - 1}.(\sin n\alpha).\cos(\theta - \alpha) + (\cos n\alpha)^{1/n}.\sin(\theta - \alpha) = 0,$$

or, on division by $(\cos n\alpha)^{1/n}.\cos(\theta - \alpha)$,

$$\tan (\theta - \alpha) = \tan n \alpha \Rightarrow \theta - \alpha = m\pi + n\alpha, \text{ for any integer } m$$

$\Rightarrow$

$$\alpha = (\theta - m\pi) / (n + 1).$$

Eliminating α from Eq. (24.15), we get the desired envelope:

$$r = a. \cos \{n (\theta - m\pi)/(n + 1) \}^{1/n}. \cos \{(n\,\theta + m\pi) / (n + 1) \}. \,//$$

§ 25. Involutes and evolutes

Definition 25.1. If tangents to a curve, say C, be normals to another curve C_5 then C_5 is called an *involute* of C and C is called the *evolute curve* of C_5.

If P $(\mathbf{r})$ is a point on C the corresponding point P_5 on C_5 is then given by

$$\mathbf{r}_5 = \mathbf{r} + u\hat{\mathbf{t}}, \qquad (25.1)$$

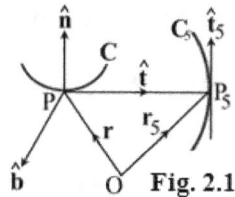

Fig. 2.1

where u = PP_5 and $\hat{\mathbf{t}}$ is the unit tangent vector to C at P. Thus, normals to C_5 are the tangents to C. So, C becomes the envelope of the family of normal to C_5. This gives rise to the alternate definition of evolute of a curve.

Definition 25.2. Evolute of a curve (say C_5) is the envelope of family of its normal.

Example 25.1. Evolute of the ellipse in Eq. (1.1.3) is represented by

$$(a\,x)^{2/3} + (b\,y)^{2/3} = (a^2 - b^2)^{2/3} . \qquad (25.2)$$

Solution. Let P $(a \cos t, b \sin t)$ be a point on the ellipse, where t is a parameter. The slope of the tangent to ellipse at P is given by

$$dy / dx = \frac{dy / dt}{dx / dt} = (- b / a) \cot t \Rightarrow \text{slope of normal} = (a / b) \tan t.$$

Hence, the equation of normal at P is

$$y - b \sin t = (a/b) \tan t. (x - a \cos t),$$

or,

$$a\,x/\cos t - by/\sin t \;=\; a^2 - b^2. \tag{25.3}$$

Differentiating Eq. (25.3) partially w.r.t. there results

$$ax \sin t \,/\, \cos^2 t + by \cos t \,/\sin^2 t \,=\, 0 \Rightarrow \tan^3 t \,=\, -(by\,/\,ax).$$

Therefore,

$$\sin t \;=\; -(by)^{1/3} \,/\, \sqrt{\{(ax)^{2/3} + (by)^{2/3}\}},$$

$$\cos t = (ax)^{1/3} \,/\, \sqrt{\{(ax)^{2/3} + (by)^{2/3}\}}.$$

Putting for these values in Eq. (25.3), the envelope of normals is derived:

$$\{(a\,x)^{2/3} + (by)^{2/3}\}\sqrt{\{(ax)^{2/3} + (by)^{2/3}\}} \;=\; a^2 - b^2.$$

Raising the power by 2/3 it assumes the desired form. //

§ 26. Series and expansion of functions

26.1. Power series

Definition 26.1. Let $\{a_o\}$, $n = 0, 1, 2, \ldots \infty$, be a sequence of real numbers. The series of the form (in ascending powers of a variable say x)

$$\sum_{n=0}^{\infty} a_n \,(x - x_o)^n \;\equiv\; a_o + a_1 (x - x_o) + a_2 (x - x_o)^2 + \ldots$$

$$+\, a_n (x - x_o)^n + \ldots \infty, \tag{26.1}$$

where x_0 is some fixed real number and the coefficients a_i's are independent of x, is called a *power series* in the variable x. It may converge for some or all values of x.

Shifting origin to x_0, above series reduces to

$$\sum u_n \equiv \sum_{n=0}^{\infty} a_n x^n \;=\; a_o + a_1 x + a_2 x^2 + \ldots + a_n x^n + \ldots \infty, \tag{26.2}$$

Associated to every power series there is a real number r, $0 \le r \le \infty$, called the *radius of convergence* of the series.

26.2. Interval of convergence of a power series

The power series $\Sigma\, u_n$ converges (to a finite definite sum) if

$$\lim{}_{n \to \infty} u_n / u_{n+1} = \lim{}_{n \to \infty} a_n / a_{n+1}.\, x = 1/x \text{ (say)} \qquad (26.3)$$

is numerically greater than 1, i.e. when $|x| < l$. Thus, the *interval of convergence* of $\Sigma\, u_n$ is $(-l,\, l)$ within which the series converges. It diverges for the values of x outside this interval. In other words, the radius of convergence of the series is

$$r = l = \lim{}_{n \to \infty} a_n / a_{n+1}. \qquad (26.4)$$

Further, the series (26.2) is called *absolutely convergent* if

$$\sum_{n=1}^{\infty} |a_n| \equiv |a_1| + |a_2| + \dots + |a_n| + \dots \infty \qquad (26.5)$$

is convergent. Since $|a_n|$ is greater or equal to a_n, an *absolutely convergent series is always convergent.*

Note 26.1. If a power series $\Sigma\, a_n x^n$ converges for $x = x_1$, then it converges absolutely for $|x| < |x_1|$.

Theorem 26.1. Let the power series in Eq. (26.2) be of radius of convergence r then both the power series

$$\Sigma\, u_n' \equiv \Sigma\, n.\, a_n x^{n-1} \quad \text{and} \quad \Sigma\, v_n \equiv \Sigma\, a_n x^{n+1} / (n+1),$$

obtained on term-wise differentiation and integration respectively are of the same radius of convergence.

Proof. The radius of convergence of the series $\Sigma\, u_n'$ is

$$\lim{}_{n \to \infty} n.\, a_n / (n+1)\, a_{n+1} = \lim{}_{n \to \infty} a_n / (1 + 1/n)\, a_{n+1},$$

whereas that of $\Sigma\, v_n$ is

$$\lim{}_{n \to \infty} \{a_n / (n+1) \div a_{n+1} / (n+2)\}$$

$$= \lim{}_{n \to \infty} (1 + 2/n)\, a_n / (1 + 1/n)\, a_{n+1}.$$

Both these limits attain the value r as given by Eq. (26.4). //

§ 27. Maclaurin's series

Consider the power series $f(x) = \sum a_n x^n$ of *radius of convergence r*. Differentiating it successively with respect to x within its radius of convergence, we find

$$f'(x) = a_1 + 2a_2 x + 3a_3 x^2 + 4a_4 x^3 + \dots ;$$

$$f''(x) = 2.1.a_2 + 3.2.a_3 x + 4.3.a_4 x^2 + \dots ;$$

$$f'''(x) = 3.2.1.a_3 + 4.3.2.\, a_4 x + \dots ; \dots\dots ;$$

$$f^{(k)}(x) = n(n-1)(n-2)\dots(n-k+1)\, a_n x^{n-k} + \dots$$

$$\Rightarrow \qquad\qquad f^{(n)}(x) = n!\, a_n + \dots$$

Evaluating $f(x)$ and its derivatives at origin, we find

$$f(0) = a_0,\ f'(0) = a_1,\ f''(0) = 2!\, a_2,\ f'''(0) = 3!\, a_3, \dots ,$$

$$f^{(n)}(0) = n!\, a_n ,\ \dots \tag{27.1}$$

Thus, the coefficients a_i's are determined and the power series in Eq. (26.2) assumes the form

$$f(x) = f(0) + x f'(0) + (x^2/2!) f''(0) + (x^3/3!) f'''(0) + \dots$$

$$+ (x^n/n!) f^{(n)}(0) + \dots \tag{27.2}$$

This form of the power series is called the Maclaurin's series and the result expressed by Eq. (27.2) is also known as Maclaurin's theorem.

Example 27.1. Expand the following functions by Maclaurin's theorem:

(i) e^x,   (ii) $\sin x$,   (iii) $\cos x$, (iv) $\tan^{-1} x$, (v) $\ln(1+x)$.

Solution. (i) The function e^x remains the same on derivation: $f^{(n)}(x) = e^x$. Hence, the coefficients are

$$a_0 = e^0 = 1,\ a_1 = 1, \qquad a_2 = 1/2!, \qquad a_3 = 1/3!, \dots ,\ a_n = 1/n!, \dots$$

Accordingly, the Eq. (27.2) becomes

$$e^x = 1 + x + x^2/2! + x^3/3! + \dots + x^n/n! + \dots \infty. \tag{27.3}$$

(ii) Derivatives of $f(x) = \sin x$ are given by

$$f'(x) = \cos x, \quad f''(x) = -\sin x, \quad f'''(x) = -\cos x, \dots,$$

$$f^{(n)}(x) = \sin(x + n\pi/2), \dots \tag{27.4}$$

Hence,

$$f(0) = 0, \quad f'(0) = 1, \quad f''(0) = 0, \quad f'''(0) = -1, \dots,$$

$$f^{(n)}(0) = \sin(n\pi/2), \dots$$

$\Rightarrow$

$$a_o = 0, \quad a_1 = 1, \quad a_2 = 0, \quad a_3 = -1/3!, \quad a_4 = 0, \quad a_5 = 1/5!, \dots$$

and the function expands as

$$\sin x = x - x^3/3! + x^5/5! - \dots \infty. \tag{27.5}$$

(iii) Derivatives of $f(x) = \cos x$ are given by

$$f'(x) = -\sin x, \quad f''(x) = -\cos x, \quad f'''(x) = \sin x, \dots,$$

$$f^{(n)}(x) = \cos(x + n\pi/2), \dots$$

$\Rightarrow$

$$f(0) = 1, \quad f'(0) = 0, f''(0) = -1, f'''(0) = 0, \quad f^{(iv)}(0) = 1, \dots,$$

$$f^{(n)}(0) = \cos(n\pi/2), \dots$$

$\Rightarrow$

$$a_o = 1, \quad a_1 = 0, \quad a_2 = -1/2!, \quad a_3 = 0, \quad a_4 = 1/4!, \dots$$

and the function expands as

$$\cos x = 1 - x^2/2! + x^4/4! - \dots \infty. \tag{27.6}$$

(iv) Derivatives of $f(x) = \tan^{-1} x$ are given by

$$f'(x) = 1/(1 + x^2), \quad f''(x) = -2x/(1 + x^2)^2,$$

$$f'''(x) = -2/(1 + x^2)^2 + 8x^2/(1 + x^2)^3 = 2(3x^2 - 1)/(1 + x^2)^3,$$

$$f^{(iv)}(x) = 24x/(1 + x^2)^3 - 48x^3/(1 + x^2)^4,$$

$\Rightarrow$

$$f^{(v)}(x) = 24 / (1 + x^2)^3 - 288x^2 / (1 + x^2)^4 + 384x^4 / (1 + x^2)^5, \ \ldots$$

$$f(0) = 0, \qquad f'(0) = 1, \ f''(0) = 0, \qquad f'''(0) = -2,$$

$$f^{iv}(0) = 0, \qquad f^{(v)}(0) = 24, \ \ldots$$

$\Rightarrow$

$$a_0 = 0, \ a_1 = 1, \qquad a_2 = 0, \qquad a_3 = -1/3, \ \ a_4 = 0, \ \ a_5 = 1/5, \ldots$$

The desired expansion is

$$\tan^{-1} x = x - x^3/3 + x^5/5 - \ \ldots \ \infty. \tag{27.7}$$

(v) Derivatives of $f(x) = \ln(1 + x)$ are given by

$$f'(x) = 1 / (1 + x), \ f''(x) = -1 / (1 + x)^2,$$

$$f'''(x) = 2 / (1 + x)^3, \qquad f^{(iv)}(x) = -6 / (1 + x)^4, \ \ldots,$$

$$f^{(n)}(x) = (-1)^{n-1}(n - 1)! / (1 + x)^n, \ \ldots$$

$\Rightarrow$

$$f(0) = 0, \ f'(0) = 1, \ f''(0) = -1, \ f'''(0) = 2, \ f^{iv}(0) = -6, \ \ldots,$$

$$f^{(n)}(0) = (-1)^{n-1}(n - 1)!, \ \ldots$$

$\Rightarrow$

$$a_0 = 0, \ \ a_1 = 1, \qquad a_2 = -1/2, \qquad a_3 = 1/3, \ a_4 = -1/4, \ \ldots,$$

$$a_n = (-1)^{n-1} / n, \ \ldots$$

The desired expansion is

$$\ln(1 + x) = x - x^2/2 + x^3/3 - x^4/4 + \ldots + (-1)^{n-1} x^n / n, \ \ldots \ \infty. \tag{27.8}$$

§ 28. Taylor's series

Starting with the general form of a power series as given by Eq. (26.1) and proceeding similarly (as in the previous section), the successive derivatives of the function $f(x) = \sum a_n (x - x_0)^n$ at the point x_0 are evaluated:

$$f(x_0) = a_0, \ f'(x_0) = a_1, \ f''(x_0) = 2!a_2, \ \ f'''(x_0) = 3!a_3, \ \ldots,$$

$$f^{(n)}(x_0) = n! \, a_n, \ldots \tag{28.1}$$

Accordingly, the power series in Eq. (26.1) assumes the form

$$f(x) = f(x_0) + f'(x_0)(x - x_0) + f''(x_0)(x - x_0)^2/2! + \ldots$$

$$f^{(n)}(x_0)(x - x_0)^n/n! + \ldots \, \infty. \tag{28.2}$$

Above series gives a generalization of Maclaurin's series in Eq. (27.2) and is called Taylor's series for the function $f(x)$ expanded around the point x_0. Putting $x - x_0 = h$ (say), the Eq. (28.2) reduces to an alternate form:

$$f(x_0 + h) = f(x_0) + h f'(x_0) + (h^2/2!) f''(x_0) + \ldots$$

$$+ (h^n/n!) \, f^{(n)}(x_0) + \ldots \, \infty, \tag{28.3a}$$

or, more generally (when x_0 is replaced by x),

$$f(x + h) = f(x) + h f'(x) + (h^2/2!) f''(x) + \ldots$$

$$+ (h^n/n!) f^{(n)}(x) + \ldots \, \infty. \tag{28.3b}$$

Further, interchanging x and h, above series may also be written as

$$f(x + h) = f(h) + x f'(h) + (x^2/2!) f''(h) + \ldots$$

$$+ (x^n/n!) f^{(n)}(h) + \ldots \, \infty. \tag{28.3c}$$

Note 28.1. Shifting origin to the point x_0, the Taylor's series (28.2) clearly reduces to Maclaurin's series in Eq. (27.2).

Example 28.1. Expand the functions $\ln(x + h)$ and $\sin^{-1}(x + h)$ in powers of x.

Solution. (i) Differentiating the function $f(x) = \ln(x + h)$ successively with respect to x:

$$f'(x) = 1/(x + h), \, f''(x) = -1/(x + h)^2, f'''(x) = 2/(x + h)^3, \ldots,$$

$$f^{(n)}(x) = (-1)^{n-1}(n - 1)! \, (x + h)^{-n}, \ldots;$$

and evaluating them at the point $x = 0$, we find

$$f(0) = \ln h, \ f'(0) = 1 / h, \ f''(0) = -1 / h^2, \ f'''(0) = 2 / h^3, \ ...,$$

$$f^{(n)}(0) = (-1)^{n-1} (n-1)! \ h^{-n}, \ ... \qquad (28.4)$$

Hence, Eq. (27.1) determines

$$a_0 = \ln h, \ a_1 = 1/h, \ a_2 = -1/2h^2, \ a_3 = 1/3h^3, \ ... , \ a_n = (-1)^{n-1} / nh^n, \ ...$$

The desired expansion is

$$\ln(x + h) = \ln h + x/h - x^2/2h^2 + x^3/3h^3 - ...$$

$$+ (-1)^{n-1} x^n / nh^n + ... \ \infty. \qquad (28.5)$$

(ii) Differentiating the function $f(x) = \sin^{-1}(x + h)$ successively with respect to x:

$$f'(x) = 1 / \{1 - (x+h)^2\}, \ f''(x) = (x+h) / \{1 - (x+h)^2\}^{3/2},$$

$$f'''(x) = 1 / \{1 - (x+h)^2\}^{3/2} + 3(x+h)^2 / \{1 - (x+h)^2\}^{5/2}, \ ...$$

and evaluating them at the point $x = 0$, we find

$$f(0) = \sin^{-1} h, \ f'(0) = 1 / \sqrt{(1 - h^2)}, \ f''(0) = h / (1 - h^2)^{3/2},$$

$$f'''(0) = (1 + 2h^2 / (1 - h^2)^{5/2}, \ ...$$

$\Rightarrow$

$$a_0 = \sin^{-1} h, \quad a_1 = 1 / \sqrt{(1 - h^2)}, \quad a_2 = h / 2! \ (1 - h^2)^{3/2},$$

$$a_3 = (1 + 2h^2) / 3! \ (1 - h^2)^{5/2}, \ ...$$

Therefore, the function expands as per Eq. (27.2):

$$\sin^{-1}(x + h) = \sin^{-1} h + x / \sqrt{(1 - h^2)} + x^2 h / 2! \ (1 - h^2)^{3/2}$$

$$+ x^3 (1 + 2h^2) / 3! \ (1 - h^2)^{5/2} + ... \ \infty. \qquad (28.6)$$

Alternately, taking $f(x) = \ln x$, its derivatives with respect to x at $x = h$ are in accordance with Eq. (28.4):

$$f'(h) = 1 / h, \qquad f''(h) = -1 / h^2, \qquad f'''(h) = 2 / h^3, \ ...$$

Correspondingly, Eq. (28.3c) assumes the form of Eq. (28.5). Similarly, the series in Eq. (28.6) can also be derived by Eq. (28.3c). //

Example 28.2. Expand the functions ln sin $(x + h)$ in powers of h by Taylor's theorem.

Solution. Taking $f(x) = \ln \sin x$ its derivatives with respect to x are:

$$f'(x) = \cot x, \qquad f''(x) = -\operatorname{cosec}^2 x, \qquad f'''(x) = 2 \operatorname{cosec}^2 x. \cot x, \ \ldots$$

hence, the Taylor's series given by Eq. (28.3b) becomes

$$\ln \sin (x + h) = \ln \sin x + h. \cot x - (h^2/2) \operatorname{cosec}^2 x$$

$$+ (h^3/3) \operatorname{cosec}^2 x. \cot x - \ldots \infty. //$$

Example 28.3. Expand sin x in powers of $x - \pi/2$.

Solution. Taking $f(x) \equiv \sin x$ as $\sin(\pi/2 + x - \pi/2)$ and putting $x_0 = \pi/2$, the Eq. (28.2) becomes

$$\sin x = f(\pi/2) + (x - \pi/2)f'(\pi/2) + \{(x - \pi/2)^2/2!\} f''(\pi/2)$$

$$+ \{(x - \pi/2)^3/3!\} f'''(\pi/2) + \ldots \infty. \tag{28.7}$$

Derivatives of sin x are obtained vide Eq. (27.4). Evaluating them at the point $x = \pi/2$:

$$f(\pi/2) = 1, \quad f'(\pi/2) = 0, \quad f''(\pi/2) = -1, \quad f'''(\pi/2) = 0,$$

$$f^{(iv)}(\pi/2) = 1, \ldots$$

Hence, Eq. (28.7) reduces to

$$\sin x = 1 - (x - \pi/2)^2/2! + (x - \pi/2)^4/4! - \ldots \infty. //$$

Example 28.4. Using Taylor' theorem prove that

$$\tan^{-1}(x + h) = \tan^{-1}x + (h. \sin z)(\sin z)/1 - (h. \sin z)^2 (\sin 2z)/2$$

$$+ (h. \sin z)^3 (\sin 3z)/3 - \ldots \infty; \text{ where } z = \cot^{-1}x.$$

Solution. For $\cot z = x$, we have

$$\sin z = 1 / \sqrt{(1 + x^2)}, \qquad \sin 2z = 2x / (1 + x^2),$$

$$\sin 3z = 3 \sin z - 4.\sin^3 z = \{3 - 4 / (1 + x^2)\} / \sqrt{(1 + x^2)}$$

$$= (3x^2 - 1) / (1 + x^2)^{3/2}, \ldots$$

Hence, the derivatives of $f(x) = \tan^{-1} x$, obtained in Example 27.1 (iv), reduce as

$$f'(x) = \sin^2 z, \ f''(x) = -(\sin^2 z) \sin 2z, \ f'''(x) = 2 (\sin^3 z) \sin 3z, \ldots$$

Accordingly, the Eq. (28.3b) assumes the desired form. //

CHAPTER 6

CALCULUS (INTEGRAL)

§ 1. Introduction

Let $y = f(x)$ defines a curve C in the xOy-plane. Let A and B be two fixed points on C whose ordinates AD, BE, the x-axis and the curve itself bound the area ADEB. If P (x, y) and Q $(x + \delta x, y + \delta y)$ are two neighbouring points on C the area of the elementary strip PLMQ lies between the areas of two rectangles PLMI and JLMQ:

$y\, \delta x < \text{Area (PLMQ)} < (y + \delta y)\, \delta x.$ (1.1)

Making Q sufficiently close to P the area of the strip PLMQ becomes

Fig. 1.1

$$\delta M = \lim_{\delta x \to 0} (y\, \delta x).$$

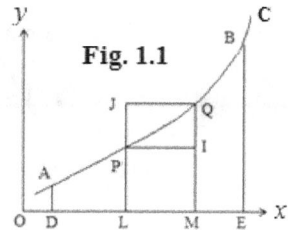

Treating the whole area ADEB as a collection of such strips PLMQ of small width δx we, thus, find

$$\text{Area ADEB} = \Sigma \lim_{\delta x \to 0} (y\, \delta x) = \lim_{\delta x \to 0} \Sigma f(x)\, \delta x. \qquad (1.2)$$

As per B. Riemann, the last limit defines the integral of the function $f(x)$ within the limits $x = a\, (= OD)$ to $x = b\, (= OE)$:

$$\int_a^b f(x)\, dx. \qquad (1.3)$$

Later on, it was discovered that the integration is a process reverse to that of differentiation; hence called as an *anti-derivation*.

§ 2. Indefinite integrals of some standard functions

Let a function $f(x)$ posses its derivative $g(x)$ with respect to x then an integral of $g(x)$ with respect to x is $f(x)$. Symbolically it is written as

$$\int g(x)\, dx = f(x). \qquad (2.1)$$

Since the derivatives of $f(x)$ and $f(x) + c$, where c is an arbitrary constant of integration (i.e. a fixed real number) are the same, the right member in Eq. (2.1) may be more appropriately replaced by $f(x) + c$. Because of this property, the above integral is called an *indefinite integral* of $g(x)$. If, however, the integration is carried out within some region bounded by two values of the variable x, say $x = a$ to $x = b$, then the integral is called a *definite integral* and it attains a definite value $f(b) - f(a)$. We then write

$$\int_a^b g(x)\, dx = \left[f(x)\right]_a^b = f(b) - f(a). \qquad (2.2)$$

In the following we obtain indefinite integrals of certain functions:

Theorem 2.1. We have

$$\int x^n\, dx = x^{n+1}/(n+1), \quad \text{if } n \ne -1; \qquad (2.3)$$

$$\int (1/x)\, dx = \ln|x| \qquad (2.4); \qquad \int e^x\, dx = e^x; \qquad (2.5)$$

$$\int a^x\, dx = a^x/\ln a \qquad (2.6); \qquad \int \sin x\, dx = -\cos x \qquad (2.7);$$

$$\int \cos x\, dx = \sin x; \qquad (2.8)$$

$$\int \tan x\, dx = \ln|\sec x| = -\ln|\cos x|; \qquad (2.9)$$

$$\int \cot x\, dx = -\ln|\operatorname{cosec} x| = \ln|\sin x|; \qquad (2.10)$$

$$\int \sec x.\tan x\, dx = \sec x, \qquad (2.11)$$

$$\int \operatorname{cosec} x.\cot x\, dx = -\operatorname{cosec} x; \qquad (2.12)$$

$$\int \sec^2 x\, dx = \tan x \qquad (2.13); \int \operatorname{cosec}^2 x\, dx = -\cot x; \qquad (2.14)$$

$$\int \sinh x\, dx = \cosh x \qquad (2.15); \int \cosh x\, dx = \sinh x. \qquad (2.16)$$

$$\int \tanh x\, dx = \ln|\cosh x| \qquad (2.17); \quad \int \coth x\, dx = \ln|\sinh x|. \qquad (2.18)$$

Corollary 2.1. $\int 1\, dx = x \quad (2.19); \quad \int (1/x^2)\, dx = -1/x; \qquad (2.20)$

$$\int \tan^2 x\, dx = \int (\sec^2 x - 1)\, dx = \tan x - x; \qquad (2.21)$$

$$\int \cot^2 x\, dx = \int (\operatorname{cosec}^2 x - 1)\, dx = -\cot x - x. \qquad (2.22)$$

Theorem 2.2. We also have

$$\int \operatorname{cosec} x \, dx = \ln| \operatorname{cosec} x - \cot x | = \ln| \tan (x/2)|, \qquad (2.23)$$

and

$$\int \sec x \, dx = \ln| \sec x + \tan x | = \ln| \tan (x/2 + \pi/4) |. \qquad (2.24)$$

Theorem 2.3. We have

$$\int \{1 / \surd (a^2 - x^2)\} \, dx = \sin^{-1} (x/a), \qquad (2.25)$$

$$-\int \{1 / \surd (a^2 - x^2)\} \, dx = \cos^{-1} (x/a), \qquad (2.26)$$

$$\int \{1 / (a^2 + x^2)\} \, dx = (1/a) \tan^{-1} (x/a), \qquad (2.27)$$

$$-\int \{1 / (a^2 + x^2)\} \, dx = (1/a) \cot^{-1} (x/a), \qquad (2.28)$$

$$\int \{1 / (a^2 - x^2)\} \, dx = (1/2a) \ln| (a + x) / (a - x) |, \qquad (2.29)$$

$$\int \{1 / (x^2 - a^2)\} \, dx = (1/2a) \ln| (x - a) / (x + a) |, \qquad (2.30)$$

$$\int \{1 / \surd (a^2 + x^2)\} \, dx = \sinh^{-1} (x/a), \qquad (2.31)$$

$$\int \{1 / \surd (x^2 - a^2)\} \, dx = \cosh^{-1} (x/a). \qquad (2.32)$$

Theorem 2.4. We have

$$\int \surd (a^2 - x^2) \, dx = (x/2) \surd (a^2 - x^2) + (a^2/2) \sin^{-1} (x/a); \qquad (2.33)$$

$$\begin{aligned} \int \surd (a^2 + x^2) \, dx &= (x/2) \surd (a^2 + x^2) + (a^2/2) \sinh^{-1} (x/a), \\ &= (x/2) \surd (a^2 + x^2) + (a^2/2). \ln \{x + \surd (a^2 + x^2)\}; \end{aligned} \qquad (2.34)$$

$$\begin{aligned} \int \surd (x^2 - a^2) \, dx &= (x/2) \surd (x^2 - a^2) - (a^2/2) \cosh^{-1} (x/a); \\ &= (x/2) \surd (x^2 - a^2) - (a^2/2). \ln \{x + \surd (x^2 - a^2)\}. \end{aligned} \qquad (2.35)$$

Theorem 2.5. Let u be a differentiable function and v easily integrable then

$$\int uv \, dx = u \int v \, dx - \int \{(du/dx) (\int v \, dx)\} dx; \qquad (2.36)$$

Note 2.1. Above method of integration is called the *method of integration by parts.*

Theorem 2.6. We have

$$\int (e^{ax} \sin bx)\, dx = (a^2 + b^2)^{-1}\, e^{ax}\,(a \sin bx - b \cos bx), \quad (2.37)$$

and

$$\int (e^{ax} \cos bx)\, dx = (a^2 + b^2)^{-1}\, e^{ax}\,(a \cos bx + b \sin bx). \quad (2.38)$$

Example 2.1. Integrate $\sin^2 x$ with respect to x.

Solution. From trigonometry, $\sin^2 x = (1 - \cos 2x)\,/\,2$. Therefore,

$$\int \sin^2 x\, dx \; = \; (1/2)\int (1 - \cos 2x)\, dx = \; (1/2)\{x - (\sin 2x)\,/\,2\}. \; //$$

Example 2.2. Integrate $\cos^4 x$ with respect to x.

Solution. We have

$$\cos^4 x = \; (1 + \cos 2x)^2\,/\,4 = (1 + 2 \cos 2x + \cos^2 2x)\,/\,4$$

$$= \{1 + 2 \cos 2x + (1 + \cos 4x)\,/\,2\}/\,4.$$

Therefore,

$$\int \cos^4 x\, dx \; = (1/4)\,[x + \sin 2x + (1/2)\,\{x + (\sin 4x)\,/\,4\}]. \; //$$

Example 2.3. Integrate $\ln |x|$.

Solution. Integrating it by method of parts taking unity as the second function we get

$$\int \ln |x|\, dx = x.\ln |x| - \int x.(1/x)\, dx = \; x.\ln |x| - x. \; // \quad (2.39)$$

§ 3. Different methods of integration

3.1. Substitution method: If the integrand contains a function, say f (x), and its derivative $f'\,(x)$ appears as a multiple of dx then we use the substitution

$$f(x) \; = \; t \quad \text{so that} \quad f'(x)\, dx \; = \; dt. \quad (3.1)$$

Hence,

$$\int [f'(x)\,/\,g\,\{f(x)\}]\, dx \; = \int \{1\,/\,g\,(t)\}\, dt.$$

3.2. $\int \{1\,/\,(ax^2 + bx + c)\}\, dx$

$$ax^2 + bx + c = a\,(x^2 + bx/a + c/a) = a\{(x + b/2a)^2 + (c/a - b^2/4a^2)\}. \quad (3.2)$$

(i) If $c/a - b^2/4a^2 > 0$, i. e. when $4ac > b^2$, above expression reduces to $a(t^2 + k^2)$, where

$$t \equiv x + b/2a \quad \text{and} \quad k^2 = c/a - b^2/4a^2 = (4ac - b^2)/4a^2$$

so that

$$ak = \sqrt{(4ac - b^2)}/2 \quad \text{and} \quad t/k = (2ax + b)/\sqrt{(4ac - b^2)}.$$

Therefore,

$$\int dx/(ax^2 + bx + c) = (1/a)\int dt/(t^2 + k^2)\}$$

$$= (1/ak)\tan^{-1}(t/k), \qquad \text{by Eq. (2.28)}$$

$$= \{2/\sqrt{(4ac - b^2)}\}.\tan^{-1}\{(2ax + b)/\sqrt{(4ac - b^2)}\}. \qquad (3.3)$$

(ii) On the other hand, if $c/a - b^2/4a^2 < 0$, i.e. when $4ac < b^2$, Eq. (3.2) reduces to $a(t^2 - k^2)$, where

$$t \equiv x + b/2a \quad \text{and} \quad k^2 = b^2/4a^2 - c/a = (b^2 - 4ac)/4a^2.$$

Therefore,

$$ak = \sqrt{(b^2 - 4ac)}/2 \quad \text{and} \quad t \pm k = \{2ax + b \pm \sqrt{(b^2 - 4ac)}\}/2a.$$

Hence, the integral reduces to

$$(1/a)\int dt/(t^2 - k^2) = (1/2ak)\ln|(t - k)/(t + k)|, \qquad \text{by Eq. (2.31)}$$

$$= \{1/\sqrt{(b^2 - 4ac)}\}\ln|\{2ax + b - \sqrt{(b^2 - 4ac)}\}/\{2ax + b + \sqrt{(b^2 - 4ac)}\}|. \qquad (3.4)$$

(iii) When $c/a - b^2/4a^2 = 0$, RHS of Eq. (3.2) reduces to $a(x + b/2a)^2$ and the integral becomes

$$(1/a)\int(x + b/2a)^{-2}\,dx = -1/a(x + b/2a) = -1/(ax + b/2). \,//$$

3.3. $\int\{(px + q)/(ax^2 + bx + c)\}\,dx$: Since

$$(px + q)/(ax^2 + bx + c) = \{(p/2a)(2ax + b) + (q - bp/2a)\}/(ax^2 + bx + c),$$

the integral splits as

$$(p/2a)\int\{(2ax + b)/(ax^2 + bx + c)\}\,dx + (q - bp/2a)\int\{1/(ax^2 + bx + c)\}\,dx. \qquad (3.5)$$

Putting $ax^2 + bx + c = t$ so that $(2ax + b)\,dx = dt$, the first integral becomes

$$(p/2a) \int (1/t) \, dt = (p/2a) \ln |t| = (p/2a) \ln |ax^2 + bx + c|, \quad \text{by Eq. (2.4)};$$

and the second integral can be evaluated by methods of previous Subsection.

3.4. Some reduction formulae

Theorem 3.1. Find a reduction formula for the integral

$$I_n = \int x^n \, e^{ax} \, dx.$$

Proof. Integrating it by the method of parts taking e^{ax} as the second function

$$I_n = \{x^n \, e^{ax} - n \int x^{n-1} \, e^{ax} \, dx\}/a = x^n \, e^{ax}/a - (n/a) I_{n-1}. \quad (3.6)$$

Replacing n by $n-1$ above formula also yields

$$I_{n-1} = x^{n-1}. e^{ax}/a - \{(n-1)/a\} I_{n-2}.$$

Hence, Eq. (3.6) reduces to

$$I_n = x^n \, e^{ax}/a - (n/a) [x^{n-1} \, e^{ax}/a - \{(n-1)/a\} I_{n-2}]$$

$$\Rightarrow$$

$$a^2 I_n = ax^n \, e^{ax} - n x^{n-1} \, e^{ax} + n (n-1) I_{n-2},$$

or,

$$a^2 I_n - n (n-1) I_{n-2} = x^{n-1} \, e^{ax} (ax - n). \quad (3.7)$$

Both Eqs. (3.6) and (3.7) represent reduction formulae for the given integral. //

Theorem 3.2. Reduction formula for the integral $I_n = \int x^n \sin ax \, dx$ is

$$a^2 I_n + n (n-1) I_{n-2} = x^{n-1}\{- ax \cos ax + n.\sin ax). \quad (3.8)$$

§ 4. Definite integrals

As seen in § 1, the integral in Eq. (1.3) has specific range of integration from $x = a$ to $x = b$. So, it defines a *definite integral*. In the following we discuss some of the basic properties of such integrals.

Theorem 4.1. We have

$$\int_{x=a}^{b} f(x)dx = \int_{t=a}^{b} f(t)dt, \tag{4.1}$$

$$\int_{x=a}^{b} f(x)dx = -\int_{x=b}^{a} f(x)dx, \tag{4.2}$$

$$\{\int_{x=a}^{c} + \int_{x=c}^{b} \} f(x)\, dx = \int_{a}^{b} f(x)\, dx, \quad \text{if } a < c < b; \tag{4.3}$$

$$\int_{x=0}^{a} f(x)\, dx = \int_{x=0}^{a} f(a-x)\, dx; \tag{4.4}$$

$$\left.\begin{array}{c} \int_{-a}^{a} f(x)dx = 2\int_{0}^{a} f(x)dx, \text{ if } f(x) \text{ is an even function of } x, \\ \text{i.e. } f(-x) = f(x), \\ = 0, \text{ if } f(x) \text{ is an odd function of } x: f(-x) = -f(x); \end{array}\right\} \tag{4.5}$$

$$\int_{x=0}^{2a} f(x)dx = 2\int_{x=0}^{a} f(x)dx, \text{ or } \qquad 0, \tag{4.6}$$

if $f(2a-x) = f(x)$ or $-f(x)$ respectvely.

Example 4.1. Evaluate the integral $\int_{x=0}^{2} |x-1|\, dx$.

Solution. The integrand $|x-1|$ is $1-x$ in the range from $x = 0$ to 1 and $x - 1$ in the range from $x = 1$ to 2. Thus, breaking the range of integration as above we get

$$\int_{x=0}^{2} |x-1|\, dx = \int_{x=0}^{1} (1-x)dx + \int_{x=1}^{2} (x-1)dx$$

$$= \left[x - x^2/2\right]_{0}^{1} + \left[x^2/2 - x\right]_{1}^{2} = (1 - 1/2) + (2 - 2) - (1/2 - 1) = 1. \text{ //}$$

Example 4.2. Evaluate the integral $\int_{x=-3}^{3} 5|x+2|\, dx$.

Solution. Breaking the range of integration from $x = -3$ to -2 and from $x = -2$ to 3, the integrand is

$|x+2| = -x-2$ when $-3 < x < -2$, and $x+2$ when $-2 < x < 3$.

Thus, the integral splits as

$$5\int_{x=-3}^{3}|x+2|\,dx = 5\int_{x=-3}^{-2}(-x-2)dx + 5\int_{x=-2}^{3}(x+2)dx =$$

$$5\left[-x^2/2-2x\right]_{-3}^{-2} + 5\left[x^2/2+2x\right]_{-2}^{3}$$

$$= 5\{(-2+4)+(9/2-6)+(9/2+6)-(2-4)\} = 65. \; //$$

Example 4.3. Prove that

$$I \equiv \int_{x=0}^{\pi/2} \ln(\sin x)\,dx = -(\pi/2)\ln 2. \tag{4.7}$$

Solution. Applying the fourth property of definite integrals given by Eq. (4.4):

$$I \equiv \int_{x=0}^{\pi/2} \ln\{\sin(\pi/2-x)\}dx = \int_{x=0}^{\pi/2} \ln(\cos x)\,dx .$$

Adding the two integrals we find

$$2I \equiv \int_{x=0}^{\pi/2} \ln(\sin x.\cos x)\,dx = \int_{x=0}^{\pi/2} \{\ln(\sin 2x)/2\}\,dx$$

$$= \int_{x=0}^{\pi/2} \ln(\sin 2x)\,dx - \int_{x=0}^{\pi/2} (\ln 2)\,dx. \tag{4.8}$$

Putting $2x = t$ so that $2dx = dt$, the first integral in (4.8) reduces to

$$(1/2)\int_{t=0}^{\pi} \ln(\sin t)\,dt = \int_{t=0}^{\pi/2} \ln(\sin t)\,dt, \qquad \text{by Eq. (4.6).}$$

This is same as I in view of Eq. (4.1). Hence, Eq. (4.8) reduces to

$$I = -(\ln 2)\left[x\right]_{0}^{\pi/2} = -(\pi/2)\ln 2. \; //$$

§ 5. Some theorems on definite integrals

Theorem 5.1. Let $f(x)$ and $g(x)$ be two functions integrable over a closed interval $[a, b]$ and k is any constant (i.e. a fixed real number) then kf, $f \pm g$ are also integrable over the same interval:

$$\int_{x=a}^{b} k f(x)\, dx = k \int_{x=a}^{b} f(x)\, dx, \qquad (5.1)$$

$$\int_{x=a}^{b} \{f(x) \pm g(x)\}\, dx = \int_{x=a}^{b} f(x)\, dx \pm \int_{x=a}^{b} g(x)\, dx. \quad (5.2)$$

Theorem 5.2. Let $f(x)$ be a function integrable over a closed interval $[a, b]$. If

$$f(x) \geq 0 \quad \forall\, x\, \varepsilon\, [a, b], \quad \text{then} \quad \int_{x=a}^{b} f(x)\, dx \geq 0; \qquad (5.3)$$

and if

$$f(x) \leq 0 \quad \forall\, x\, \varepsilon\, [a, b], \quad \text{then} \quad \int_{x=a}^{b} f(x)\, dx \leq 0. \qquad (5.4)$$

Theorem 5.3. Let $f(x)$ be a function integrable over a closed interval $[a, b]$ and m and M be two bounds of the function: $m \leq f(x) \leq M \quad \forall\, x\, \varepsilon\, [a, b]$, then

$$m(b-a) \leq \int_{x=a}^{b} f(x)\, dx \leq M(b-a). \qquad (5.5)$$

Theorem 5.4. (*Mean value theorem for integrals*) If $f(x)$ is continuous on the interval $[a, b]$, then there exists a point $x_o\, \varepsilon\, (a, b)$ such that

$$\int_{x=a}^{b} f(x)\, dx = (b-a) f(x_o). \qquad (5.6)$$

Theorem 5.5. (*First fundamental theorem of calculus*) Let $f(t)$ be a continuous function in $[a, b]$ and

$$F(x) \equiv \int_{x=a}^{x} f(t)\, dt, \quad \text{for some } x\, \varepsilon\, [a, b], \qquad (5.7)$$

then

$$F'(x) = f(x). \qquad (5.8)$$

Theorem 5.6. (*Second fundamental theorem of calculus*) Let $f(t)$ be a continuous function in $[a, b]$ with its anti-derivative $g(t)$ on the same interval, then

$$\int_{t=a}^{b} f(t)\, dt = g(b) - g(a). \qquad (5.9)$$

Note 5.1. In order to Apply above theorem continuity of the integrand between the limits of integration must be ensured.

Example 5.1. Check if the integral $\int_{x=-1}^{1} (-2/x^3)dx$ can be evaluated?

Solution. Since $\lim_{x\to 0} (-2/x^3) = \lim_{\varepsilon\to 0} \{-2/(-\varepsilon)^3\} = \infty$,
whereas

$$\lim_{x\to 0} (-2/x^3) = \lim_{\varepsilon\to 0} \{-2/\varepsilon^3\} = -\infty;$$

the integrand is not continuous at $x = 0$ ε $[-1, 1]$. Hence, the integral
cannot be evaluated. //

Example 5.2. Find the derivative of the function

$$f(x) = \int_0^x (t^4 + 1)\, dt.$$

Solution. By Theorem 5.5, $f'(x) = x^4 + 1$. Alternately,

$$f(x) = \left[t^5/5 + t\right]_0^x = x^5/5 + x \quad \Rightarrow \quad f'(x) = x^4 + 1. //$$

Example 5.3. Find the derivative of the function

$$f(x) = \int_x^2 \{1/(15 - t)\}dt.$$

Solution. Changing the order of integration:

$$f(x) = \int_2^x \{1/(t - 15)\}dt$$

and applying Theorem 5.5, we have

$$f'(x) = 1/(x - 15). //$$

Example 5.4. Find the derivative of the function $f(x) = \int_0^x \sqrt{(\sin t)}\, dt$.

Solution. By Theorem 5.5, we have $f'(x) = \sqrt{(\sin x)}$. //

Example 5.5. Find the derivatives of the function $f(x) = \int_{-1}^{\pi^2} (3^t)\, dt$.

Solution. By Theorem 5.5, $df(x)/dx^2 = 3^x$. But,

$$df(x)/dx^2 = \{df(x)/dx\}/(dx^2/dx) = \{df(x)/dx\}/2x.$$

$\Rightarrow$

$$df(x)/dx = 2x \cdot 3^x. //$$

Example 5.6. Find x_0 by Mean value theorem for integrals when

$$f(x) = 3 + 2x, \qquad [a, b] = [1, 3].$$

Solution. By Theorem 5.4, we have

$$(3 - 1)f(x_0) = \int_1^3 (3 + 2x)\, dx = \left[3x + x^2\right]_1^3 = (9 + 9) - (3 + 1) = 14,$$

or,

$$2(3 + 2x_0) = 14 \qquad \Rightarrow \qquad x_0 = 2.\ //$$

§ 6. Area of a region by single integration

Let a continuous function $f(x)$ be defined in a closed interval $[a, b]$ such that $f(x) \geq 0 \ \forall x \ \varepsilon \ [a, b]$ then area of the region R bounded by the graph of the function, two parallels of y-axis ($x = a$ and $x = b > a$) and x-axis has been defined by Eq. (1.3):

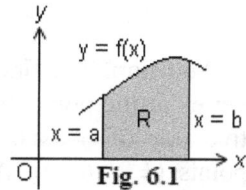

$$A(R) = \int_{x=a}^{b} f(x)\, dx. \qquad (6.1)$$

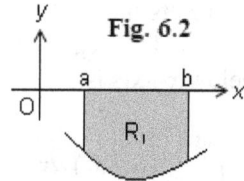

On the other hand, if $f(x) \leq 0 \ \forall x \ \varepsilon \ [a, b]$ then area of the region R bounded by the graph of the function, two parallels of y-axis ($x = a$ and $x = b > a$) and x-axis is

$$A(R_1) = -\int_{x=a}^{b} f(x).\ dx. \qquad (6.2)$$

Example 6.1. Find the area of the region bounded by the lines $x = 1$, $x = 2$, x-axis and the graph of the function $f(x) = x^2 + 1$.

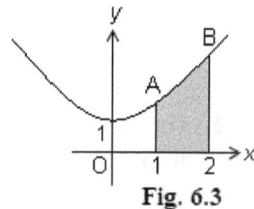

Solution. Area of the shaded region, by Eq. (1.1), is

$$\int_{x=1}^{2} (x^2 + 1)\, dx = \left[x^3/3 + x\right]_1^2 = (8/3 + 2) - (1/3 + 1) = 10/3 \text{ sq. units. } //$$

Example 6.2. Find the area of the region bounded by the lines $x = \pi$, $x = 2\pi$, x-axis and the graph of the function $f(x) = \sin x$.

Solution. Since $\sin x \leq 0 \ \forall x \ \varepsilon \ [\pi, 2\pi]$, the area of the shaded region,

by Eq. (1.2), is

$$-\int_{x=\pi}^{2\pi} \sin x \, dx = \left[\cos x\right]_{\pi}^{2\pi} = 1 + 1 = 2 \text{ sq.}$$

units. //

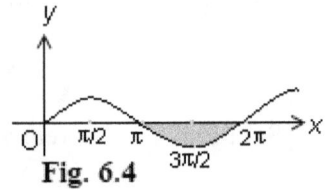

Fig. 6.4

Example 6.3. Find area of the region bounded by the lines $x = -1$, $x = 1 \cdot 5$ and graphs of the functions

$$f(x) = 6 - x^2 \text{ and } g(x) = x^2 - 2.$$

Solution. The first function represents a parabola $x^2 = -y + 6$ with vertex at the point $(0, 6)$, axis along (the negative side of) y-axis. It intersects the x-axis in points $(\pm \sqrt{6}, 0)$. Also, the second function represents a parabola $x^2 = y + 2$ with vertex at $(0, -2)$, axis along the positive side of y-axis. It intersects the x-axis in points $(\pm \sqrt{2}, 0)$. Area of the shaded regions: R above x-axis and R_1 below x-axis is

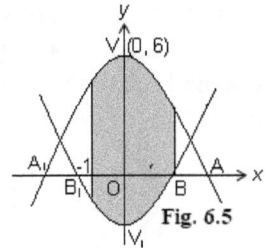

Fig. 6.5

$$\int_{x=-1}^{1\cdot5} (6 - x^2) \, dx - \int_{x=-1}^{1\cdot5} (x^2 - 2) \, dx = \left[6x - x^3/3 - x^3/3 + 2x\right]_{-1}^{1\cdot5}$$

$$= 8 \, (1 \cdot 5 + 1) - (2/3) \, \{(1 \cdot 5)^3 + 1\} = 20 - (2/3)(27/8 + 1)$$

$$= 20 - 35/12 = 205/12. \, //$$

Example 6.4. Find area of the region bounded by x-axis, the line $x = e$ and graph of the function $f(x) = \ln x$.

Solution. Giving different (real) values to x the function $y = f(x)$ possess the following values:

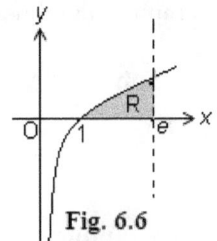

$$x = \quad 0, \quad <1, \quad 1, \quad >1, \quad e, \quad >e;$$

$$y = -\infty, \quad <0, \quad 0, \quad >0, \quad 1, \quad >1.$$

Fig. 6.6

So, the negative side of y-axis is a vertical asymptote of the graph that lies to the right of y-axis. The area as bounded above is the shaded region R above x-axis. So, it is given by Eqs. (1.1) and (2.40):

$$\int_{x=1}^{e} (\ln x)\, dx = \left[x.\ln x - x\right]_{0}^{e} = e.\ln e - e - (1.\ln 1 - 1) = 1. \text{ //}$$

Example 6.5. Find area of the region bounded by y-axis, lines $x = 1$, $y = x$ and graph of the function $f(x) = e^x$.

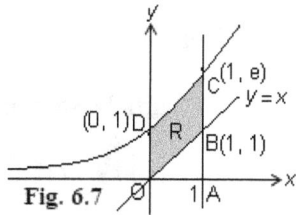

Fig. 6.7

Solution. Giving different (real) values to x the function $y = f(x)$ possess the following values:

$$x = -\infty, \quad 0, \quad > 0, \quad 1, \quad > 1.$$

$$y = \quad 0, \quad 1, \quad > 1, \quad e, \quad > e.$$

So, the negative side of x-axis is a horizontal asymptote of the graph that lies above the x-axis. The area as bounded above is the shaded region R above x-axis. So, it is given by Eq. (1.1):

$$\int_{x=0}^{1} e^x dx - \text{area of } \triangle OAB = \left[e^x\right]_{0}^{1} - 1/2 = e - 3/2. \text{ //}$$

Example 6.6. Find area of the region bounded by the lines $x = \pm 1$, $y = |x|$ and graph of the function $f(x) = x^2 - 1$.

Solution. The graph represents a parabola $x^2 = y + 1$ with vertex $(0, -1)$ and axis along the positive side of y-axis. It intersects the x-axis in points $(\pm 1, 0)$. Also, $y = |x|$ gives the lines $y = \pm x$. The area bounded above is the shaded portion R comprising of triangles OAB and OA_1B_1

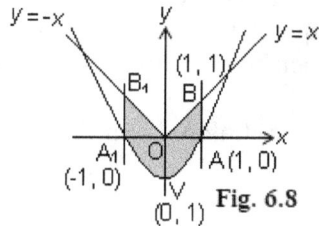

Fig. 6.8

lying above x-axis, and the region A_1VAA_1 lying below x-axis. So, it is given by Eq. (1.2):

$$\text{Area OAB} + \text{area OA}_1\text{B}_1 - \int_{x=-1}^{1} (x^2 - 1)dx = 1/2 + 1/2 - \left[x^3/3 - x\right]_{-1}^{1}$$

$$= 1 - (1/3 - 1) + (-1/3 + 1) = 3 - 2/3 = 7/3.$$

Alternately, above integral can be evaluated by using the property in Eq. (4.5) of definite integrals. The integrand $x^2 - 1$ being an even function of x we, therefore, have

$$-\int_{-1}^{1}(x^2-1)dx = -2\int_{0}^{1}(x^2-1)dx = -2\Big[x^3/3-x\Big]_{0}^{1} = -2\,(1/3-1) = 4/3.\ //$$

§ 7. Displacement

Let a particle start moving in a straight line from a point O and describes two neighbouring positions P ($OP = s$) and Q ($s + \delta s$) in time intervals t and $t + \delta t$ (both measured from O). Its velocity at P is defined by

$$\lim\nolimits_{\delta t \to 0} (\delta s)\,/\,\delta t = ds\,/\,dt = \dot{s} = v\,(t). \qquad (7.1)$$

On the other hand, if the particle acquires the velocity v at a certain time t. The distance s travelled by it from time t_1 to t_2 is then calculated by

$$s = \int_{t_1}^{t_2} v\,(t)\,dt. \qquad (7.2)$$

Example 7.1. If a moving particle acquires velocity $v\,(t) = t^2$ ft./sec. at time t secs. How far does the particle move from time $t = 1$ sec. to 3 secs?

Solution. By Eq. (7.2), we have $s = \int_{t=1}^{3} t^2 dt = \Big[t^3/3\Big]_{1}^{3} = 26/3$ ft. //

Example 7.2. An α–particle enters an accelerator at time $t = 0$. It acquires a velocity $v\,(t) = (10)^9\,t^3$ ft./sec. after t secs. How far does the particle move during the first $(10)^{-2}$ sec.?

Solution. From Eq. (7.2), we have

$$s = \int_{t=0}^{1/100} (10)^9\, t^3 dt = (10)^9\Big[t^4/4\Big]_{0}^{1/100} = 2{\cdot}5\ \text{ft.}\ //$$

§ 8. Work done by a force

Definition 8.1. If a constant force f acting upon a particle displaces the particle through a distance, say a, in its direction then the work done by the force is

$$w = f.a. \qquad (8.1)$$

If, however, the force is variable i.e. a function of a variable $x\ \varepsilon\ [a, b]$ acting along x-axis, work done by the force in displacing a particle from $x = a$ to b (along x-axis) is defined by

$$w = \int_{x=a}^{b} f(x)\, dx. \tag{8.2}$$

Example 8.1. A spring is stretched from the length 1/4 to 5/2 units under a force $f(x) = 6x$. Find the work done.

Solution. Work done, by Eq. (8.2), is

$$w = \int_{x=1/4}^{5/2} 6x\,dx = 3\left[x^2 \right]_{1/4}^{5/2} = 3(25/4 - 1/16) = 297/16 \text{ units. //}$$

Example 8.2. A repulsive force created by an electric charge e_1 to repel another charge e_2 by distance x is

$$f(x) = k.e_1.e_2 / x^2.$$

Find the work done by $f(x)$ in moving the charge e_2 from a point A $(0, a)$ to the point B $(0, b)$ if the charge e_1 is located at the origin.

Solution. Putting $k.e_1.e_2 = h$ (const.), the work done, by Eq. (8.2), is

$$w = \int_{x=a}^{b} (h/x^2)\, dx = h\left[-1/x \right]_{a}^{b} = h\,(-1/b + 1/a). \text{ //}$$

Example 8.3. The earth attracts an object of weight m pounds distant x miles from its centre by the force $f(x) = (4000)^2.m/x^2$. How much work is required to lift a 200 pounds load into 500 mile high orbit accounting the radius of the earth as 4000 miles?

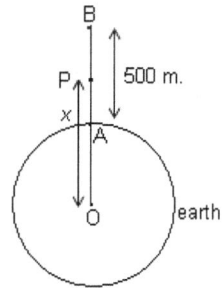

Solution. Let the load of 200 lbs. be lifted from a point A (on the surface of the earth) to a point B for 500 miles. The work required , by (3.2), is

$$w = \int_{x=4000}^{4500} f(x)dx = (4000)^2.200 \int_{x=4000}^{4500} (1/x^2)dx$$

$$= (4000)^2.200 \left[-1/x \right]_{4000}^{4500} = (4000)^2.200\,(-1/4500 + 1/4000)$$

$$= 4000 \times 200 \times 500 / 4500 = (8/9).(10)^5 . \text{ //}$$

Fig. 8.1

§ 9. Volume of solids of revolution

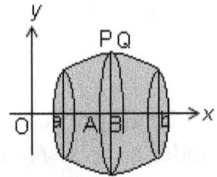

Fig. 9.1

Let a solid be obtained by revolving the graph of a function $f(x)$ about x-axis. Let P (x, y) and Q $(x + \delta x, y + \delta y)$ be two neighbouring points on the surface of the solid. Sections of solid through these points and parallel to y-axis describe a circular disc of radius AP $= y$ and thickness AB $= \delta x$. Therefore, volume of the disc being the product of its area and thickness is $\delta V = \pi y^2.\delta x$. Treating whole solid as the totality of such elementary discs volume of the solid between the lines $x = a$ and $x = b$ may be obtained by

$$V = \lim_{\delta x \to 0} \sum_{x=a} \pi y^2.\delta x = \int_{x=a}^{b} \pi y^2.dx = \pi \int_{x=a}^{b} \{f(x)\}^2 \, dx. \quad (9.1)$$

Example 9.1. Find the volume of the solid of revolution of the region enclosed by x-axis and the curve $y = \sqrt{(16 - x^2)}$ about x-axis.

Solution. The curve is a semi-circle of radius 4, centre at origin and lying above x-axis. Revolution of this semi-circle about x-axis generates a sphere of same radius 4 and centre at origin. Its volume, by Eq. (9.1), is

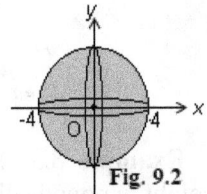

Fig. 9.2

$$V = \pi \int_{x=-4}^{4} y^2 dx = \pi \int_{-4}^{4} (16 - x^2) dx$$

$$= 2\pi \int_{0}^{4} (16 - x^2) dx = 2\pi \left[16x - x^3/3 \right]_{0}^{4}$$

$$= 2\pi (64 - 64/3) = 256\pi / 3. \; //$$

Example 9.2. Find the volume of the solid of revolution obtained by revolving the graph of function $f(x) = x^2$ from $x = 1$ to $x = 3$ about the x-axis.

Solution. The graph is a parabola with vertex at origin, axis along the positive side of y-axis. The desired volume is given by Eq. (9.1):

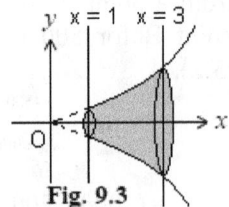

Fig. 9.3

$$V = \pi \int_{x=1}^{3} y^2.dx = \pi \int_{1}^{3} x^4 \, d$$

$$= \pi \left[x^5 / 5 \right]_1^3 = \pi (243 - 1) /5 = 242 \, \pi /5. \, //$$

§ 10. Area of a region by double integration

10.1. In rectangular Cartesian coordinates

Let AB be any arc of a curve C: $y = f(x)$ between the lines $x = a$ and $x = b$. The area enclosed by these lines, x-axis and the curve in the xOy-plane is to be found. Let P (x, y) and Q $(x + \delta x, y + \delta y)$ be two neighbouring points

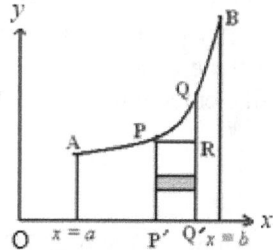
Fig. 10.1

on C. Area of any rectangular element of dimensions δx and δy on the elementary strip PLMQ is $\delta x.\delta y$. For this strip, δx remains constant while δy varies from 0 to LP = $f(x)$. Thus, the total area of this strip is the sum of all such elements:

$$\text{Area} = \lim_{\delta y \to 0} \sum \delta x.\delta y$$

$$= \int_{y=0}^{f(x)} \delta x. dy = \delta x. \int_{y=0}^{f(x)} 1. dy = \delta x. \left[y \right]_0^{f(x)} = \delta x. f(x). \quad (10.1)$$

Further, the area ADEB is the sum of all such elementary strips where x varies from a to b:

$$\text{Area ADEB} = \lim_{\delta x \to 0} \sum f(x). \delta x$$

$$= \int_{x=a}^{b} f(x). dx = \int_{x=a}^{b} \left\{ \int_{y=0}^{f(x)} 1. dy \right\}. dx. \quad (10.2)$$

10.2. In polar coordinates

Let BC be any arc of a curve $r = f(\theta)$ and the area enclosed by this arc and two radii vectors OB and OC is desired. Let P (r, θ) and Q $(r + \delta r, \theta + \delta \theta)$ be two neighbouring points on this arc. Arc PQ being small, the lamina OPQ is almost triangular. Let DEE'D' be an element of this lamina. Its area is the difference of areas of triangles OD'E' and ODE:

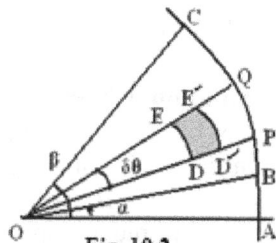
Fig. 10.2

$$(1/2).\{(r + \delta r)^2.\sin \delta \theta - r^2.\sin \delta \theta\} = (1/2).\{2r.\delta r + (\delta r)^2\}.\sin \delta \theta$$

$$= (1/2).\{2r.\delta r + (\delta r)^2\}.\delta\theta = r.\,\delta r.\,\delta\theta,$$

where, $\delta\theta$ being small, $\sin\delta\theta = \delta\theta$ is accounted and approximations are applied up to the second order. Thus, the term $(\delta r)^2.\delta\theta$, being smaller, is dropped. Throughout the triangular lamina OPQ, $\delta\theta$ remains constant while r varies from 0 to OP $= f(\theta)$. The area of this lamina is the sum of all elements like DEE'D':

$$\text{Area OPQ} = \sum_{r=0}^{f(\theta)}(r.\delta r.\delta\theta) = (\delta\theta).\int_{r=0}^{f(\theta)} r.\,dr.$$

Area of the sector OBC, where the vectorial angle θ varies from $\angle$ AOB $= \alpha$ to $\angle$ AOC $= \beta$, say, is the sum of areas of all lamina like OPQ:

$$\text{Area OBC} = \sum_{\theta=\alpha}^{\beta}\left\{\int_{r=0}^{f(\theta)} r.\,dr\right\}.\delta\theta = \int_{\theta=\alpha}^{\beta}\left\{\int_{r=0}^{f(\theta)} r.\,dr\right\}.d\theta. \quad (10.3)$$

§ 11. Change of order of integration

When a function $f(x, y)$ is continuous in a rectangular region R (a, b, c, d) it is seen [cf. 22], Theorem 13.3.1, that the order of integration w.r.t. two variables (x varying from a to b and y varying from c to d) can be interchanged:

$$\int_{x=a}^{b}\left\{\int_{y=c}^{d} f(x, y).\,dy\right\}.dx = \int_{y=c}^{d}\left\{\int_{x=a}^{b} f(x, y).\,dx\right\}.dy. \quad (11.1)$$

Example 11.1. Change the order of integration in the double integral

$$\int_{x=0}^{a}\left\{\int_{y=0}^{\sqrt{a^2-x^2}} f(x, y).\,dy\right\}.dx.$$

Fig. 11.1

Solution. Observing the range of integration of the variable y we notice that it varies from $y = 0$ (i.e. starting from x-axis) and rising up to the boundary of circle $x^2 + y^2 = a^2$. Together with that x varies from 0 (i.e. from y-axis) to $x = a$ (i.e. moves up to above circle). Since both variables are non-negative the area of integration is a quarter of the circle and lies in the first quadrant. The process of integration suggests that first the summation is performed vertically and then horizontally. In order to change the order of integration, the summation process is to be done first horizon-

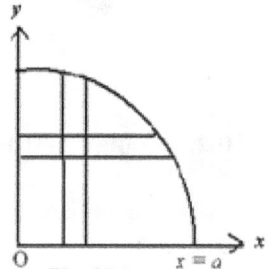

tally: for which x will vary from 0 up to the circle, i.e. up to $\sqrt{(a^2 - y^2)}$; next vertically: for which y will vary from 0 up to the boundary of circle, i.e. up to a. Therefore, the integral transforms to

$$\int_{y=0}^{a} \left\{ \int_{x=0}^{\sqrt{a^2 - y^2}} f(x, y) . dx \right\} . dy. \; //$$

Example 11.2. Change the order of integration in the double integral

$$\int_{\theta=0}^{\pi/2} \left\{ \int_{r=0}^{2a.\cos\theta} f(r,\theta).r.dr \right\}.d\theta.$$

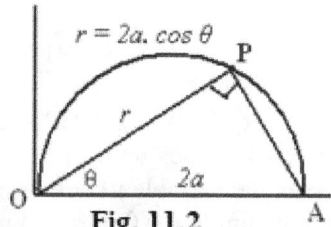

$r = 2a. \cos\theta$

Fig. 11.2

Solution. The variable r varies from $r = 0$ (i.e. starting from pole and rising up to the boundary of semi-circle $r = 2a.\cos\theta$. Together with that θ varies from 0 (i.e. from the initial line) and increases up to a line perpendicular to the initial line. Thus, the region of integration is throughout a semi-circle of diameter $2a$ described on the initial line and lying in the first quadrant of the real plane. In the changed order of integration, the variable θ is to vary from the initial line (where it is 0) and rises up to the semi-circle, i.e. $\cos^{-1}(r/2a)$. On the other hand, the radius vector r varies from 0 to $2a$ describing the semi-circle. As such, the integral transforms to

$$\int_{r=0}^{2a} \left\{ \int_{\theta=0}^{\cos^{-1}(r/2a)} f(r,\theta).r.d\theta \right\}.dr. \; //$$

Example 11.3. Change the order of integration in the double integral

$$\int_{x=0}^{2a} \left\{ \int_{y=x^2/4a}^{3a-x} f(x, y).dy \right\}.dx. \qquad (11.2)$$

Solution. The variable y varies from a point of parabola $x^2 = 4ay$ to the straight line $x + y = 3a$. Solving these equations we get the point of intersection P $(2a, a)$ of these paths, which lies in the first quadrant. Together with that, the variable x varies from 0 (i.e. from the y-axis) to $2a$, which is the abscissa of the point P. Thus, the range of integration is the shaded region en-

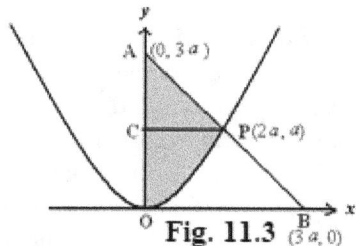

A $(0, 3a)$

$P(2a, a)$

Fig. 11.3 $_{(3a, 0)}^{B}$

closed by y-axis, the parabola and the straight line. We divide this region into two parts: right triangle ACP and the region OCP lying within y-axis, the parabola and the line PC drawn parallel to x-axis through P. The coordinates of the point of intersection of the line and the y-axis, i.e. A are $(0, 3a)$. Changing the order of integration, we notice that the variable x varies from y-axis to the line APB and y starting from the line CP rises up to the point A in order to describe the $\triangle$ ACP. Thus, considering a horizontal strip ST of the $\triangle$ ACP its area, by Eq. (11.1), is

$$\int_{y=a}^{3a} \left\{ \int_{x=0}^{3a-y} f(x, y).\, dx \right\}.\, dy.$$

Next, the region OCP is described when the variable x starting from y-axis stretches up to the parabola. Thus, its limits are $x = 0$ to $2\sqrt{(ay)}$. Also, the variable y starts from the origin and rises up to the line CP having its limits $y = 0$ to a. Taking a horizontal strip UV of this region OCP its area is found by Eq. (11.1):

$$\int_{y=0}^{a} \left\{ \int_{x=0}^{2\sqrt{ay}} f(x, y).\, dx \right\}.\, dy.$$

Thus, the integral (11.2), in changed order of integration, splits as the sum of above two integrals. //

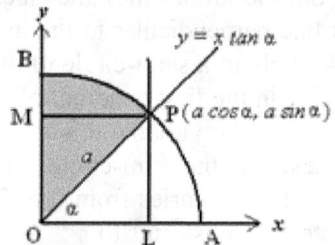

Fig. 11.4

Example 11.4. Change the order of integration

$$\int_{x=0}^{a\cos\alpha} \left\{ \int_{y=x.\tan\alpha}^{\sqrt{a^2 - x^2}} f(x, y).\, dy \right\} dx,$$

and evaluate the integral when $f(x, y) = 1$. Also, verify the result geometrically.

Solution. (i) The variable y starting from the straight line $y = x.\tan\alpha$ rises up to the circle $x^2 + y^2 = a^2$. The point of intersection of these paths is P $(a.\cos\alpha, a.\sin\alpha)$ and it lies in the first quadrant. Also, the variable x starts from y-axis and stretches up to the point P. Thus, the range of integration is the shaded region enclosed by y-axis, the circle and the straight line OP. Dividing this region into two parts: right triangle OMP and the region MBP lying within y-axis, the circle and the line PM drawn parallel to x-axis through P. The coordinates of the point of intersection of this line and the y-axis, i.e. M are $(0, a.\sin\alpha)$. In the changed order of integration, for the Δ OMP the variable x varies from y-axis to the line OP, while y starting from the origin rises up to the line MP. Thus, for their limits: $x = 0$ to $x = y$. cot α and $y = 0$ to $y = a$. sin α, the area of this triangle, by Eq. (11.1), is

$$\int_{y=0}^{a\sin\alpha} \left\{ \int_{x=0}^{y\cot\alpha} f(x, y).\,dx \right\}.\,dy.$$

Next, in the changed order of integration, the region MBP is described when x starts from y-axis and stretches up to the circle. Thus, its limits are $x = 0$ to $\sqrt{(a^2 - y^2)}$. Also, the variable y starts from the line MP and rises up to the circle having its limits $y = a$. sin α to a. Hence, its area, by Eq. (11.1), is

$$\int_{y=a.\sin\alpha}^{a} \left\{ \int_{x=0}^{\sqrt{a^2-y^2}} f(x, y).\,dx \right\}.\,dy.$$

Thus, the given integral, in changed order of integration, splits as the sum of above two integrals.

(ii) For particular case, when the integrand is 1, the sum of above integrals simplifies to

$$\int_{y=0}^{a\sin\alpha} y\cot\alpha.\,dy + \int_{y=a\sin\alpha}^{a} \sqrt{a^2 - y^2}.\,dy$$

$$= (1/2).\left\{ a^2 \sin^2\alpha.\cot\alpha + \left[y.\sqrt{a^2 - y^2} + a^2 \sin^{-1}(y/a) \right]_{a\sin\alpha}^{a} \right\}$$

$$= (a^2/2).(\sin\alpha.\cos\alpha + \pi/2 - \sin\alpha.\cos\alpha - \alpha) = (a^2/2).(\pi/2 - \alpha).$$

(iii) When $f(x, y) = 1$, the original integral represents area of the shaded portion OPB, which is a sector of the circle subtending angle $\pi/2 - \alpha$. Evidently, the area evaluated above is the area of this sector obtainable by the formula:

(square of radius).(angle subtended by the circular arc at the centre)/2. //

§ 12. Changing order of integration when limits are defined by an inequality

Many a times we encounter with the problems when variables assume limits defined by certain inequalities. For instance, we wish to evaluate the integral

$$I = \iint f(x, y). \, dx. \, dy, \qquad (12.1)$$

when the region of integration is extended to all positive values of x and y subject to the condition $x + y \leq h$. Thus, here the region of integration is bounded by both the coordinate axes and the straight line $x + y = h$. Taking an elementary strip LP parallel to y-axis the area is described by limits $y = 0$ to $h - x$ and $x = 0$ to h. Thus, in view of Eq. (11.1), the area of the region is described by the integral

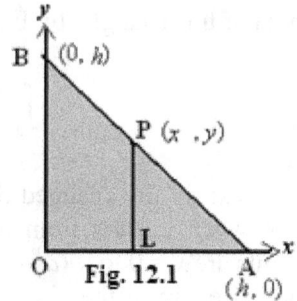

Fig. 12.1

$$\int_{x=0}^{h} \left\{ \int_{y=0}^{h-x} f(x, y). \, dy \right\}. \, dx.$$

Next, in the changed order of integration, considering an elementary horizontal strip MP, the same region can also be described by the limits $x = 0$ to $h - y$ and $y = 0$ to h. So, the area of the region, by Eq. (11.1), is also equal to

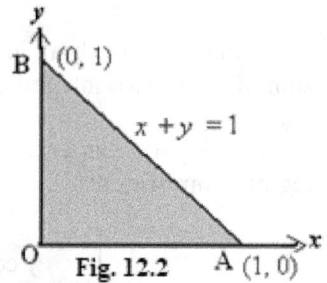

Fig. 12.2

$$\int_{y=0}^{h} \left\{ \int_{x=0}^{h-y} f(x, y). \, dx \right\}. \, dy.$$

Example 12.1. (*Dirichlet's integral*) Evaluate the double integral

$$\iint_{R} x^{m-1}. \, y^{n-1}. \, dx. \, dy$$

over the region R bounded by the condition $x + y \leq 1$, where x, y, m, n are all positive.

Solution. As described above in the beginning of this Section, we have the integral

$$\int_{x=0}^{1} \left\{ \int_{y=0}^{1-x} x^{m-1}. \, y^{n-1} \right). \, dy \right\} dx = (1/n). \int_{x=0}^{1} x^{m-1}. \left[y^{n} \right]_{y=0}^{1-x} . \, dx$$

$$= (1/n).\int_{x=0}^{1} x^{m-1}.(1-x)^n.dx$$

$$= (1/n).B(m,n+1) = (1/n).\frac{\Gamma(m).\Gamma(n+1)}{\Gamma(m+n+1)} = \frac{\Gamma(m).\Gamma(n)}{\Gamma(m+n+1)},$$

cf. [25], Eq. (19.2.2) //

§ 13. Triple integration

Triple integrals are generally used to compute volume of any enclosed space in three dimensions. Let $z = f(x, y)$ represents a closed surface in 3-dimensional Euclidean space. We wish to find the volume enclosed within this curved surface. Let us consider an elementary cuboid of sides δx, δy, δz. The volume of this elementary cuboid is $\delta x.\delta y.\delta z$. To get the volume of whole space enclosed within the solid we first integrate it with respect to z varying from $z = 0$ to $f(x, y)$. In order to integrate it further w.r.t. y and x, we consider the curve of intersection of this surface with the plane $z = 0$ (i.e. the coordinate plane

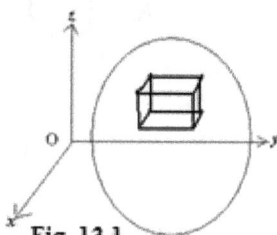
Fig. 13.1

xOy). This curve will facilitate for the limits of x and y. Thus, the total volume will be found by the integral

$$V = \int_x \int_y \int_{z=0}^{f(x,y)} dx \, . \, dy \, . \, dz. \qquad (13.1)$$

Note 13.1. An nth order integral for the volume of a solid in n-dimensional space can also be similarly defined.

Dirichlet's (double) integral was evaluated in Example 12.1 under certain conditions. Here, we consider a similar triple integral in 3 variables x, y and z.

Theorem 13.1. The value of triple integral

$$I_3 \equiv \iiint x^{l-1}.y^{m-1} z^{n-1}.dx.dy.dz,$$

where x, y, z, l, m, n are all positive, subject to the condition

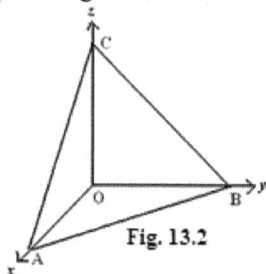
Fig. 13.2

$$x + y + z \leq 1, \qquad\qquad (13.2)$$

is

$$I_3 = (1/n). \, B\,(m, \, n + 1). \, B\,(l, \, m + n + 1)$$

$$= \Gamma(l) . \, \Gamma(m) . \, \Gamma(n) \, / \Gamma(l + m + n + 1). \qquad\qquad (13.3)$$

Proof. The bounding relation represents the plane region ABC making equal intercepts of length 1 on either of coordinate axes. Thus, the triple integration is to be carried over the space contained within the tetrahedron OABC, wherein the variable z varies from 0 to $1 - x - y$. This plane intersects the coordinate plane $z = 0$ in the line $x + y = 1$. Hence, the variable y within above space varies from 0 to $1 - x$, while x varies from 0 to 1. Thus, the integral is evaluated over these limits:

$$I_3 = \int_{x=0}^{1} \int_{y=0}^{1-x} \int_{z=0}^{1-x-y} x^{l-1} . \, y^{m-1} . \, z^{n-1} . \, dx. \, dy. \, dz$$

$$= \int_{x=0}^{1} x^{l-1} . \left\{ \int_{y=0}^{1-x} y^{m-1} . \left(\int_{z=0}^{1-x-y} z^{n-1} . \, dz \right) . \, dy \right\} . \, dx$$

$$= (1/n). \int_{x=0}^{1} x^{l-1} . \left\{ \int_{y=0}^{1-x} y^{m-1} . (1 - x - y)^n . \, dy \right\} . \, dx. \qquad (13.4)$$

The integral w.r.t. y, under the substitution $y = (1 - x).t$, so that $dy = (1 - x).dt$, transforms as

$$\int_{t=0}^{1} (1 - x)^{m-1} t^{m-1} . (1 - x)^n . (1 - t)^n . (1 - x). \, dt$$

$$= (1 - x)^{m+n} . \int_{t=0}^{1} t^{m-1} . (1 - t)^n . \, dt = (1 - x)^{m+n} .B\,(m, n + 1),$$

cf. [25], Eq. (19.1.1). It may also be noted that while carrying integration w.r.t. y, the variable x is treated as a constant. Therefore, the integral in Eq. (13.4) reduces to

$$I_3 = (1/n). \, B\,(m, \, n + 1). \int_{x=0}^{1} x^{l-1} . (1 - x)^{m+n} . \, dx$$

$$= (1/n). \, B\,(m, \, n + 1).B\,(l, \, m + n + 1),$$

again [op cit. Eq. (19.1.1)]. Putting from [25], Eqs. (19.2.2) and (19.2.9) in terms of Gamma functions, it assumes the desired form. //

Note 13.1. Similarly, a Dirichlet's integral of pth order can be evaluated:

$$I_p = \Gamma(l_1).\,\Gamma(l_2).\,\Gamma(l_3) \ldots \Gamma(l_p) / \Gamma(l_1 + l_2 + l_3 + \ldots + l_p + 1). \quad (13.5)$$

Corollary 13.1. The value of triple integral discussed above under the condition

$$x + y + z \leq h \quad (13.6)$$

is

$$I_3 = h^{l+m+n}.\,\Gamma(l).\,\Gamma(m).\,\Gamma(n) / \Gamma(l + m + n + 1). \quad (13.7)$$

Corollary 13.2. The triple integral discussed above under the condition

$$(x/a)^p + (y/b)^q + (z/c)^r \leq 1,$$

has value

$$I_3 = (a^l.b^m.c^n/p.q.r).\Gamma(l/p).\Gamma(m/q).\Gamma(n/r) / \Gamma(l/p + m/q + n/r + 1). \quad (13.8)$$

Example 13.1. Evaluate the nth order integral

$$I_n = \underbrace{\iiint \ldots \int}_{n} dx_1.dx_2.dx_3 \ldots dx_n,$$

extended to all positive values of the variables x_i's subject to the condition

$$x_1^2 + x_2^2 + x_3^2 + \ldots + x_n^2 \leq R^2.$$

Solution. Taking the integrand $x_1^{1-1}.x_2^{1-1}.x_3^{1-1} \ldots x_n^{1-1}$, and re-writing the condition as

$$(x_1/R)^2 + (x_2/R)^2 + (x_3/R)^2 + \ldots + (x_n/R)^2 \leq 1,$$

Corollary 13.2 is applicable. Generalizing the result of the Corollary to nth order we, thus, have

$$I_n = (R^n/2^n).\,\Gamma(1/2).\,\Gamma(1/2) \ldots \Gamma(1/2) / \Gamma(1/2 + 1/2 + \ldots + 1/2 + 1)$$

$$= (R^n/2^n).(\sqrt{\pi})^n / \Gamma(n/2 + 1) = (R\sqrt{\pi}/2)^n / \Gamma(n/2 + 1). \,//$$

§ 14. Transformation of multiple integrals

In the following we discuss the transformation of an nth order integral

$$I_n = \int\int\int .....\int f(x_1, x_2, x_3, ..., x_n) \cdot dx_1 \cdot dx_2 \cdot dx_3 .... dx_n, \quad (14.1)$$

by substitution of variables from x_i's to new variables y_i's, $i = 1, 2, 3, ...,$ n:

$$x_i = x_i (y_1, y_2, y_3, ... , y_n). \quad (14.2)$$

Before we discuss the main integral, we present here a process of elimination of $n - 1$ (new) variables $y_1, y_2, ... , y_{n-1}$ from the last relation of the set of n equations (14.2) in the following steps:

Step 1. Considering three linear equations

$$a_1 x_1 + a_2 x_2 + a_3 x_3 = 0, \qquad b_1 x_1 + b_2 x_2 + b_3 x_3 = 0,$$
$$\left.\begin{array}{c} \\ \\ \end{array}\right\} \quad (14.3)$$
$$c_1 x_1 + c_2 x_2 + c_3 x_3 = \lambda \, (\neq 0).$$

Solving the first two of these in x_1, x_2, x_3:

$$\frac{x_1}{\begin{vmatrix} a_2 & a_3 \\ b_2 & b_3 \end{vmatrix}} = \frac{x_2}{\begin{vmatrix} a_3 & a_1 \\ b_3 & b_1 \end{vmatrix}} = \frac{x_3}{\begin{vmatrix} a_1 & a_2 \\ b_1 & b_2 \end{vmatrix}} = k, \quad \text{say} \cdot \quad (14.4)$$

Putting these values of x_1, x_2, x_3 in the third of Eqs. (14.3), we get

$$c_1 \cdot \begin{vmatrix} a_2 & a_3 \\ b_2 & b_3 \end{vmatrix} + c_2 \cdot \begin{vmatrix} a_3 & a_1 \\ b_3 & b_1 \end{vmatrix} + c_3 \cdot \begin{vmatrix} a_1 & a_2 \\ b_1 & b_2 \end{vmatrix} = \frac{\lambda}{k}, \; \Delta_3 \equiv \begin{vmatrix} a_1 & a_2 & a_3 \\ b_1 & b_2 & b_3 \\ c_1 & c_2 & c_3 \end{vmatrix} = \frac{\lambda}{k}.$$

$$\Rightarrow \quad \lambda = k. \Delta_3 = x_3. \Delta_3 / \Delta_2, \quad \text{where} \quad \Delta_2 = \begin{vmatrix} a_1 & a_2 \\ b_1 & b_2 \end{vmatrix}. \quad (14.5)$$

Step 2. Now, we considering four linear equations in four variables x_i's, $i = 1, 2, 3, 4$:

$$a_1 x_1 + a_2 x_2 + a_3 x_3 + a_4 x_4 = 0,$$
$$b_1 x_1 + b_2 x_2 + b_3 x_3 + b_4 x_4 = 0,$$
$$\left.\begin{array}{c} \\ \\ \\ \\ \end{array}\right\} \quad (14.6)$$
$$c_1 x_1 + c_2 x_2 + c_3 x_3 + c_4 x_4 = 0,$$
$$d_1 x_1 + d_2 x_2 + d_3 x_3 + d_4 x_4 = \lambda \, (\neq 0).$$

Solving the first three of these and eliminating the variables from the fourth equation, analogous to Eq. (14.5), we derive $\lambda = x_4 . \Delta_4 / \Delta_3$, where Δ_3 is as above but

$$\Delta_4 \equiv \begin{vmatrix} a_1 & a_2 & a_3 & a_3 \\ b_1 & b_2 & b_3 & b_4 \\ c_1 & c_2 & c_3 & c_3 \\ d_1 & d_2 & d_3 & d_4 \end{vmatrix} .$$

Step 3. The problem can be generalized to n linear equations, and we similarly derive

$$\lambda = x_n . \Delta_n / \Delta_{n-1}.$$

Step 4. Returning back to the set of n equations represented by Eq. (14.2), it is possible to eliminate $n - 1$ variables $y_1, y_2, \ldots , y_{n-1}$ from the last relation of the set of equations (14.2) reducing x_n as a function of x_1, $x_2, \ldots , x_{n-1}$ and y_n :

$$x_n = \varphi (x_1, x_2, \ldots , x_{n-1}, y_n). \tag{14.7}$$

Further, applying Taylor's theorem, we can find the differentials of x_i's given by Eqs. (14.2):

$$dx_i = (\partial x_i / \partial y_1).dy_1 + (\partial x_i / \partial y_2).dy_2 + \ldots + (\partial x_i / \partial y_n).dy_n.$$

Thus,

$$dx_n = (J_n / J_{n-1}). \, dy_n ,$$

where

$$J_r = \partial (x_1, x_2, \ldots , x_r) / \partial (y_1, y_2, \ldots , y_r).$$

Hence, the integral (14.1) reduces to

$$I_n = \iiint \ldots \int f_n (x_1, x_2, \ldots , x_{n-1}, \, y_n) . \frac{J_n}{J_{n-1}} . dx_1 . dx_2 \ldots dx_{n-1} . dy_n . \tag{14.8}$$

Step 5. Next, taking first $n - 1$ equations of the set of Eqs. (14.2), and eliminating $n - 2$ variables $y_1, y_2, \ldots , y_{n-2}$ from the $(n - 1)^{\text{th}}$ equation, we may get:

$$x_{n-1} = \psi (x_1, x_2, \ldots , x_{n-2}, y_{n-1}, y_n) \Rightarrow dx_{n-1} = (J_{n-1} / J_{n-2}). \, dy_{n-1}.$$

Accordingly, the integral vide Eq. (14.8) further reduces to

$$I_n = \iiint \ldots \int f_{n-1} (x_1, x_2, \ldots , x_{n-2}, y_{n-1}, y_n).$$

$$\cdot \frac{J_n}{J_{n-1}} \cdot \frac{J_{n-1}}{J_{n-2}} . dx_1 . dx_2 ... dx_{n-2} . dy_{n-1} . dy_n .$$

Continuing this process of elimination of variables, finally, we derive $dx_1 = J_1 . dy_1$, reducing the integral to

$$I_n = \iiint ... \int f_1(y_1, y_2, ..., y_{n-1}, y_n) . J_n . dy_1 . dy_2 ..... dy_{n-1} . dy_n. \quad (14.9)$$

§ 15. Transformation into polar coordinates

15.1. Double integrals

The rectangular Cartesian coordinates x and y are connected to 2-dimensional polar coordinates r and θ by the relations

$$x = r. \cos \theta \qquad \text{and} \qquad y = r. \sin \theta, \qquad (15.1)$$

together with

$$r = \sqrt{(x^2 + y^2)} \quad \text{and} \quad \theta = \tan^{-1}(y/x). \qquad (15.2)$$

Accordingly, their partial derivatives found by

$$\partial x/\partial r = \cos \theta, \ \partial x/\partial \theta = -r.\sin \theta, \ \partial y/\partial r = \sin \theta, \ \partial y/\partial \theta = r.\cos \theta, \quad (15.3)$$

determine the Jacobian

$$J = \frac{\partial (x, y)}{\partial (r, \theta)} = \begin{vmatrix} \partial x/\partial r & \partial x/\partial \theta \\ \partial y/\partial r & \partial y/\partial \theta \end{vmatrix} = \begin{vmatrix} \cos \theta & -r.\sin \theta \\ \sin \theta & r.\cos \theta \end{vmatrix} = r. \quad (15.4)$$

Hence, the double integral in Eq. (12.1) transforms as

$$\iint f(x, y) . dx . dy = \iint f_1(r, \theta) . r . d\theta . dr. \qquad (15.5)$$

15.2. Triple integrals

The rectangular Cartesian coordinates x, y and z are connected to 3-dimensional polar coordinates r, θ and φ by the relations

$$x = r. \sin \theta. \cos \varphi, y = r. \sin \theta. \sin \varphi, \quad z = r. \cos \theta. \qquad (15.6)$$

together with

$$r = \sqrt{(x^2 + y^2 + z^2)}, \quad \theta = \tan^{-1}\sqrt{\{(x^2 + y^2)/z^2\}}, \quad \varphi = \tan^{-1}(y/x). \quad (15.7)$$

Accordingly, their partial derivatives found by

$$\partial x/\partial r = \sin\theta . \cos\varphi, \quad \partial x/\partial\theta = r.\cos\theta.\cos\varphi, \quad \partial x/\partial\varphi = -r.\sin\theta.\sin\varphi;$$

$$\partial y/\partial r = \sin\theta.\sin\varphi, \quad \partial y/\partial\theta = r.\cos\theta.\sin\varphi, \quad \partial y/\partial\varphi = r.\sin\theta.\cos\varphi;$$

$$\partial z/\partial r = \cos\theta, \quad \partial z/\partial\theta = -r.\sin\theta, \quad \partial z/\partial\varphi = 0; \quad (15.8)$$

determine the Jacobian

$$J = \frac{\partial(x,y,z)}{\partial(r,\theta,\varphi)} = \begin{vmatrix} \sin\theta.\cos\varphi & r.\cos\theta.\cos\varphi & -r.\sin\theta.\sin\varphi \\ \sin\theta.\sin\varphi & r.\cos\theta.\sin\varphi & r.\sin\theta.\cos\varphi \\ \cos\theta & -r.\sin\theta & 0 \end{vmatrix}$$

$$= r^2.\sin\theta. \begin{vmatrix} \sin\theta.\cos\varphi & \cos\theta.\cos\varphi & -\sin\varphi \\ \sin\theta.\sin\varphi & \cos\theta.\sin\varphi & \cos\varphi \\ \cos\theta & -\sin\theta & 0 \end{vmatrix}$$

$$= r^2.\sin\theta.\{\sin\varphi\,(\sin^2\theta.\sin\varphi + \cos^2\theta.\sin\varphi)$$

$$+ \cos\varphi\,(\sin^2\theta.\cos\varphi + \cos^2\theta.\cos\varphi)\}$$

$$= r^2.\sin\theta.(\sin^2\varphi + \cos^2\varphi) = r^2.\sin\theta. \quad (15.9)$$

Hence, the triple integral transforms as

$$\iiint f(x,y,z).dx.dy.dz = \iiint f_1(r,\theta,\varphi).r^2.\sin\theta.dr.d\theta.d\varphi.$$
$$(15.10)$$

§ 16. Dirichlet's and Liouville's integrals

In the following we discuss the transformation of the integral

$$\iiint f(x+y+z).x^{l-1}.y^{m-1}.z^{n-1}.dx.dy.dz, \quad (16.1)$$

where x, y, z, l, m, n are all positive and the variables are connected by Eq. (13.2). It may be noted that the integrand here is more general than the one considered in Theo. 13.1. Setting

$$x+y+z = u, \quad y+z = uv, \quad z = uvw$$

$\Rightarrow$ $\qquad\qquad x = u.\,(1-v),\ y\ =\ u\,v.\,(1-w);$ $\qquad\qquad$ (16.2)

and considering the partial derivatives of x, y, z, w.r.t. u, v, w, the Jacobian may be found:

$$J = \frac{\partial\,(x,\,y,\,z)}{\partial\,(u,\,v,\,w)} = \begin{vmatrix} \partial x/\partial u & \partial x/\partial v & \partial x/\partial w \\ \partial y/\partial u & \partial y/\partial v & \partial y/\partial w \\ \partial z/\partial u & \partial z/\partial v & \partial z/\partial w \end{vmatrix} = \begin{vmatrix} 1-v & -u & 0 \\ v\,(1-w) & u\,(1-w) & -u\,v \\ v\,w & w\,u & u\,v \end{vmatrix}$$

$$= u^2 v. \begin{vmatrix} 1-v & -1 & 0 \\ v\,(1-w) & 1-w & -1 \\ v\,w & w & 1 \end{vmatrix} = u^2 v. \begin{vmatrix} 1-v & -1 & 0 \\ v & 1 & 0 \\ vw & w & 1 \end{vmatrix} = u^2 v.$$

Accordingly, the integral (16.1) transforms as

$$\iiint f(u).\{u\,(1-v)\}^{l-1}.\{uv\,(1-w)\}^{m-1}.(uvw)^{n-1}.u^2\,v\,.du\,.\,dv\,.\,dw$$

$$= \iiint \{f\,(u).\,u^{l+m+n-1}.v^{m+n-1}.(1-v)^{l-1}\}\{w^{n-1}.(1-w)^{m-1}\}\,du.dv.dw.$$

The variables x, y, z being positive, the condition (1.2) implies $y + z \leq 1$ and $z \leq 1$. Therefore, Eqs. (16.2) conclude $u \leq 1$, $v \leq 1$ and $w \leq 1$. Thus, all the three new variables u, v, w vary from 0 to 1. Hence, above integral reduces to

$$\int_{u=0}^{1} f(u).u^{l+m+n-1}.\left\{\int_{v=0}^{1} v^{m+n-1}.(1-v)^{l-1}.\left(\int_{w=0}^{1} w^{n-1}.(1-w)^{m-1}.\,dw\right).\,dv\right\}du$$

$$= B\,(m+n,l)\,B\,(n,m)\int_{u=0}^{1} f\,(u).u^{l+m+n-1}\,.\,du$$

$$= \frac{\Gamma\,(l).\Gamma\,(m).\Gamma\,(n)}{\Gamma\,(l+m+n)}.\int_{u=0}^{1} f\,(u).u^{l+m+n-1}\,.\,du, \qquad (16.3)$$

cf. [25], Eq. (19.2.9).

16.1. Particular cases

(i) If the bounding condition in Eq. (13.2) is replaced by the one given in Eq. (13.6), the triple integral in Eq. (16.1) transforms to

$$\frac{\Gamma(l).\Gamma(m).\Gamma(n)}{\Gamma(l+m+n)}.\int_{u=0}^{h} f(u).u^{l+m+n-1}.du. \qquad (16.4)$$

(ii) If $f(x, y, z) = 1$, the integral in Eq. (16.1) reduces to the form given in Theo. 13.1. In such case, the value of the integral obtained in Eq. (16.3) is easily evaluated:

$$\frac{\Gamma(l).\Gamma(m).\Gamma(n)}{\Gamma(l+m+n)}.\left[\frac{u^{l+m+n}}{l+m+n}\right]_{u=0}^{1} = \frac{\Gamma(l).\Gamma(m).\Gamma(n)}{(l+m+n).\Gamma(l+m+n)}$$

$$= \frac{\Gamma(l).\Gamma(m).\Gamma(n)}{\Gamma(l+m+n+1)},$$

cf. [25], Eq. (19.2.2). As per expectation, it is the same result as in Theo. 13.1.

CHAPTER 7

COMPLEX VARIABLE

§ 1. Functions of a complex variable

If for each value of a complex variable $z = x + iy$ in a given region R, we have one or more values of $w = u + iv$ then w is called a function of z and we write

$$w = u(x, y) + iv(x, y) = f(z), \tag{1.1}$$

where $u = u(x, y)$ and $v = v(x, y)$ are real functions of x, y. The function w is called *single-valued* if to each value of z there corresponds one and only one value of w; otherwise it is a multi-valued function of z.

§ 2. Limit of $f(z)$

A function $w = f(z)$ is said to have a limit l as $z \to z_0$ if for every real number ε there exists a positive real number δ such that

$$|f(z) - l| < \varepsilon \text{ and } |z - z_0| < \delta$$

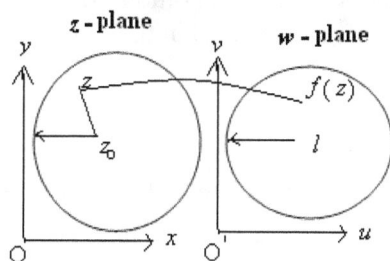

Fig. 2.1

are consistent; i.e. for every $z \neq z_0$ in the δ-disc (of z-plane), $f(z)$ has a value in the ε-disc (of w-plane). We write

$$\lim_{z \to z_0} f(z) = l.$$

Definition 2.1. The function $f(z)$ is said to be *continuous* at a point $z = z_0$, if

$$\lim_{z \to z_0} f(z) = f(z_0).$$

Further, $f(z)$ is said to be continuous in any region R if it is continuous at every point of R. Also, if the function $w = f(z) = u + iv$ is continuous at $z = z_0$ then both $u(x, y)$ and $v(x, y)$ are continuous at $z = z_0$, i.e. at $x = x_0$ and $y = y_0$. Conversely, if u, v are continuous at (x_0, y_0) then so will be $f(z)$ at $z = z_0$.

§ 3. Derivative of $f(z)$ with respect to z

Let $w = f(z)$ be a single-valued function of the variable $z = x + iy$ then the derivative of $f(z)$ with respect to z is defined by

$$dw / dz \equiv f'(z) = \lim_{\delta z \to 0} \{f(z + \delta z) - f(z)\}/ \delta z, \qquad (3.1)$$

if the limit exists and has the same value for all the different ways in which $\delta z \to 0$.

Suppose P (x, y) is a fixed point with its neighbouring point Q $(x + \delta x, y + \delta y)$ in a given region R. The point Q may approach to P along any path: straight or curved in R to cause $\delta z \to 0$ in any manner. The necessary and sufficient conditions for the existence of dw/dz are deduced in the following.

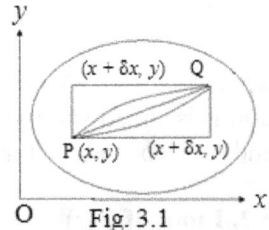

Fig. 3.1

Theorem 3.1. The derivative of $w = f(z)$ exists for all values of z iff:

(i) the first order partial derivatives: $\partial u/\partial x$, $\partial u/\partial y$, $\partial v/\partial x$ and $\partial v/\partial y$ of u, v are continuous functions of x, y in R;

(ii) $u_x \equiv \partial u /\partial x = \partial v /\partial y \equiv v_y, \quad u_y \equiv \partial u /\partial y = - \partial v /\partial x \equiv - v_x.$ \qquad (3.2)

Note 3.1. The equations (3.2) are called the *Cauchy-Riemann* (or briefly C-R) conditions.

§ 4. Analytic functions

A single-valued function $f(z)$ possessing a unique derivative with respect to z at all points of region R is called an *analytic* or *regular* function of z in that region. A point where $f(z)$ does not possess a derivative is called a *singular point* of the function.

Note 4.1. The Cauchy-Riemann Eqs. (3.2) are necessary and sufficient conditions for an analytic function $f(z) = u + iv$ in R.

The real and imaginary parts (namely u and v) of an analytic function are called *conjugate* functions.

Example 4.1. The function $w = \ln z$ is analytic at all points except at origin. Find dw/dz.

Solution. Let

$$w = u + iv = \ln(x + i y) = (1/2) \ln(x^2 + y^2) + i \tan^{-1}(y/x)$$

$\Rightarrow$

$$u = (1/2) \ln(x^2 + y^2) \quad \text{and} \quad v = \tan^{-1}(y/x). \qquad (4.1)$$

Differentiating u, v partially with respect to x, y we get

$$\left. \begin{array}{ll} u_x = x/(x^2 + y^2), & u_y = y/(x^2 + y^2), \\[2mm] v_x = - y/(x^2 + y^2), & v_y = x/(x^2 + y^2); \end{array} \right\} \qquad (4.2)$$

which clearly satisfy C-R equations (3.2) establishing the analytic nature of the function. However,

$$dw / dz = \partial u / \partial x + i \, \partial v / \partial x,$$

$$= x/(x^2 + y^2) - i y/(x^2 + y^2), \qquad \text{by Eq. (4.2)}$$

$$= (x - i y)/(x^2 + y^2) = (x - i y)/(x + i y)(x - i y) = 1/(x + i y) = 1/z$$

exists at all points except at $z = 0$. Thus, the origin is a singular point of the analytic function w. //

Example 4.2. The polar form of Cauchy-Riemann equations is

$$\partial u / \partial r = (1/r) \, \partial v / \partial \theta, \quad \partial u / \partial \theta = - r \, \partial v / \partial r. \qquad (4.3)$$

Also, deduce that

$$\partial^2 u / \partial r^2 + (1/r) \, \partial u / \partial r + (1/r^2) \, \partial^2 u / \partial \theta^2 = 0. \qquad (4.4)$$

Solution. (i) For the polar coordinates (r, θ) of a point P (x, y) the complex coordinate z of P is

$$z = x + i y = r(\cos \theta + i \sin \theta) = r e^{i \theta}. \qquad (4.5)$$

Accordingly, the function

$$w = f(z) \quad \text{is} \quad u + i v = f(r e^{i \theta}). \qquad (4.6)$$

Differentiating Eq. (4.6) partially with respect to r, θ:

$$\partial u \,/\, \partial r \,+ i\, \partial v \,/\, \partial r \,= \, f'(r\,e^{\,i\theta})\,e^{\,i\theta}, \tag{4.7}$$

$$\partial u/\partial\theta + i\,\partial v/\partial\theta = f'(r.e^{\,i\theta})\,ir.e^{\,i\theta} = ir\,(e^{\,i\theta}f') = ir\,(\partial u/\partial r + i\,\partial v/\partial r),$$

by Eq. (4.7). Equating real and imaginary parts in the last equation we derive Eq. (4.3). //

(ii) Differentiating Eq. (4.3) further with respect to r, θ we also obtain

$$\partial^2 u\,/\partial r^2 = -\,(1/r^2)\,\partial v/\partial\theta \,+ (1/r)\,\partial^2 v/\partial r\,\partial\theta, \quad \partial^2 u\,/\,\partial\theta^2 = -\,r.\partial^2 v/\partial\theta\,\partial r.$$

Putting from Eq. (4.3) and above relations in the first member of Eq. (4.4), we derive

$$\partial^2 u\,/\,\partial r^2 + (1/r)\,\partial u\,/\,\partial r + (1/\,r^2)\,\partial^2 u\,/\,\partial\theta^2$$

$$= -\,(1/r^2)\,\partial v/\partial\theta + (1/r)\,\partial^2 v/\partial r\,\partial\theta + (1/r^2)\,\partial v\,/\partial\theta - (1/r)\,\partial^2 v\,/\partial\theta\,\partial r = 0.\ //$$

§ 5. Applications of C-R conditions to Laplace equation

Let $f(z)$ be an analytic function of z in some region R of the z-plane. So, there hold the Cauchy-Riemann conditions given by Eqs. (3.2). Differentiating the first of these w.r.t. x and the second one w.r.t. y we get

$$u_{xx} \equiv \partial^2 u\,/\,\partial x^2 = \partial^2 v\,/\,\partial x\,\partial y \quad \text{and} \quad u_{yy} \equiv \partial^2 u\,/\partial y^2 = -\,\partial^2 v\,/\,\partial y\,\partial x.$$

Adding them and using the result $\partial^2 v\,/\,\partial x\,\partial y = \partial^2 v\,/\,\partial y\,\partial x$, we get the Laplace equation

$$\nabla^2 u \equiv \partial^2 u\,/\partial x^2 + \partial^2 u\,/\partial y^2 = 0. \tag{5.1}$$

Similarly, differentiating the second of equations (3.2) w.r.t. x and the first one w.r.t. y we derive

$$\partial^2 u\,/\,\partial y\,\partial x = \partial^2 v\,/\,\partial y^2, \qquad \partial^2 u\,/\,\partial x\,\partial y = -\,\partial^2 v\,/\,\partial x^2$$

$$\Rightarrow$$

$$\nabla^2 v \equiv \partial^2 v\,/\,\partial x^2 + \partial^2 v\,/\,\partial y^2 = 0. \tag{5.2}$$

Thus, both the functions u and v satisfying C-R conditions also satisfy the Laplace equation. Hence, they are called as the *harmonic functions*. The theory of harmonic functions will be called as the *potential theory*.

Note 5.1. The function $u = 1/r$, where $r^2 = x^2 + y^2 + z^2$, is harmonic in the three-dimensional Euclidean space E_3. The Laplace equation in E_3 is

$$\nabla^2 u \equiv \partial^2 u / \partial x^2 + \partial^2 u / \partial y^2 + \partial^2 u / \partial z^2 = 0.$$

Example 5.1. Find an analytic function whose real part is

$$u(x, y) = \sin 2x / (\cosh 2y - \cos 2x). \tag{5.3}$$

Solution. Let $f(z) = u + iv$ so that

$$f'(z) = \partial u / \partial x + i \partial v / \partial x = \partial u / \partial x - i \partial u / \partial y, \quad \text{by C-R Eqs. (3.2)}$$

or,

$$f'(z) = \{2 \cos 2x \, (\cosh 2y - \cos 2x)$$

$$- \sin 2x.(2 \sin 2x)\} / (\cosh 2y - \cos 2x)^2$$

$$+ i \sin 2x \, (2 \sinh 2y) / (\cosh 2y - \cos 2x)^2$$

$$= 2\{(\cos 2x.\cosh 2y - 1) / (\cosh 2y - \cos 2x)^2$$

$$+ 2i \sin 2x.\sinh 2y / (\cosh 2y - \cos 2x)^2,$$

For $y = 0$, $z = x$ (along the real line), by Milne-Thomson method, we get

$$f'(z) = 2 \, (\cos 2z - 1) / (1 - \cos 2z)^2 + i.0 = -2/(1 - \cos 2z)$$

$$= -2/\{1 - (1 - 2 \sin^2 z)\} = -1/\sin^2 z = -\csc^2 z.$$

Integrating it w.r.t. z, we get $f(z) = \cot z + i \, C$, where C is some constant of integration. //

Example 5.2. If $f(z) = u(x, y) + i \, v(x, y)$ is a regular function prove that

$$\{\partial^2 / \partial x^2 + \partial^2 / \partial y^2\}.|f(z)|^2 = 4 \, |f'(z)|^2. \tag{5.4}$$

Solution. $|f(z)|^2 = u^2 + v^2 = \varphi(x, y)$, say. Differentiating it partially w.r.t. x, y successively, we obtain

$$\varphi_x \equiv \partial\varphi / \partial x = 2 \, (u.u_x + v.v_x), \qquad \varphi_y \equiv \partial\varphi / \partial y = 2 \, (u.u_y + v.v_y),$$

$$\varphi_{xx} \equiv \partial^2 \varphi / \partial x^2 = 2 \, (u_x^2 + u.u_{xx} + v_x^2 + v.v_{xx}),$$

$$\varphi_{yy} \equiv \partial^2 \varphi / \partial y^2 = 2 \, (u_y^2 + u. \, u_{yy} + v_y^2 + v. \, v_{yy}),$$

$\Rightarrow$

$$\varphi_{xx} + \varphi_{yy} \equiv 2\{ \, (u_x^2 + u_y^2) + (v_x^2 + v_y^2) \, \}$$

$$+ 2\{ \, u \, (u_{xx} + u_{yy}) + v \, (v_{xx} + v_{yy}) \}. \tag{5.5}$$

By hypothesis, $f(z)$ is regular, so the functions u, v satisfy C-R conditions (3.2) as well as Laplace Eqs. (5.1) and (5.2). Hence, Eq. (5.5) simplifies to

$$\varphi_{xx} + \varphi_{yy} = 2\{(u_x^2 + v_x^2) + (v_x^2 + u_x^2)\} = 4 \, (u_x^2 + v_x^2) \, = \, 4 \, |f'(z)|^2,$$

i.e.

$$\{\partial^2 / \partial x^2 + \partial^2 / \partial y^2\}\varphi \equiv \{\partial^2 / \partial x^2 + \partial^2 / \partial y^2\}.|f(z)|^2 = 4 \, |f'(z)|^2. \, //$$

§ 6. Integral of a complex function along a curve

Let $f(z)$ be a continuous function of a complex variable $z = x + i \, y$ defined at all points of a curve C with end points A and B. Divide C into n parts at the points

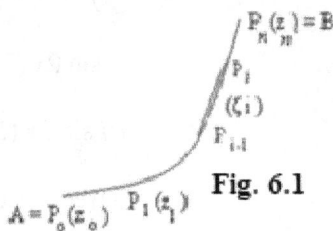

Fig. 6.1

$$A \equiv P_0 \, (z_0), \, P_1(z_1), \, P_2 \, (z_2), \ldots, \, P_i \, (z_i), \ldots, \, P_n \, (z_n) \equiv B.$$

Let the arc–interval $P_{i-1} \, P_i$ be denoted by δz_i and ζ_i be any point of this arc. Then the limit of the sum $\displaystyle\sum_{i=1}^{n} f(\zeta_i) \, \delta z_i$, as $n \to \infty$, in such a way that the length of the chord $\delta z_i = P_{i-1} P_i$ approaches to zero, defines the line–integral of $f(z)$ taken along the path C:

$$\int_C f(z) \, dz \equiv \lim_{\delta z \to 0} \sum_{i=1}^{n} f(\zeta_i) \, \delta z_i. \tag{6.1}$$

Writing $f(z) \equiv u \, (x, \, y) + i \, v \, (x, \, y)$ and noting $dz = dx + i \, dy$, above integral becomes

$$\int_C f(z) \, dz = \int_C \, (u \, dx - v \, dy) + i \int_C \, (v dx + u \, dy). \tag{6.2}$$

Thus, the line integral of a complex function $f(z)$ reduces to two line–integrals of real functions.

Example 6.1. For an integer $n \neq -1$ and a circle C

$$|z - a| = r. \qquad (6.3),$$

prove that

(i) $$\int_C dz/(z-a) = 2\pi i,$$

(ii) $$\int_C (z-a)^n \, dz = 0.$$

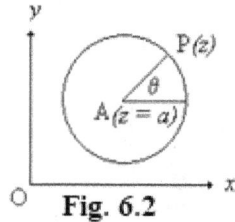

Fig. 6.2

Solution. The equation of the circle C may be written in a parametric form:

$$z - a = r e^{i\theta} \quad \Rightarrow \quad dz = r e^{i\theta}. i \, d\theta,$$

where θ varies from 0 to 2π along C. Hence,

(i) $$\int_C dz/(z-a) = \int_0^{2\pi} (r e^{i\theta}. i / r e^{i\theta}) \, d\theta = i \int_0^{2\pi} d\theta = 2\pi i,$$

(ii) $$\int_C (z-a)^n \, dz = \int_0^{2\pi} r^n e^{ni\theta}. ir e^{i\theta} \, d\theta = i r^{n+1} \int_0^{2\pi} e^{(n+1)i\theta} \, d\theta$$

$$= \{r^{n+1}/(n+1)\}. \left[e^{(n+1)i\theta} \right]_0^{2\pi} = \{r^{n+1}/(n+1)\}[e^{2(n+1)\pi i} - e^0] = 0,$$

if $n \ne -1$. We have also used $\exp\{2(n+1)\pi i\} = 1 = e^0$ therein. //

Example 6.2. Evaluate the integral $\int_0^{2+i} (\bar{z})^2 \, dz$ along:

(i) the line $y = x/2$,

(ii) the real axis up to $x = 2$, and then vertically up to $z = 2 + i$.

Solution. (i) Along the line OA we have

$$x = 2y, \qquad z \equiv x + iy = (2 + i) y, \qquad (6.4)$$

$$dz = (2 + i) \, dy, \quad \bar{z} = x - iy = (2 - i) y.$$

Since

$$z = 0 \Rightarrow y = 0 \text{ and } z = 2 + i \quad \Rightarrow \quad y = 1, \text{ by Eq. (6.4)}$$

we, therefore, have

$$I = \int_0^{2+i} (\bar{z})^2 \, dz = \int_0^1 (2-i)^2 y^2 (2+i) \, dy = 5(2-i) \int_0^1 y^2 \, dy$$

$$= 5(2-i)\left[\; y^3/3 \;\right]_0^1 = 5\,(\,2-i)/3. \; //$$

(ii) $\displaystyle I = \int_{OB} (\bar{z})^2\, dz + \int_{AB} (\bar{z})^2\, dz. \quad (6.5)$

Along OB: $z = x, \bar{z} = x, dz = dx, y = 0$; and along BA: $z = 2 + i\,y, \bar{z} = 2 - iy, \; dz = i\, dy, x = 2$. Therefore, Eq. (1.5) reduces to

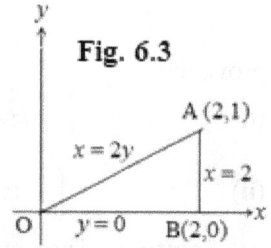

$$I = \int_0^2 x^2\, dx + \int_0^1 (2 - i\,y)^2\, i\, dy = \left[\; x^3/3 \;\right]_0^2 + i\int_0^1 (4 - y^2 - 4i\,y)\, dy$$

$$= 8/3 + i\left[\; 4y - y^3/3 - 2i\,y^2 \;\right]_0^1 = (14 + 11i)/3. \; //$$

§ 7. Cauchy's integral theorem

Theorem 7.1. If $f(z)$ is an analytic function and $f'(z)$ is continuous at each point within or on a closed curve C, then

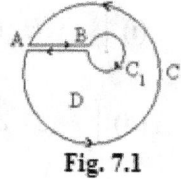

$$\int_C f(z)\, dz = 0. \qquad (7.1)$$

Theorem 7.2. (*Extension of Cauchy's theorem*) If $f(z)$ is analytic in the region D bounded by two simple closed curves C and C_1, then

$$\int_C f(z)\, dz = \int_{C_1} f(z)\, dz. \qquad (7.2)$$

§ 8. Cauchy's integral formula

Theorem 8.1. If $f(z)$ is an analytic function within and on a closed curve C, then

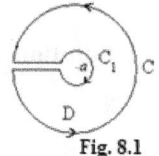

$$f(a) = (1/2\pi\, i) \int_C \{ f(z)\,/\,(z - a)\}\, dz. \qquad (8.1)$$

Corollary 8.1. The derivatives of Eq. (3.1) with respect to a yield

$$f'(a) \equiv \partial f/\partial a = (1/\,2\pi i) \int_C f(z)\,/\,(z - a)^2\}\, dz,$$

$$f''(a) \equiv (2!/\,2\pi i) \int_C \{ f(z)\,/\,(z - a)^3\}\, dz,$$

. . . .

$$f^{(n)}(a) \equiv (n!/2\pi i) \int_C \{f(z)/(z-a)^{n+1}\} \, dz. \qquad (8.2)$$

Example 8.1. Evaluate the integral $\int_C \{z^2 - z + 1)/(z-1)\} \, dz$ along the circle: **(i)** $|z| = 1$, **(ii)** $|z| = 1/2$.

Solution. (i) Comparing the given integral with that in Eq. (3.1) we note $a = 1$ and $f(z) = z^2 - z + 1$ implies $f'(z) = 2z - 1$, which exists within and on C. Therefore, $f(z)$ is analytic within and on the given circle $C: |z| = 1$. Hence, by Eq. (3.1),

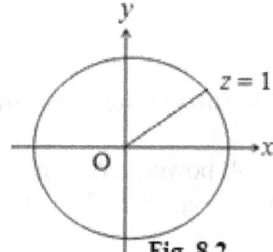

Fig. 8.2

$$f(a) = f(1) = 1 = (1/2\pi i)\int_C \{f(z)/(z-1)\}.dz$$

$\Rightarrow \qquad \int_C \{f(z)/(z-1)\}dz = 2\pi i. \; //$

(ii) In case of the circle $C' : |z| = 1/2$, the point $z = 1$ lies outside C'. Hence, $(z^2 - z + 1)/(z-1)$ as a whole is analytic everywhere within and on C. So, by Cauchy's theorem 2.1,

$$\int_{C''} \{(z^2 - z + 1)/(z - 1)\}. \, dz = 0. \; //$$

Example 8.2. Evaluate the integrals:

(i) $\int_C \{e^{2z}/(z-1)(z-2)\} \, dz$, C being a circle $|z| = 3$;

(ii) $\int_C \{\sin^2 z/(z - \pi/6)^3\} \, dz$, C being a circle $|z| = 1$.

Solution. (i) For $1/(z-1)(z-2) = -1/(z-1) + 1/(z-2)$, the integral splits as

$$\int_C \{e^{2z}/(z-1)(z-2)\}. \, dz$$

$$= -\int_C \{e^{2z}/(z-1)\}dz + \int_C \{e^{2z}/(z-2)\}dz. \qquad (8.3)$$

The function $f(z) \equiv e^{2z}$ is analytic within the circle $|z| = 3$ with the points $z = 1, 2$ as the singular points within C. Hence, by Cauchy's integral formula in Eq. (3.1), above integral results as

$$- 2\pi i\, f(1) + 2\pi i\, f(2) = 2\pi i\, (- e^2 + e^4).\ //$$

(ii) The function $f(z) \equiv \sin^2 z$, with $f'(z) \equiv \sin 2z$ is analytic within and on the circle C: $|z| = 1$ and the point $z = \pi/6 \approx 0.524$ lies within C. Hence, by Cauchy's integral formula in Eq. (3.2),

$$\int_C \{\sin^2 z\, /(z - \pi/6)^3\}dz \;=\; \pi i\, f''(\pi/6) \;=\; \pi i\, [2 \cos 2z]_{z=\pi/6} = \pi i.\ //$$

§ 9. Integration of a power series and Laurent's series

A power series in ascending powers of a (real) variable x is define dvide Eq. (5.26.2). For

$$u_n \equiv a_n x^n, \qquad u_{n+1} / u_n = (a_{n+1} / a_n)\, x,$$

the series $\Sigma\, u_n$ converges (to a finite definite sum) if

$$\lim_{n \to \infty} u_{n+1} / u_n \;=\; \lim_{n \to \infty} (a_{n+1} / a_n)\, x \;=\; l\, x\ \text{(say)}$$

is numerically less than 1, i.e. when $|x| = 1/\,l$. Thus, the *interval of convergence* of $\Sigma\, u_n$ is $-1/\,l < x < 1/\,l$ within which it converges. It diverges for the values of x outside this interval.

9.1. Series of complex terms

Replacing the real coefficients a's by complex numbers $a_m + ib_m$, $m = 1, 2, \ldots, n, \ldots, \infty$, Eq. (5.26.2), a power series in complex terms may be written as

$$\sum_{n=1}^{\infty} (a_n + i\, b_n) \equiv (a_1 + i\, b_1) + (a_2 + i\, b_2) + \ldots + (a_n + i\, b_n) + \ldots \infty. \quad (9.1)$$

If two series $\Sigma\, a_n$ and $\Sigma\, b_n$ converge to the sums A and B respectively, then series in Eq. (9.1) is said to converge to the sum $A + iB$. Thus, for a convergent series, there should hold

$$\lim_{n \to \infty} (a_n + i\, b_n) \;=\; 0.$$

Further, the series in Eq. (9.1) is called *absolutely convergent* if

$$\sum_{n=1}^{\infty} |a_n + i\, b_n| \equiv |a_1 + i\, b_1| + |a_2 + i\, b_2| + \ldots + |a_n + i\, b_n| + \ldots \infty \quad (9.2)$$

is convergent. Since $|a_n|$ and $|b_n|$ are both less or equal to $|a_n + i\, b_n| \equiv \sqrt{\{a_n^2 + b_n^2\}}$, an *absolutely convergent series* is *convergent*.

Note 9.1. If a power series $\Sigma\, a_n\, z^n$ converges for $z = z_1$, then it converges absolutely for $|z| < |z_1|$.

Theorem 9.1. (*Taylor's series*) If $f(z)$ is analytic within a circle C having centre at a, then for z inside C, $f(z)$ can be expanded in an infinite series containing powers of $z - a$:

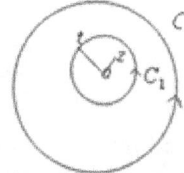

$$f(z) = f(a) + f'(a)\,(z - a) + (1/2!)\,f''(a)\,(z - a)^2 + \ldots$$

$$+ (1/n!)\,f^{(n)}(a)\,(z - a)^n + \ldots \infty. \qquad (9.3)$$

Fig. 9.1

Note 9.2. (Complex) analytic functions can always be expressed in terms of a power series of the form in Eq. (9.4).

Theorem 9.2. (*Laurent's series*) Let $f(z)$ be an analytic function of z in a ring shaped region D bounded by two concentric circles: C_1 (of radius r_1) and C (of radius $r > r_1$) each of them having their centre at a. Then, for all values of z in D we have

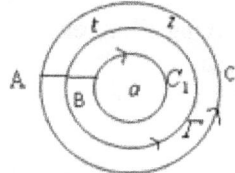

$$f(z) = a_0 + a_1\,(z - a) + a_2\,(z - a)^2 + \ldots \infty$$

$$+ a_{-1}\,(z - a)^{-1} + a_{-2}\,(z - a)^{-2} + \ldots \infty, \qquad (9.4)$$

where

Fig. 9.2

$$a_n \equiv (1/2\pi i) \int_{\Gamma} \{f(t)/(t - a)^{n+1}\}\, dt. \qquad (9.5)$$

Γ being any curve in D encircling C_1.

Example 9.1. Expand $f(z) = 1/(z - 1)(z - 2)$ in the region:

(i) $|z| < 1$, **(ii)** $1 < |z| < 2$, **(iii)** $|z| > 2$, **(iv)** $0 < |z - 1| < 1$.

Solution. (i) $f(z) = 1/(z - 1)(z - 2) = 1/(z - 2) - 1/(z - 1)$

$$= -(1/2)\,(1 - z/2)^{-1} + (1 - z)^{-1}. \qquad (9.6)$$

For $z < 1$, both $z/2$ and z are less than 1. Hence, both the terms in above sum can be expanded into convergent infinite series:

$$f(z) = -(1/2)(1 + z/2 + z^2/4 + z^3/8 + \dots \infty) + (1 + z + z^2 + z^3 + \dots \infty)$$

$$= 1/2 + (3/4)z + (7/8)z^2 + (15/16)z^3 + \dots \infty),$$

which is a Taylor's series for $f(z)$. //

(ii) For $1 < |z| < 2$, we write Eq. (9.6) as:

$$f(z) = -1/2(1 - z/2) - 1/z(1 - z^{-1})$$

$$= -(1/2)(1 - z/2)^{-1} - (1/z)(1 - z^{-1})^{-1};$$

and note that both $|z/2|$ and $|z^{-1}|$ are less than 1. Hence, above sums expand into convergent series as

$$f(z) = -(1/2)(1 + z/2 + z^2/4 + z^3/8 + \dots \infty)$$

$$- (1/z)(1 + z^{-1} + z^{-2} + z^{-3} + \dots \infty)$$

$$= \dots - z^{-4} - z^{-3} - z^{-2} - z^{-1} - 1/2 - z/4 - z^2/8 - z^3/16 - \dots,$$

giving a Laurent's series for $f(z)$. //

(iii) For $|z| > 2$, we write Eq. (9.6) as:

$$f(z) = 1/z(1 - 2z^{-1}) - 1/z(1 - z^{-1})$$

$$= z^{-1}(1 - 2z^{-1})^{-1} - z^{-1}(1 - z^{-1})^{-1}.$$

As $|z| > 2$, both $|z^{-1}|$ and $|2z^{-1}|$ are less than 1. So, above sums expand into convergent (infinite) series:

$$f(z) = z^{-1}(1 + 2z^{-1} + 4z^{-2} + 8z^{-3} + \dots \infty)$$

$$- z^{-1}(1 + z^{-1} + z^{-2} + z^{-3} + \dots \infty) = z^{-2} + 3z^{-3} + 7z^{-4} + \dots \infty. //$$

(iv) For $0 < |z - 1| < 1$, we write Eq. (9.6) as

$$f(z) = 1/\{(z-1) - 1\} - 1/(z - 1) = -1/\{1 - (z-1)\} - (z-1)^{-1}$$

$$= -\{1 - (z-1)\}^{-1} - (z-1)^{-1}.$$

Since $z - 1 < 1$ the first term in above function expands into a convergent infinite series:

$$f(z) = -\{1 + (z-1) + (z-1)^2 + (z-1)^3 + \ldots \infty\} - (z-1)^{-1}. \; //$$

§ 10. Singular point of an analytic function and residue of the function

We have already defined a singular point of the function $f(z)$ in the § 4 as a point where the function ceases to be analytic. Let $z = a$ be such a point of the function $f(z)$ that there exists a small circle centered at a which has no other singular points of $f(z)$, the point $z = a$ is then called an *isolated singular point*. In that case the function $f(z)$ can be expanded in a Laurent's series around $z = a$:

$$f(z) = \{c_0 + c_1(z-a) + c_2(z-a)^2 + \ldots \infty\}$$

$$+ \{c_{-1}(z-a)^{-1} + c_{-2}(z-a)^{-2} + \ldots \infty\}. \tag{10.1}$$

If $c_{-n}(z-a)^{-n}$ is the last non-vanishing term in the second series in Eq. (10.1) then the singular point is called a *pole of nth order*. A first order pole is called a *simple pole*.

However, if there is no last term in above series of negative powers of $(z-a)$, i.e. the series extends up to infinity, the point $z = a$ is called an *essential singularity*.

Definition 10.1. The coefficient of $(z-a)^{-1}$ in the expansion of $f(z)$ around an isolated singularity (irrespective of its order) is called the *residue* of $f(z)$ at that point:

$$c_{-1} \equiv \text{Res.} f(a) = (1/2\pi i).\int_C f(z).\,dz \tag{10.2}$$

$$\Rightarrow \int_C f(z)\,dz = 2\pi i.\,\text{Res.} f(a). \tag{10.3}$$

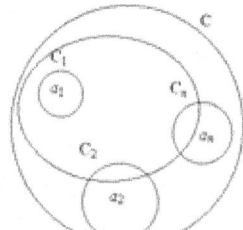

Theorem 10.1. (*Residue theorem*) If $f(z)$ is analytic within a closed curve C save for a finite number of singular points $a_1, a_2, \ldots, a_n$ within C, then

Fig. 10.1

$$\int_C f(z)\,dz = 2\pi i.\{\text{sum of residues of } f(z) \text{ at singular points } a_1, a_2, \ldots, a_n\}.$$

$$\tag{10.4}$$

Theorem 10.2. Residue of a function $f(z)$ at its simple pole $z = a$ is calculated by:

$$\text{Res. } f(a) = \lim_{z \to a} \{(z - a)f(z)\}, \tag{10.5}$$

or, alternately

$$\text{Res. } f(a) = \varphi(a) / \psi'(a), \tag{10.6}$$

when

$$f(z) = \varphi(z) / \psi(z), \text{ and } \psi(z) = (z - a) F(z), F(a) \neq 0. \tag{10.7}$$

Theorem 10.3. The residue of a function $f(z)$ at its nth order pole $z = a$ is

$$\text{Res. } f(a) = \{1/(n-1)!\}\{d^{n-1}(z-a)^n f(z) / dz^{n-1}\}_{z \to a}. \tag{10.8}$$

Example 10.1. Determine the poles of the function

$$f(z) = z^2 / (z - 1)^2 (z + 2),$$

and the residue at each pole. Hence, evaluate $\int_C f(z)\, dz$, where C is the circle $|z| = 2.5$.

Solution. (i) The function has a simple pole at $z = -2$ where its residue, by Eq. (10.5), is

$$\text{Res. } f(-2) = \lim_{z \to -2} \{(z + 2)f(z)\}$$

$$= \lim_{z \to -2} \{z^2 / (z - 1)^2\} = 4/9. \text{ //}$$

(ii) The function also has a pole of second order at $z = 1$ and its residue there, by Eq. (10.8), is

$$\text{Res. } f(1) = \lim_{z \to 1} d\{(z - 1)^2 f(z)\}/dz = \lim_{z \to 1} d\{z^2 /(z + 2)\}/dz$$

$$= \lim_{z \to 1} \{2z /(z + 2) - z^2 /(z + 2)^2\} = 5/9. \text{ //}$$

(iii) Clearly, $f(z)$ is analytic on the circle C: $|z| = 2.5$ and at all points within C except at the poles $z = 1, -2$. Hence, by (Residue) Theorem 10.1,

$$\int_C f(z)\, dz = 2\pi i\{\text{Res. } f(1) + \text{Res. } f(-2)\}$$

$$= 2\pi i\,(4/9 + 5/9) = 2\pi i. \text{ //}$$

§ 11. Evaluation of real definite integrals by contour integration

Many real definite integrals can be more easily evaluated by means of Residue Theorem. The contours chosen will consist of straight lines and circular arcs.

11.1. Integration around a unit circle

An integral of the type $\int_0^{2\pi} f(\sin\theta, \cos\theta)\, d\theta$, where the integrand is a rational function of $\sin\theta$ and $\cos\theta$, can be evaluated along a unit circle $z = e^{i\theta}$. For

$$\left.\begin{array}{l} \sin\theta = (e^{i\theta} - e^{-i\theta})/2i = (z - 1/z)/2i, \\[2mm] \cos\theta = (e^{i\theta} + e^{-i\theta})/2 = (z + 1/z)/2, \end{array}\right\} \qquad (11.1)$$

the given integral takes the form $\int_C f(z)\, dz$, where $f(z)$ is a rational function of z and C is a unit circle $|z| = 1$. Hence, the integral equals $2\pi i$ times the sum of its residues at the poles lying within C.

Example 11.1. Show by contour integration that

$$I \equiv \int_0^{2\pi} \{\cos 2\theta / (1 - 2a\cos\theta + a^2)\}\, d\theta = 2\pi a^2 / (1 - a^2), \quad a^2 < 1.$$

Solution. Putting $z = e^{i\theta}$ so that $dz = ie^{i\theta}\, d\theta \Rightarrow d\theta = dz / iz$ and $\cos\theta$ is given by Eq. (11.1). Also,

$$\cos 2\theta = (e^{2i\theta} + e^{-2i\theta})/2 = (z^2 + 1/z^2)/2,$$

we have

$$I = \int_C [(z^2 + 1/z^2)/2iz\{1 - a(z + 1/z) + a^2\}]\, dz$$

$$= (1/2i) \int_C \{(z^4 + 1)/z^2 (z - az^2 - a + a^2 z)\}\, dz$$

$$= (1/2i) \int_C \{(z^4 + 1)/z^2 (z - a)(1 - az)\}\, dz \equiv \int_C f(z)\, dz,$$

where C is the unit circle $|z| = 1$ and

$$f(z) \equiv (1/2i)(z^4 + 1) / z^2 (z - a)(1 - az)$$

has simple poles at $z = a$, $1/a$ and a pole of second order at $z = 0$. Of these only $z = 0$, a lie within the unit circle for given condition: $a^2 < 1$. Therefore,

$$\text{Res.} f(a) = \lim_{z \to a} \{(z - a) f(z)\}$$

$$= (1/2i) \lim_{z \to a} (z^4 + 1) / z^2 (1 - az) = (a^4 + 1) / 2i.a^2 (1 - a^2),$$

and

$$\text{Res.} f(0) = \lim_{z \to 0} (d/dz) \{z^2 f(z)\}$$

$$= (1/2i) \lim_{z \to 0} (d/dz) \{(z^4 + 1) / (z - a)(1 - az)\}$$

$$= (1/2i) \lim_{z \to 0} [4z^3/(z - a)(1 - az) - (z^4 + 1)/(z - a)^2 (1 - az)$$

$$+ a (z^4 + 1)/(z - a)(1 - az)^2]$$

$$= (1/2i)\{- 1/a^2 + a /(- a)\} = (- 1/2i)(1 + 1/a^2) = - (a^2 + 1)/2ia^2.$$

Hence,

$$I = 2\pi i. \{\text{Res.} f(a) + \text{Res.} f(0)\}$$

$$= \pi. \{(a^4 +1) / a^2 (1 - a^2) - (a^2 +1) / a^2\} = 2\pi a^2 / (1 - a^2). \text{ //}$$

11.2. Integration around a small semi-circle

The integral $\int_{-\infty}^{\infty} f(x)\, dx$ can be evaluated along a contour consisting of a semi-
circle $C_R : |z| = R$ together with the diameter
closing the semi-circle. If $f(z)$ has no singular point on the real axis, by Residue Theorem we have,

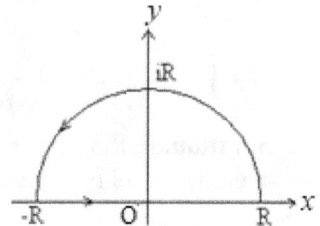
Fig. 11.1

$$\int_{C} f(z)\, dz + \int_{-r}^{-R} f(x)\, dx = 2\pi i \sum \text{Res.} f(a),$$

where $z = a$, etc. are some poles of $f(z)$ within C_R. Finally, letting R approach infinity, we find

$$\int_{-\infty}^{\infty} f(x)\, dx = \lim_{R \to \infty} 2\pi i \sum \text{Res.} f(a), \qquad (11.2)$$

provided $\int_{C} f(z)\, dz \to 0$ as $R \to \infty$.

Example 11.2. Evaluate $\int_{-\infty}^{\infty} \{x^2 / (x^2 + 1)(x^2 + 4)\}\, dx$.

Solution. Let us consider a complex integral $\int_C f(z)\, dz$, where C is the contour consisting of the semi-circle C_R of radius R together with the part of the real axis from $x = -R$ to R (see Fig. 11.1). The integrand $f(z) \equiv z^2 / (z^2 + 1)(z^2 + 4)$ has simple poles at $z = \pm i, \pm 2i$ of which $z = i, 2i$ lie within C. Therefore,

$$\text{Res.}\, f(i) = \lim_{z \to i} \{(z - i) f(z)\}$$

$$= \lim_{z \to i} z^2 / (z + i)(z^2 + 4) = -1/6i = i/6,$$

$$\text{Res.}\, f(2i) = \lim_{z \to 2i} \{(z - 2i) z^2 /(z^2 + 1)(z - 2i)(z + 2i)\} = 1/3i = -i/3,$$

Hence, by Residue Theorem,

$$\int_C f(z)\, dz = 2\pi i\, (i/6 - i/3) = \pi/3. \tag{11.3}$$

Also,
$$\int_C f(z)\, dz = \int_{C_R} f(z)\, dz + \int_{-R}^{-R} f(x)\, dx. \tag{11.4}$$

Letting $R \to \infty$ so that any point z on C_R satisfies $|z| \to \infty$,

$$f(z)\} \equiv z^2 / (z^2 + 1)(z^2 + 4) = 1 / z^2 (1 + z^{-2})(1 + 4z^{-2}) \to 0.$$

Consequently, $\lim_{|z|=R \to \infty} \int_{C_R} f(z)\, d = 0$, so that Eqs. (11.3) and (11.4) yield

$$\int_{-\infty}^{\infty} \{x^2 / (x^2 + 1)(x^2 + 4)\}\, dx = \pi/3. //$$

11.3. Integration around rectangular contours

Example 11.3. Evaluate $I = \int_{-\infty}^{\infty} \{e^{ax} / (e^x + 1)\} dx,\ 0 < a < 1$.

Solution. We consider the integral

$$\int_C f(z)\, dz \equiv \int_C \{e^{az} / (e^z + 1)\} dz,$$

where C is the rectangle with vertices at A $(R, 0)$, B $(R, 2\pi)$, C $(-R, 2\pi)$ and D $(-R, 0)$, $R > 0$. The function $f(z) \equiv e^{az} /(e^z + 1)$ has infinitely large number of poles given by

$$e^z = -1 = e^{(2n+1)\pi i}$$

$\Rightarrow$

$z = (2n+1)\pi i = \pm \pi i, \pm 3\pi i, \ldots;$

$n = 0, \pm 1, \pm 2, \pm 3, \ldots;$

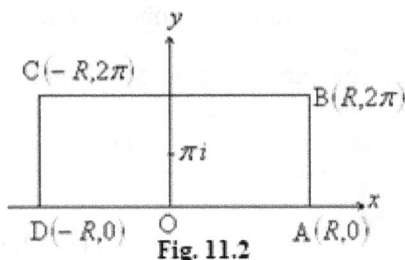

Fig. 11.2

of which only $z = \pi i$ lies within the rectangle. By Residue Theorem,

$$\int_C f(z)dz = 2\pi i.\text{Res}.f(\pi i) = 2\pi i.\lim_{z \to \pi i} e^{az}/\{d(e^z+1)/dz\}, \text{ by Eq.(10.6)}$$

$$= 2\pi i.\lim_{z \to \pi i} e^{az}/e^z = -2\pi i\, e^{a\pi i}. \tag{11.5}$$

Also,

$$\int_C f(z)\,dz = (\int_{AB} + \int_{BC} + \int_{CD} + \int_{DA})f(z)\,dz = \int_0^{2\pi} f(R+iy)\, i\, dy$$

$$+ \int_{x=-R}^{-R} f(x+2\pi i)\,dx + \int_{y=2\pi}^{0} f(-R+iy)\, i\, dy + \int_{x=-R}^{-R} f(x)\,dx, \tag{11.6}$$

where we have used:

$z = R + iy \Rightarrow dz = i\, dy$ along AB, $z = x + 2\pi i \Rightarrow dz = dx$ along BC,

$z = -R + iy \Rightarrow dz = i\, dy$ along CD, and $z = x \Rightarrow dz = dx$ along DA.

Putting for $f(z)$ in Eq. (11.6) and inverting the direction of the second and third integrals we have

$$\int_C f(z)\,dz = i\int_0^{2\pi} [e^{a(R+iy)}/\{e^{(R+iy)}+1\}]dy$$

$$- \int_{-R}^{-R} [e^{a(x+2\pi i)}/\{e^{(x+2\pi i)}+1\}]\,dx - i\int_0^{2\pi} [e^{a(-R+iy)}/\{e^{(-R+iy)}+1\}]\,dy$$

$$+ \int_{-r}^{-R} \{e^{ax}/(e^x+1)\}\,dx. \tag{11.7}$$

For any two complex numbers z_1 and z_2 there holds

$$|z_1| \geq |z_2| \Rightarrow |z_1+z_2| \geq |z_1| - |z_2|. \tag{11.8}$$

Hence,

$$|e^{(R+iy)}+1| \geq e^R - 1 \text{ and } |e^{a(R+iy)}| = e^{aR}, \text{ as } |e^{iay}| = 1;$$

the integrand in the first integral in right hand side of Eq. (11.7) satisfies

$$\left| e^{a(R+iy)} / \{ e^{(R+iy)} + 1 \} \right| \le e^{aR} / (e^R - 1) \to 0 \text{ as } R \to \infty, \text{ for } a < 1.$$

Also, the third integrand in RHS of Eq. (11.7) satisfies

$$\left| e^{a(-R+iy)} / \{ e^{(-R+iy)} + 1 \} \right| \le e^{-aR} / (1 - e^{-R}) \to 0 \text{ as } R \to \infty, \text{ for } 0 < a.$$

Thus, letting $R \to \infty$, Eq. (11.7) reduces to

$$f(z)\, dz = - e^{2a\pi i} \int_{-\infty}^{\infty} \{ e^{ax} / (e^x + 1) \} dx + \int_{-\infty}^{\infty} \{ e^{ax} / (e^x + 1) \} d = (1 - e^{2a\pi i})\, I,$$

which together with Eq. (11.5) determines

$$I = 2\pi i\, e^{a\pi i} / (e^{2a\pi i} - 1) = 2\pi i / (e^{a\pi i} - e^{-a\pi i}) = \pi / \sin a\pi. \; //$$

11.4. Indenting the contours having poles on the real axis

So far the integrands considered before did not have any pole(s) on the real axis. But, now we consider the case when they have some simple poles on the real axis. Such poles may be deleted by encircling them by semi-circles of sufficiently small radii.

Example 11.4. Evaluate $I \equiv \int_0^\infty \{ (\sin mx)/x \}\, dx, \; m > 0.$

Solution. Consider the integral

$$\int_C f(z)\, dz \equiv \int_C (e^{imz} / z)\, dz,$$

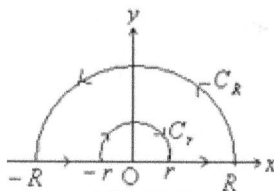

Fig. 11.3

where C consists of (i) the real axis from $x = r$ to R, (ii) the upper half of the circle C_R: $|z| = R$, (iii) the real axis from $x = -R$ to $-r$, (iv) the upper half of the circle C_r: $|z| = r$. Since the function $f(z)$ has no singular points inside C (the only singular point at $z = 0$ has been deleted by indenting it by a semi-circle C_r), by (Cauchy's) Theorem 7.1, there follows

$$\int_r^R f(x)\, dx + \int_{C_R} f(z)\, dz + \int_{-r}^{-R} f(x)\, dx + \int_C f(z)\, dz = 0. \quad (11.9)$$

Now, for $z = R e^{i\theta}$ along C_R and therefore $dz = iR e^{i\theta}.d\theta$, we have

$$\int_{C_R} f(z)\, dz \;=\; i \int_0^{2\pi} e^{imR\,(\cos\theta + i\sin\theta)}\, d\theta.$$

Since

$$\left| e^{imR\,(\cos\theta + i\sin\theta)} \right| \;=\; \left| e^{-mR\sin\theta + imR\cos\theta} \right| \;=\; e^{-mR\sin\theta},$$

$$\left| \int_{C_R} f(z)\, dz \right| \le \int_0^{\pi} e^{-mR\sin\theta}\, d\theta = 2\int_0^{\pi/2} e^{-mR\sin\theta}\, d\theta, \text{ as } F\,(\pi - \theta) = F\,(\theta),$$

$$\le 2\int_0^{\pi/2} e^{-mR\,(2\theta/\pi)}\, d\theta, \qquad \text{for } \theta \,\varepsilon\, [0,\, \pi/2], \text{ and } 2\theta/\pi \le \sin\theta;$$

$$= (\pi/mR)\Big[\, -e^{-2mR\theta/\pi}\,\Big]_0^{\pi/2} = (\pi/mR)[1 - e^{-mR}] \to 0, \text{ as } R \to \infty, \text{ for } m > 0.$$

Also, for $z = r\, e^{i\theta}$ along C_r and therefore $dz = ir\, e^{i\theta}.\, d\theta$, we have

$$\int_{C_R} f(z)\, dz = i\int_\pi^0 e^{imr\,(\cos\theta + i\sin\theta)}\, d\theta \le i\int_\pi^0 e^{-m r \sin\theta}\, d\theta \to i\int_\pi^0 d\theta = -\pi\, i,$$

as $r \to 0$. Hence, letting $R \to \infty$ and $r \to 0$, (11.9) reduces to

$$\{ \int_0^\infty + \int_{-\infty}^{-0} \}\, f(x)\, dx - \pi\, i = 0$$

$$\Rightarrow \qquad \int_{-\infty}^\infty f(x)\, dx \equiv \int_{-\infty}^\infty (e^{imx}/x)\, dx = \pi i. \qquad\qquad (11.10)$$

Equating the imaginary parts on both sides, we get

$$\int_{-\infty}^\infty \{(\sin mx)/x\}\, dx = 2\int_0^\infty \{(\sin mx)/x\}\, dx \;=\; \pi$$

$$\Rightarrow \qquad I = \pi/2 \quad \text{as } f(-x) = f(x).\; //$$

Note 11.1. Equating the real parts in (11.10) we can also derive

$$\int_{-\infty}^\infty \{(\cos mx)/x\}\, dx \;=\; 0.$$

Example 11.5. Show by contour integration

$$\int_0^\infty \{x^{p-1}/(1+x)\}\, dx \;=\; \pi/\sin p\pi, \qquad\qquad 0 < p < 1.$$

Solution. Putting $x = e^t$ so that $dx = e^t\, dt$ the given integral assumes the form

$$I = \int_{-\infty}^{\infty} \{e^{(p-1)t} / (1 + e^t)\} \, e^t \, dt = \int_{-\infty}^{\infty} \{e^{pt} / (1 + e^t)\} dt,$$

which is the same as in Ex. 11.3. //

§ 12. Conformal transformation and geometrical interpretation of $w = f(z)$

Let $f(z)$ be a function of the complex variable $z = x + iy$ in the z-plane (determined by Ox- and Oy-coordinate axes) and $w = u + iv$ be another complex variable in the w-plane (determined by u- and v-axes). All the four variables x, y, u, v are real. The transformation of a real variable, say x, onto another real variable (say y) effected by some functional equation $y = g(x)$, may be plotted as a curve on the two-dimensional xOy-plane. But, in case of function w:

$$w = u + iv = f(x + iy) = f(z), \qquad (12.1)$$

involving four real variables u, v, x, y a 4-dimensional complex region is required. Since 4-dimensional graph paper is not possible we make use of two complex regions: z-plane (for the variables x and y) and w-plane (for the variables u and v). Thus, we understand that when the complex variable z describes a curve C on the z-plane, the point w correspondingly describes another curve C′ on the w-plane. Since for each point P (x, y) on C in z-plane there corresponds a pint P′(u, v) on C′ in w-plane we say that the curve C (of z-plane) is transformed onto C′ (of w-plane) by Eq. (12.1). This defines a mapping or transformation of z-plane onto the w-plane.

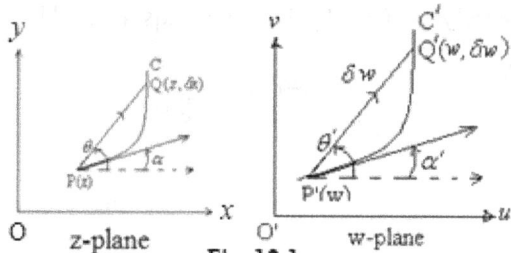

Fig. 12.1

Definition 12.1. Let C, D be two curves in z-plane intersecting each other at the point P and the corresponding curves (transformed curves under the transformation (12.1)) in w-plane be C′ and D′ intersecting at P′. If the angle of intersection of the curves at P is the same as that at P′ in magnitude as well as in the sense of direction then the transformation in Eq. (12.1) becomes *conformal*.

Theorem 12.1. The transformation effected by an analytic function $w = f(z)$ is conformal at every point of the z-plane where $f'(z) \neq 0$.

Note 12.1. A point, where $f'(z) = 0$, is called a *critical point* of the transformation (1.1).

Corollary 12.1. The tangent to the curve C gets rotated through an angle

$$\varphi \equiv \text{amp} \, |f'(z)| \tag{12.2}$$

under the given conformal transformation vide Eq.(12.1).

Corollary 12.2. The arc of a curve C gets magnified by the factor $|f'(z)|$ under a conformal transformation given by Eq. (12.1).

Corollary 12.3. The infinitesimal areas of the z-plane get magnified by the factor $|f'(z)|^2$.

Lemma 12.1. The Jacobian $J\{(u, v)/(x, y)\} = |f'(z)|^2$.

Proof. $\qquad\qquad$ LHS $= \begin{vmatrix} \partial u/\partial x & \partial u \partial y \\ \partial v/\partial x & \partial v/\partial y \end{vmatrix}$.

The function $f(z)$ being analytic its real and imaginary parts: u and v satisfy the Cauchy-Riemann Eqs. (3.2). Hence, the Jacobian reduces to

$$\begin{vmatrix} \partial u/\partial x & -\partial v/\partial x \\ \partial v/\partial x & \partial u/\partial x \end{vmatrix} = (\partial u/\partial x)^2 + (\partial v/\partial x)^2 = |f'(z)|^2. \; //$$

Note 12.2. The angle preserving property of a conformal transformation has many applications. For instance, consider the flow of an incompressible fluid in a plane with velocity potential $\varphi(x, y)$ and the stream function $\psi(x, y)$. Both of these functions are real and imaginary parts of some analytic function $w = f(z)$. As $\varphi = \text{const.}$ and $\psi = \text{const.}$ represent a system of orthogonal curves these are transformed into a set of orthogonal lines in the w-plane under the conformal transformation in Eq. (12.1).

Thus, the conjugate functions φ and ψ satisfy the Laplace equation and remain conjugate even after the transformation in Eq. (12.1).

§ 13. Some standard conformal transformations

13.1. Translation. We consider a transformation

$$w = z + c, \qquad (13.1)$$

of z-plane onto the w-plane, where

$$c = c_1 + i c_2 \qquad (13.2)$$

is some complex constant: c_1, c_2 being real. Writing z and w in terms of x, y and u, v respectively the transformation in Eq. (13.1) becomes

$$u + iv = x + iy + c_1 + i c_2 \implies u = x + c_1 \text{ and } v = y + c_2. \qquad (13.3)$$

Thus, the transformation in Eq. (13.1) maps a point P (x, y) of z-plane onto P′ $(x + c_1, \ y + c_2)$ of w-plane. Every point of z-plane maps onto a corresponding point of w-plane in the same way. In other words, a region of z-plane gets shifted into a region of w-plane along the vector $c = (c_1, c_2)$ and the shape and size of the region are preserved. Such transformations are called *translations*.

Theorem 13.1. A translation maps circles of z-plane onto circles of w-plane.

13.2. Rotation. We consider a transformation

$$w = cz, \qquad (13.4)$$

where

$$c = \rho e^{i\alpha} = \rho (\cos \alpha + i \sin \alpha) \implies |c| = \rho, \qquad (13.5)$$

is a complex constant. Taking $z = re^{i\theta}$ and $w = Re^{i\varphi}$, Eq. (13.4) becomes

$$w = \rho e^{i\alpha} . re^{i\theta} = (\rho r) e^{i(\theta + \alpha)} \qquad (13.6)$$

$\implies$

$$R = \rho r \quad \text{and} \quad \varphi = \theta + \alpha; \qquad (13.7)$$

thus, mapping a point P (r, θ) of the z-plane onto the point P′ $(\rho r, \theta + \alpha)$ in the w-plane. Hence, the transformation in Eq. (13.4) consists of magnification (respectively contraction) of the radius vec or OP by $\rho = |c|$, if $\rho > 1$ (respectively $\rho < 1$); and its rotation through an angle $\alpha = \text{amp } c$. Accordingly, any figure in the z-plane gets transformed in a geometrically similar figure in the *w*-plane.

Theorem 13.2. A rotation maps circles of *z*-plane onto circles of *w*-plane.

Proof. The circle $z = re^{i\theta}$ of *z*-plane transforms onto the circle in Eq. (13.6), which also represents a circle in the *w*-plane. //

13.3. Inversion and reflection

Now we consider a transformation

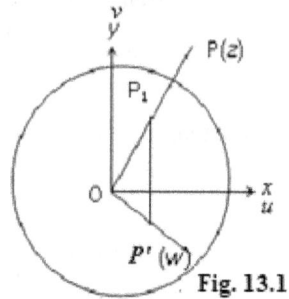
Fig. 13.1

$$w = 1/z. \tag{13.8}$$

Taking $z = re^{i\theta}$ and $w = Re^{i\varphi}$ above transformation becomes

$$Re^{i\varphi} = (1/r) e^{-i\theta} \quad (8.9) \quad \Rightarrow \quad R = 1/r \quad \text{and} \quad \varphi = -\theta.$$

It is convenient here to consider the two planes (*z*-plane and *w*-plane) superimposed over each other. Thus, for any point P (r, θ) of *z*-plane P_1 $(1/r, \theta)$ is its inverse with respect to a unit circle centered at O. Further, reflection of P_1 through the real axis is a point P' $(1/r, -\theta)$ in the *w*-plane representing the transformed point of P (*z*) in the *w*-plane under the transformation (13.8).

As such, the transformation (13.8) is an inversion of *z* with respect to the unit circle $|z| = 1$ followed by a reflection of the inverse of *z* through the real axis.

Theorem 13.3. The transformation (13.8) maps the interior (respectively exterior) of the unit circle $|z| = 1$ onto the exterior (resp. interior) of the unit circle $|w| = 1$. In particular, the origin $z = 0$ corresponds to the point at infinity and the point at infinity maps onto the origin.

Theorem 13.4. The transformation given by Eq. (13.8) maps a circle on *z*-plane onto a circle on *w*-plane. If the circle on *z*-plane passes through origin its image on *w*-plane is a straight line.

Note 13.1. Since a straight line can be regarded as a special case of a circle (of radius infinity) the above theorem can be restated as:

"The transformation given by Eq. (13.8) always maps a circle of z-plane onto a circle of w-plane".

Example 13.1. Find the transformation that maps the points:

$$z = -1, i, 1 \ (\text{of } z\text{-plane}) \quad \text{onto} \quad w = 1, i, -1 \ (\text{of } w\text{-plane})$$

respectively.

Solution. Let the transformation (12.1) maps the points of z-plane onto respective points of w-plane so that the relations:

$$f(-1) = 1, \quad f(i) = i, \quad \text{and} \quad f(1) = -1$$

hold simultaneously. But, this is possible only when $f(z) = -1/z = w$. //

Example 13.2. Find the image of

$$\left| z - 2i \right| = 2 \tag{13.9}$$

under the transformation in Eq.(13.8). Also, show that the hyperbola given by Eq. (1.1.4) (on the z-plane) maps onto the lemniscate:

$$\rho^2 = \cos 2\varphi \tag{13.10}$$

(on the w-plane).

Solution. (i) Rewriting Eq. (13.9) as

$$\left| x + iy - 2i \right| = \left| x + i(y - 2) \right| = 2,$$

and taking its modulus we get

$$x^2 + (y - 2)^2 = 4, \quad \text{or} \quad x^2 + y^2 - 4y = 0, \tag{13.11}$$

that represents a circle (through origin) on the z-plane. The same, under Eq. (13.8), maps onto

$$1 / (u^2 + v^2) + 4v / (u^2 + v^2) = 0; \quad \text{or} \quad 4v + 1 = 0,$$

which is a straight line on w-plane. //

(ii) Also, the hyperbola in Eq. (1.1.4) transforms onto

$$u^2 - v^2 = (u^2 + v^2)^2.$$

Writing $w = \rho e^{i\varphi}$ so that $u = \rho.\cos\varphi$ and $v = \rho.\sin\varphi$, above equation further reduces to

$$\rho^2(\cos^2\varphi - \sin^2\varphi) = \rho^4 \quad \Rightarrow \quad \rho^2 = \cos 2\varphi. \; //$$

§ 14. Bilinear transformation

The transformation

$$w = (az + b) / (cz + d), \tag{14.1}$$

where a, b, c, d are some complex constants satisfying

$$ad - bc \neq 0, \tag{14.2}$$

is called a *bilinear* transformation.

Note 14.1. The transformation given by Eq. (14.1) was first studied by August Ferdinand *Möbius* (1790 – 1868 A.D.) hence it is also called after him.

The condition vide Eq. (14.2) ensures that

$$dw/dz = \{a(cz + d) - c(az + b)\}/(cz + d)^2 = (ad - bc)/(cz + d)^2 \neq 0,$$

so that, by Theorem 12.1, the transformation becomes conformal. In the lack of above condition $dw/dz = 0$ causing every point of the z-plane become a critical point.

Theorem 14.1. The inverse mapping of Eq. (14.1) is also bilinear.

Note 14.2. It is clear from Eq. (14.1) that each point of z-plane except $z = -d/c$ corresponds to a unique point in the w-plane. Similarly, it is also evident that each point of w-plane except $w = a/c$ maps onto a unique point in z-plane.

Furthermore, the exceptional points $z = -d/c$ of z-plane (respectively $w = a/c$ of w-plane) map onto infinity of w-plane (resp. z-plane).

14.1. Invariant points of a bilinear transformation

If a point $z = z_1$ (of z-plane) maps onto itself in the w-plane, i.e. $w = z$, then Eq. (14.1) becomes

$$z = (az + b) / (cz + d) \quad \Rightarrow \quad cz^2 + (d - a)z - b = 0. \tag{14.3}$$

The roots of this quadratic equation are called the *invariant* or *fixed* points of the bilinear transformation in Eq. (14.1). If these roots are equal the transformation is called *parabolic*.

Dividing both the numerator and denominator in the right hand side of (14.1) by one of the non-vanishing constant coefficient, say b, and rewriting Eq. (14.1) as

$$w = (a_1 z + 1) / (c_1 z + d_1), \tag{14.4}$$

where $a_1 = a/b$, etc. we note that there are only three essential constants. Thus, only three conditions are required to determine a bilinear transformation. Consequently, three distinct points z_1, z_2, z_3 (of z-plane) can be mapped onto any three specified points w_1, w_2, w_3 (of w-plane).

Theorem 14.2. A bilinear transformation maps circles into circles.

Theorem 14.3. A bilinear transformation preserves the cross-ratio of four points.

Example 14.1. Find the bilinear transformation mapping the points $z = 1, i, -1$ (of z-plane) onto the points $w = i, 0, -i$. Hence, find the image of $z < 1$ and the invariant points of the transformation.

Solution. (i) Let a transformation vide Eq. (14.1) map the points

$$z_1 = 1, \quad z_2 = i, \quad z_3 = -1 \quad \text{and} \quad z_4 = z, \quad \text{(of } z\text{-plane)}$$

onto

$$w_1 = i, \quad w_2 = 0, \quad w_3 = -i \quad \text{and} \quad w_4 = w, \quad \text{(of } w\text{-plane)}$$

respectively. Applying Eq. (14.7) we then have

$$i(-i - w) / (i - w)(-i) = (1 - i)(-1 - z) / (1 - z)(-1 - i)$$

$$\Rightarrow$$

$$(w + i)/(w - i) = (1 - i)(1 + z) / (1 + i)(z - 1)$$

$$= \{(1 - i)z + (1 - i)\}/\{(1 + i)z - (1 + i)\}.$$

Applying the method of *componendo* and *dividendo* we derive

$$w / i = (z - i) / (-iz + 1) \quad \Rightarrow \quad w = (iz + 1) / (-iz + 1). \text{//} \tag{14.5}$$

(ii) Solving Eq. (14.5) for z:

$$i z (1 + w) = w - 1$$

$\Rightarrow$

$$z = (w - 1) / i (w + 1) = i (1 - w) / (1 + w), \qquad (14.6)$$

the inequality $|z| < 1$ reduces to

$$|z| = |1 - w| / |1 + w| < 1 \Rightarrow |1 - u - iv| < |1 + u + iv| \ ,$$

or,

$$(1 - u)^2 + v^2 < (1 + u)^2 + v^2 \Rightarrow 0 < 4u \Rightarrow u > 0.$$

Thus, the interior of the circle $|z| = 1$ transforms onto the entire half of the w-plane to the right of the imaginary axis. //

(iii) To find the invariant points of the transformation we put $w = z$ in Eq. (14.5):

$$i z^2 + (i - 1) z + 1 = 0$$

$\Rightarrow$

$$z = [1 - i \pm \sqrt{\{(i - 1)^2 - 4i\}}] / 2i = - \{1 + i \mp \sqrt{(6i)}\}/2,$$

determining the invariant points. //

Example 14.2. The transformation (14.1) transforms the circle $|w| = 1$ onto a straight line in z-plane if $|a| = |c|$.

§ 15. Some special conformal transformations

15.1. Transformation $\qquad w = z^2$ $\qquad\qquad\qquad (15.1)$

Writing $w = u + iv$ and $z = x + iy$, the given transformation becomes

$$u + iv = (x + iy)^2 = (x^2 - y^2) + i (2xy) \Rightarrow u = (x^2 - y^2) \qquad (15.2)$$

and

$$v = 2x\,y. \qquad\qquad\qquad (15.3)$$

Thus, when $u = $ const. (say a), Eq. (15.2) represents a rectangular hyperbola. Similarly, when $v = b$ (const.), Eq. (15.3) represents:

$$x\,y = b / 2, \qquad\qquad\qquad (15.3a)$$

which is also a rectangular hyperbola referred to its asymptotes as (rectangular) coordinate axes. Thus, we have the:

Theorem 15.1. A pair of lines parallel to the axes in w-plane: $u = a$ and $v = b$ map into a pair of orthogonal rectangular hyperbolas in the z-plane under the transformation vide Eq. (15.1).

On the other hand, $x = c$ (const.) reduces Eqs. (15.2) and (15.3) to

$$y^2 = c^2 - u \qquad\qquad \text{and} \qquad y = v/2c.$$

Elimination of y from these equations determines

$$v^2 = 4c^2 (c^2 - u), \tag{15.4}$$

representing a parabola in w-plane. Similarly, $y = d$ (const.) reduces Eqs. (15.2) and (15.3) to

$$x^2 = d^2 + u \qquad \text{and} \qquad x = v/2d.$$

Elimination of x from these yields

$$v^2 = 4d^2 (d^2 + u), \tag{15.5}$$

which is also a parabola in w-plane. Thus, we have proved the:

Theorem 15.2. A pair of lines: $x = c, y = d$ parallel to the coordinate axes in z-plane map onto orthogonal parabolas in the w-plane under the transformation vide Eq. (15.1).

The critical points of the transformation (15.1) are given by

$$dw / dz = 2z = 0 \qquad \Rightarrow \qquad z = 0.$$

Theorem 15.3. The upper half of the z-plane, where θ varies from 0 to π, transforms into the entire w-plane ($0 \le \varphi \le 2\pi$) under the transformation vide Eq. (15.1).

15.2. The transformation $\qquad\qquad w = z^n,$ $\qquad\qquad$ (15.6)

(n being a positive integer). The critical point of the transformation is given by

$$dw / dz = n z^{n-1} = 0 \qquad \Rightarrow \qquad z = 0.$$

Also, the polar form of the transformation is

$$Re^{i\varphi} = r^n e^{in\theta},\qquad(15.7)$$

$\Rightarrow$

$$R = r^n \qquad \text{and} \qquad \varphi = n\theta.\qquad(15.8)$$

For a sector of z-plane, where θ varies from 0 to π/n, correspondingly on w-plane φ varies from 0 to π. Thus, we have the:

Theorem 15.4. The transformation vide Eq. (15.6) is conformal everywhere except at origin and it maps a sector of z-plane of central angle π/n to cover the upper half of w-plane.

15.3. The transformation $\qquad\qquad w = e^z.\qquad$ (15.9)

Writing $w = \rho\, e^{i\varphi}$ and $z = x + iy$ the transformation reduces to

$$\rho e^{i\varphi} = \exp(x + iy) \quad\Rightarrow\quad \rho = e^x \quad \text{and} \quad \varphi = y.\quad(15.10)$$

Thus, the lines parallel to y-axis (with equation $x = a$) map onto circles $\rho = $ const. in w-plane. Also, the lines parallel to x-axis (with equation $y = b$) map into the radius vector $\varphi = b$ (const.) of the w-plane. Therefore, for any horizontal strip of height 2π in the z-plane (where y varies from 0 to 2π), Eq. (15.10) determines $\varphi = 0$ to 2π. Thus, such horizontal strip covers once the entire w-plane.

Theorem 15.5. The rectangular region ABCD of z-plane (bounded by lines

$$a_1 \le x \le a_2 \quad \text{and} \quad b_1 \le y \le b_2)$$

transforms onto the region A′B′C′D′ of w-plane bounded by the circles

$$e^a \le \rho \le e^a \quad \text{and} \quad \text{rays} \quad b_1 \le \varphi \le b_2.$$

15.4. Transformation $\qquad w = \cosh z.\qquad$ (15.11)

Writing the variables z and w in terms of their real and imaginary parts the transformation vide Eq. (15.11) becomes

$$u + iv = \cosh(x + iy) = \cosh x.\cosh(iy) + \sinh x.\sinh(iy)$$

$$= \cosh x.\cos y + i\sinh x.\sin y,$$

where we have used

$$\sin (iy) = i \sinh y, \quad \text{and} \quad \cos (iy) = \cosh y. \quad (15.12)$$

Separating the real and imaginary parts we, thus, obtain

$$u = \cosh x. \cos y \quad \text{and} \quad v = \sinh x. \sin y. \quad (15.13)$$

Elimination of x from these relations yields

$$u^2 / \cos^2 y - v^2 / \sin^2 y = \cosh^2 x - \sinh^2 x = 1, \quad (15.14)$$

that represents a *hyperbola* on w-plane for some constant value of y.

Similarly, elimination of y from Eq. (15.12) yields

$$u^2 / \cosh^2 x + v^2 / \sinh^2 x = \cos^2 y + \sin^2 y = 1, \quad (15.15)$$

which is an *ellipse* on w-plane for $x = $ const. Thus, we have the:

Theorem 15.6. The lines parallel to x-axis (having equation $y = $ const.) on z-plane map onto hyperbolas in the w-plane, while those parallel to y-axis (with equation $x = $ const.) map onto ellipses in the w-plane.

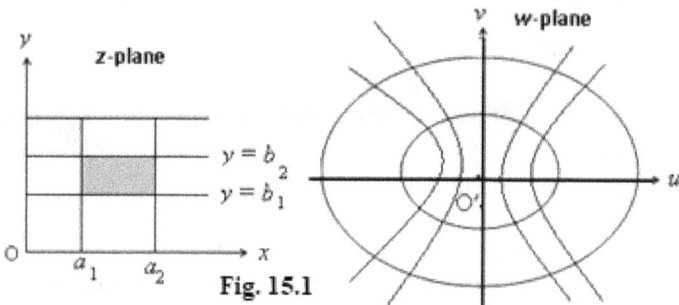

Fig. 15.1

Theorem 15.7. The rectangular region bounded by the lines $a_1 \le x \le a_2$ and $b_1 \le y \le b_2$) on the z-plane transforms into the shaded portion of the w-plane (bounded by the corresponding hyperbolas and ellipses).

15.5. Joukowski's transformation $\quad w = z + 1/z. \quad (15.16)$

Since $dw/dz = 1 - 1/z^2 = (z - 1)(z + 1)/z^2$, which vanishes at the points $z = \pm 1$, the transformation is conformal except at these points.

These critical points correspond to the points $w = \pm 2$ of the w-plane. Changing z into polar coordinates the transformation (15.16) becomes

$$w = u + iv = r (\cos \theta + i \sin \theta) + 1/r (\cos \theta + i \sin \theta)$$

$$= r (\cos \theta + i \sin \theta) + (1/r) (\cos \theta - i \sin \theta)$$

$\Rightarrow$

$$u = (r + 1/r) \cos \theta \quad \text{and} \quad v = (r - 1/r) \sin \theta. \quad (15.17)$$

Eliminating θ from these we obtain

$$u^2 / (r + 1/r)^2 + v^2 / (r - 1/r)^2 = 1. \quad (15.18)$$

Also,

$$r + 1/r = u / \cos \theta \quad \text{and} \quad r - 1/r = v / \sin \theta$$

$\Rightarrow$

$$2r = u / \cos \theta + v / \sin \theta \quad \text{and} \quad 2/r = u / \cos \theta - v / \sin \theta,$$

which, on multiplication with each other, yield

$$4 = u^2 / \cos^2 \theta - v^2 / \sin^2 \theta. \quad (15.19)$$

It is clear from Eq. (15.18) that the circles $r =$ const. of z-plane transform into a family of ellipses in the w-plane. For

$$(r + 1/r)^2 - (r - 1/r)^2 = (2r)(2/r) = 4 \text{ (i.e. a constant),}$$

these ellipses are confocal. In particular, the unit circle $r = 1$ of the z-plane maps onto the segment $u = -2$ to $u = 2$ of the real axis (having equation $v = 0$) in the w-plane.

Also, Eq. (15.19) shows that the radial lines $\theta =$ const. of the z-plane transform into a family of hyperbolas which are also confocal for $\cos^2 \theta - \sin^2 \theta =$ const. as θ is const.

Note 15.1. From Eq. (15.17) we note that

$$v = 0 \quad \text{if} \quad r = \pm 1, \quad \text{or} \quad \sin \theta = 0 \Rightarrow \theta = 0, \pi.$$

Thus, the stream line $v = 0$ (of w-plane) consists of the unit circle $r = 1$ and the x-axis (along which θ varies from 0 to π) of z-plane.

§ 16. Complex potential and application of complex analysis to flow problems

It is seen in § 5, that both real and imaginary parts of an analytic function $f(z)$ satisfy the Laplace equation. These solutions (of the Laplace equation), so-called the *harmonic* (or *conjugate*) functions provide solution to a number of field and flow problems.

We consider the motion of an incompressible fluid in two-dimensions (i.e. in a plane). Let the flow be in a plane parallel to the xOy-plane and the velocity vector **v** of the motion be irrotational, i.e. its *curl* vanishes:

$$\text{curl } \mathbf{v} \equiv \nabla \times \mathbf{v} = \mathbf{0}. \tag{16.1}$$

Since there holds a vector identity (cf. Eq. (52...) for a scalar function φ (x, y):

$$\text{curl (grad } \varphi) \equiv \nabla \times (\nabla \varphi) = (\nabla \times \nabla) \varphi = \mathbf{0} \tag{16.2}$$

the Eq. (16.1) implies

$$\mathbf{v} = \text{grad } \varphi = (\partial \varphi / \partial x)\,\hat{\mathbf{i}} + (\partial \varphi / \partial y)\,\hat{\mathbf{j}} \tag{16.3}$$

where $\hat{\mathbf{i}}$, $\hat{\mathbf{j}}$ are unit vectors along rectangular Cartesian coordinate axes Ox and Oy respectively.

Definition 16.1. The function $\varphi (x, y)$ satisfying Eq. (16.3) is called the *velocity potential* and the curves

$$\varphi (x, y) = c \text{ (const.)} \tag{16.4}$$

are called the *equi-potential lines*.

Since the flow is in a plane parallel to xOy-plane, the velocity vector **v** can be expressed as

$$\mathbf{v} = v_x\,\hat{\mathbf{i}} + v_y\,\hat{\mathbf{j}}. \tag{16.5}$$

Comparing Eqs. (16.3) and (16.5) we, therefore, have

$$v_x \equiv \partial v / \partial x = \partial \varphi / \partial x \quad \text{and} \quad v_y \equiv \partial v / \partial y = \partial \varphi / \partial y \tag{16.6}$$

$\Rightarrow$

$$\partial v_x / \partial x = \partial^2 v / \partial x^2 = \partial^2 \varphi / \partial x^2, \ \partial v_y / \partial y = \partial^2 v / \partial y^2 = \partial^2 \varphi / \partial y^2. \tag{16.7}$$

Further, as the fluid is incompressible the divergence of **v** vanishes:

$$\text{div } \mathbf{v} = \nabla .\mathbf{v} = (\hat{\mathbf{i}}\, \partial/\partial x + \hat{\mathbf{j}}\, \partial/\partial y) . (v_x\hat{\mathbf{i}} + v_y\hat{\mathbf{j}})$$

$$= \partial v_x /\partial x + \partial v_y /\partial y = 0. \qquad (16.8)$$

Putting from Eq. (16.7) in Eq. (16.8), one gets

$$\partial^2 v /\partial x^2 + \partial^2 v /\partial y^2 \equiv \partial^2 \varphi /\partial x^2 + \partial^2 \varphi /\partial y^2 = 0, \qquad (16.9)$$

proving that the *velocity potential is harmonic*. As a result, there also exists a conjugate harmonic function $\psi (x, y)$ such that

$$w = f(z) = \varphi (x, y) + i\,\psi (x, y) \qquad (16.10)$$

becomes analytic.

Also, the slope of the tangent at a point of the curve

$$\psi (x, y) = c' \qquad (16.11)$$

is given by

$$(\partial\psi /\partial x)\, dx + (\partial\psi /\partial y)\, dy = 0$$

$\Rightarrow$

$$m_1 \equiv \{dy /dx\}_\psi = -(\partial\psi/\partial x) /(\partial\psi/\partial y) = (\partial\varphi/\partial y) /(\partial\varphi/\partial x), \qquad (16.12)$$

by Cauchy-Riemann conditions given by Eqs. (3.2). Further, putting from Eq. (16.6) above slope reduces to

$$dy /dx = v_y /v_x = \tan \theta,$$

proving that the velocity of the fluid is along the tangent to the curve given by Eq. (16.11), i.e. a particle of the fluid moves along this curve. Such curves are called *stream lines* and the function $\psi (x, y)$ as the *stream function*. Similar to Eq. (16.12), the slope of equi-potential lines in Eq. (16.4) can be found as

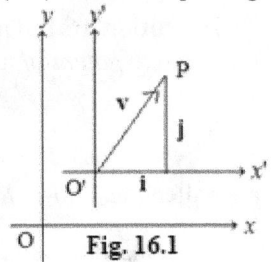

Fig. 16.1

$$m_2 \equiv \{dy /dx\}_\varphi = -(\partial\varphi/\partial x) /(\partial\varphi/\partial y). \qquad (16.13)$$

The results in Eqs. (16.12) and (16.13) imply $m_1.m_2 = -1$. Thus, we have the:

Theorem 16.1. The equi-potential lines and the stream lines cut each other orthogonally.

Differentiating Eq. (16.10) along the real line (along which $z = x$) and using C-R conditions and Eq. (16.6), we derive:

$$dw \,/\, dz = \partial\varphi \,/\, \partial x + i\,\partial\psi \,/\, \partial x = \partial\varphi \,/\, \partial x - i\,\partial\varphi \,/\, \partial y = v_x - i\,v_y.$$

Thus, the magnitude of the fluid velocity is

$$|\mathbf{v}| = \sqrt{(v_x^2 + v_y^2)} = |\,dw \,/\, dz\,|. \tag{16.14}$$

The flow pattern of the fluid is, thus, completely represented by the complex function $w\,(z)$ given by Eq. (16.1.0). This function is called the *complex potential.*

Note 16.1. In electrostatics and gravitational fields the curves given by Eqs. (16.4) and (16.11) are called *equi-potential lines* and *lines of force* respectively. On the other hand, in heat flow problems, they are known as *isothermals* and *heat flow lines*.

In the following examples, we demonstrate methods to compute functions: $\varphi\,(x, y)$ or $\psi\,(x, y)$ when one of these is given.

Example 16.1. If $w = \varphi + i\psi$ represents the complex potential for an electric field and
$$\psi = x^2 - y^2 + x \,/\, (x^2 + y^2), \tag{16.15}$$
find the function φ.

Solution. Differentiating Eq. (16.15) successively with respect to x and y we obtain

$$\partial\psi \,/\, \partial x = 2x + 1/(x^2 + y^2) - 2x^2 \,/(x^2 + y^2)^2, \tag{16.16}$$

$$\partial\psi \,/\, \partial y = -2y - 2xy \,/\, (x^2 + y^2)^2, \tag{16.17}$$

$$\partial^2\psi \,/\, \partial x^2 = 2 - 6x \,/(x^2 + y^2)^2 + 8x^3 \,/(x^2 + y^2)^3,$$

$$\partial^2\psi \,/\, \partial y^2 = -2 - 2x \,/\, (x^2 + y^2)^2 + 8xy^2 \,/\, (x^2 + y^2)^3.$$

Clearly, these derivatives satisfy the Laplace equation (5.1) making function $u = \psi$ harmonic. Also, both the conjugate functions φ and ψ will satisfy the Cauchy-Riemann conditions:

$$\partial\varphi \,/\, \partial x = \partial\psi \,/\, \partial y = -2y - 2xy \,/\, (x^2 + y^2)^2, \tag{16.18}$$

and

$$\partial \varphi / \partial y \ = \ - \partial \psi / \partial x \ = \ - 2x - 1/(x^2 + y^2) + 2x^2/(x^2 + y^2)^2, \quad (16.19)$$

where Eqs. (16.16) and (16.17) have been used.

Integrating Eq. (16.18) with respect to x partially, we derive

$$\varphi(x, y) \ = \ - 2xy + y / (x^2 + y^2) + \eta(y), \quad (16.20)$$

where $\eta(y)$ is some arbitrary function of y. Differentiating Eq. (16.20) partially w.r.t. y:

$$\partial \varphi / \partial y \ = \ - 2x + 1 / (x^2 + y^2) - 2y^2 / (x^2 + y^2)^2 + \eta'(y),$$

and putting from Eq. (16.19) we derive $\eta'(y) = 0 \Rightarrow \eta(y) = c$ (const.). Accordingly, Eq. (16.20) determines

$$\varphi(x, y) \ = \ - 2xy + y / (x^2 + y^2) + c. \ //$$

Example 16.2. If the potential function is $\log_e (x^2 + y^2)$, find the flux function and the complex potential function.

Solution. Let Eq. (16.10) represent the complex function with potential function

$$\varphi(x, y) \ = \ \log_e (x^2 + y^2), \quad (16.21)$$

and the flux function $\psi(x, y)$. Differentiating Eq. (16.21) partially w.r.t. x and y we obtain

$$\partial \varphi / \partial x = 2x / (x^2 + y^2), \qquad \partial \varphi / \partial y \ = \ 2y / (x^2 + y^2),$$

$$\partial^2 \varphi / \partial x^2 \ = \ 2 / (x^2 + y^2) - 4x^2 / (x^2 + y^2)^2,$$

$$\partial^2 \varphi / \partial y^2 \ = \ 2 / (x^2 + y^2) - 4y^2 / (x^2 + y^2)^2.$$

These derivatives satisfy the Laplace equation (5.1) making the function $u = \varphi$ harmonic. Hence, both φ and ψ satisfy the Cauchy- Riemann conditions:

$$\partial \varphi / \partial x = 2x / (x^2 + y^2) = \partial \psi / \partial y, \quad (16.22)$$

and

$$\partial \varphi / \partial y = 2y / (x^2 + y^2) = - \partial \psi / \partial x. \quad (16.23)$$

Integrating Eq. (16.22) partially w.r.t. y we get

$$\psi\,(x,\,y) = 2x \int \{1/(x^2 + y^2)\}\ dy + \xi\,(x) = 2\ \tan^{-1}\,(y/x) + \xi\,(x), \quad (16.24)$$

where $\xi\,(x)$ is an arbitrary function of x alone. Differentiating above integral equation partially w.r.t. x and putting from Eq. (16.23) we evaluate

$$- 2y/(x^2 + y^2) + \xi'\,(x) = -\,2y/(x^2 + y^2) \Rightarrow \xi'\,(x) = 0 \Rightarrow \xi\,(x) = c\ (\text{const.}).$$

Hence, Eq. (16.24) reduces to $\psi\,(x,\,y) = 2\ \tan^{-1}\,(y/x) + c$. Accordingly, Eq. (16.10) becomes

$$w\,(z) = \ln\,(x^2 + y^2) + i\{2\ \tan^{-1}\,(y/x) + c\}$$

$$= 2\{(1/2)\ \ln\,(x^2 + y^2) + i\ \tan^{-1}\,(y/x)\} + ci = 2\ \ln z + ci, \ \text{by Eq. (1.1.1).} \ //$$

Example 16.3. A two-dimensional flow has the stream function ψ find the velocity potential φ:

(i) $\psi = -\,y\,/\,(x^2 + y^2),$ (ii) $\psi = \tan^{-1}\,(y\,/\,x).$

Solution. (i) Differentiating ψ partially w.r.t. x and y we get

$$\partial\psi\,/\partial x = 2xy/(x^2 + y^2)^2, \quad \partial\psi\,/\,\partial y = -\,1/(x^2 + y^2) + 2y^2/(x^2 + y^2)^2,$$

$$\partial^2\psi\,/\,\partial x^2 = 2y\,/\,(x^2 + y^2)^2 - 8x^2\,y\,/\,(x^2 + y^2)^3,$$

$$\partial^2\psi\,/\,\partial y^2 = 6y\,/\,(x^2 + y^2)^2 - 8y^3\,/\,(x^2 + y^2)^3\ .$$

Clearly, these derivatives satisfy

$$\partial^2\psi\,/\,\partial x^2 + \partial^2\psi\,/\,\partial y^2 = 8y\,/\,(x^2 + y^2)^2 - 8y\,(x^2 + y^2)\,/\,(x^2 + y^2)^3 = 0,$$

establishing the harmonic character of ψ. Also, both φ and ψ satisfy the Cauchy-Riemann conditions:

$$\partial\varphi\,/\,\partial x = \partial\psi\,/\,\partial y = -\,1\,/\,(x^2 + y^2) + 2y^2\,/\,(x^2 + y^2)^2, \quad (16.25)$$

and

$$\partial\varphi\,/\,\partial y = -\,\partial\psi\,/\,\partial x = -\,2xy\,/\,(x^2 + y^2)^2. \quad (16.26)$$

Integrating Eq. (16.26) partially w.r.t. y we get

$$\varphi\,(x,\,y) = x\,/\,(x^2 + y^2) + \xi\,(x).$$

The same on partial differentiation w.r.t. x yields

$$1/(x^2 + y^2) - 2x^2 /(x^2 + y^2)^2 + \xi'(x) = -1/(x^2 + y^2) + 2y^2 /(x^2 + y^2)^2,$$

by Eq. (16.25); or,
$$2/(x^2 + y^2) + \xi'(x) = 2(x^2 + y^2) / (x^2 + y^2)^2$$

$\Rightarrow$

$$\xi'(x) = 0 \quad \Rightarrow \quad \xi(x) = c \text{ (const.).}$$

Therefore,
$$\varphi(x, y) = x/(x^2 + y^2) + c. \text{ //}$$

(ii) $\partial\psi / \partial x = -y/(x^2 + y^2),$ $\partial\psi / \partial y = x/(x^2 + y^2),$

$\partial^2\psi / \partial x^2 = 2xy/(x^2 + y^2)^2,$ $\partial^2\psi / \partial y^2 = -2xy/(x^2 + y^2)^2$

$\Rightarrow$
$$\partial^2\psi / \partial x^2 + \partial^2\psi / \partial y^2 = 0,$$

establishing that $\psi(x, y)$ is harmonic. Also, both φ and ψ satisfy the Cauchy-Riemann conditions:

$$\partial\varphi / \partial x = \partial\psi / \partial y = x/(x^2 + y^2)$$

$\Rightarrow$
$$\varphi(x, y) = (1/2) \ln(x^2 + y^2) + \eta(y), \tag{16.27}$$

and
$$\partial\varphi / \partial y = -\partial\psi / \partial x = y/(x^2 + y^2). \tag{16.28}$$

Differentiating Eq. (16.27) partially w.r.t. y and putting from Eq. (16.28) we evaluate $\eta(y)$:
$$\eta'(y) = 0 \Rightarrow \eta(y) = c \text{ (const.).}$$

Therefore, Eq. (16.27) determines

$$\varphi(x, y) = (1/2) \log_e(x^2 + y^2) + c. \text{ //}$$

CHAPTER 8

DETERMINANTS

§ 1. Determinant

A square matrix (cf. [25]) possesses a determinant having some value (a real or complex number). The determinant is denoted by enclosing its elements within two solidi.

Example 1.1. The square matrix

$$A = \begin{bmatrix} a & b & c \\ d & e & f \\ g & h & i \end{bmatrix}, \tag{1.1}$$

has its determinant

$$|A| = \begin{vmatrix} a & b & c \\ d & e & f \\ g & h & i \end{vmatrix}. \tag{1.2}$$

The right member in Eq. (1.2) is expansible along any row (or column) into a linear sum of the elements of that row (or column) multiplied by their cofactors. Thus, expanding $|A|$ along the first row:

$$|A| = a \begin{vmatrix} e & f \\ h & i \end{vmatrix} - b \begin{vmatrix} d & f \\ g & i \end{vmatrix} + c \begin{vmatrix} d & e \\ g & h \end{vmatrix}$$

$$= a(ei - hf) - b(di - gf) + c(dh - ge). \tag{1.3}$$

The elements of A being scalars (i.e. numbers) the right member of Eq. (1.3) has some scalar value.

Definition 1.1. The coefficients of a, b, c in Eq. (1.3) are called their *cofactors*. The signs of these cofactors are taken alternately starting with positive for the cofactor of one-one position element, that is a, in the determinant.

Example 1.2. The determinant of an identity matrix of any order is of value 1, while that of a square null matrix is always zero.

Example 1.3. The determinant of a scalar matrix $x\,I$ of order $n \times n$ is x^n.

Example 1.4. The determinant of a scalar multiple xA of a square matrix A of order $n \times n$ is $x^n \,|\, A \,|$.

Example 1.5. A square matrix and its transpose possess the same determinant. So is the case with symmetric and skew–symmetric matrices.

§ 2. Some properties of determinants

2.1. Addition and subtraction by a common multiple of any row (or column) to another row (or column) in a determinant leaves the determinant unaltered.

2.2. Interchange of any two rows (or columns) in a determinant changes its sign.

2.3. Multiplication of any row (or column) in a determinant by a scalar results in the multiplication of whole determinant by that scalar.

2.4. Interchange of rows and columns (retaining their serial order) in a determinant leaves the determinant unaltered.

(Note 2.1. It is already seen in Example 1.5 above.)

2.5. A determinant having at least two rows (or columns) proportional (or identical) is of zero value.

2.6. The linear sum of the elements of one row (or column) together with cofactors of the elements of another row (or column) in a determinant vanishes identically.

Proof. Replacing the cofactors of a, b, c in Eq. (1.3) by the cofactors of elements of, say second row, in Eq. (1.1) makes

$$- a \begin{vmatrix} b & c \\ h & i \end{vmatrix} + b \begin{vmatrix} a & c \\ g & i \end{vmatrix} - c \begin{vmatrix} a & b \\ g & h \end{vmatrix}$$

$$= - a\,(b\,i - h\,c) + b\,(a\,i - g\,c) - c\,(a\,h - g\,b) = 0. \; //$$

2.7. A determinant can be split into the sum of two (or more) determinants as follows:

$$\begin{vmatrix} a & b & c \\ d & e & f \\ g_1+g_2 & h_1+h_2 & i_1+i_2 \end{vmatrix} = \begin{vmatrix} a & b & c \\ d & e & f \\ g_1 & h_1 & i_1 \end{vmatrix} + \begin{vmatrix} a & b & c \\ d & e & f \\ g_2 & h_2 & i_2 \end{vmatrix}. \qquad (2.1)$$

2.8. A determinant of product of matrices is the product of their determinants:

$$|AB| = |A||B|.$$

Note 2.2. Multiplication of two determinants of same order can be defined in the similar manner as in case of matrices.

Example 2.1. The determinant of *Vandermonde* matrix is

$$|V_n| = \begin{vmatrix} 1 & x_1 & x_1^2 & \cdots & x_1^{n-1} \\ 1 & x_2 & x_2^2 & \cdots & x_2^{n-1} \\ 1 & x_3 & x_3^2 & \cdots & x_3^{n-1} \\ \cdot & \cdot & \cdot & \cdots & \cdot \\ 1 & x_n & x_n^2 & \cdots & x_n^{n-1} \end{vmatrix} = \prod_{i>j}(x_i - x_j), \ i,j = 1, 2, 3, \ldots, n$$

$$= \{(x_n - x_1)(x_{n-1} - x_1)\ldots(x_2 - x_1)\}.\{(x_{n-1} - x_2)\ldots(x_3 - x_2)\}\ldots(x_n - x_{n-1}).$$

Solution. Subtracting R_1 from the remaining rows and expanding the determinant along C_1, $|Vn|$ reduces to

$$\begin{vmatrix} x_2 - x_1 & x_2^2 - x_1^2 & \cdots & x_2^{n-1} - x_1^{n-1} \\ x_3 - x_1 & x_3^2 - x_1^2 & \cdots & x_3^{n-1} - x_1^{n-1} \\ \cdot & \cdot & \cdots & \cdots \\ x_n - x_1 & x_n^2 - x_1^2 & \cdots & x_n^{n-1} - x_1^{n-1} \end{vmatrix}.$$

Taking common factors from all the $n-1$ rows out it becomes

$$(x_n - x_1)(x_{n-1} - x_1)\ldots(x_2 - x_1).\begin{vmatrix} 1 & x_2 + x_1 & \cdots & \cdots \\ 1 & x_3 + x_1 & \cdots & \cdots \\ \cdot & \cdot & \cdots & \cdots \\ 1 & x_n + x_1 & \cdots & \cdots \end{vmatrix}.$$

Repeating the process of subtraction of R_1 from the remaining rows and expanding the determinant along C_1, it reduces to

$$(x_n - x_1)(x_{n-1} - x_1)\ldots(x_2 - x_1)\cdot \begin{vmatrix} x_3 - x_2 & \cdots & \cdots \\ \cdot & \cdots & \cdots \\ x_n - x_2 & \cdots & \cdots \end{vmatrix}.$$

Again, taking common factors from all the $n-2$ rows out it becomes

$$\{(x_n - x_1)(x_{n-1} - x_1)\ldots(x_2 - x_1)\}\{(x_n - x_2)(x_{n-1} - x_2)\ldots(x_3 - x_2)\} \mid V_{n-2} \mid.$$

Continuing in the same way, we get the result.

Alternately, subtracting x_1 multiple of each preceding column from the next one, $C_i - x_1 C_j$, $i > j = 1, 2, \ldots, n$, and expanding along R_1, the determinant reduces to

$$\begin{vmatrix} x_2 - x_1 & x_2(x_2 - x_1) & \cdots & x_2^{n-2}(x_2 - x_1) \\ x_3 - x_1 & x_3(x_3 - x_1) & \cdots & x_3^{n-2}(x_2 - x_1) \\ \cdot & \cdot & \cdots & \cdots \\ x_n - x_1 & x_n(x_n - x_1) & \cdots & x_n^{n-2}(x_2 - x_1) \end{vmatrix}.$$

Taking common factors from all the $n-1$ rows out it becomes

$$(x_n - x_1)(x_{n-1} - x_1)\ldots(x_2 - x_1)\cdot \begin{vmatrix} 1 & x_2 & \cdots & x_2^{n-2} \\ 1 & x_3 & \cdots & x_3^{n-2} \\ \cdot & \cdot & \cdots & \cdots \\ 1 & x_n & \cdots & x_n^{n-2} \end{vmatrix},$$

involving the $(n-1)^{\text{th}}$ order determinant in variables $x_2, x_3, \ldots, x_n$ and can be expanded by the same process. //

CHAPTER 9

DIFFERENTIAL EQUATIONS (ORDINARY)

§ 1. Introduction

Equations involving one (or more) dependent variable(s), independent variable(s) and derivatives of dependent variable(s) with respect to the independent variable(s) are called *differential equations*. Presently, we consider ordinary differential equations involving *one* independent variable, say x, *one* dependent variable, say y (which is a function of x), and derivatives of y with respect to x. When the differential equation is written in a rational form, the highest order of differentiation of y w.r.t. x appearing in the equation defines the *order* of the differential equation; while the highest degree of dy/dx defines the *degree* of the differential equation. Thus, the differential equation

$$dy \,/\, dx \;=\; f(x, y) \qquad\qquad (1.1)$$

is an *ordinary differential equation* of *first degree* and of *first order* in two variables: x (independent) and y (dependent). Also, a differential equation is said to be *linear* when both the dependent variable y and its derivative dy/dx are of the first degree. Thus,

$$P(x). (dy \,/\, dx) + Q(x). y \;=\; R(x), \qquad\qquad (1.2)$$

where P, Q, R are some functions of x alone, is a linear differential equation. Any relation between the dependent variable and the independent variable satisfying the differential equation is called a *solution* (or *integral*) of the differential equation. A solution having the same number of arbitrary constants equal to the order of the differential equation is called a *general solution* (or *complete integral* or *complete primitive*). In the next section, we deal with the differential equations of the type (1.1) and discuss different methods for their integration.

On contrary, there are differential equations of the *first* order but of degree more than one. For example, if $F(x, y, p)$ is a function of any degree in its variables: (independent) x, (dependent) y and $p \equiv dy \,/dx$, then a relation

$$F(x, y, p) \;=\; 0 \qquad\qquad (1.3)$$

expresses a differential equation of above type. The solution of such differential equations will be discussed in the Section 4.

The linear differential equations of *any order* (but of *degree one*) with constant coefficients will be discussed in the Section 5. The following equation

$$a_0 (d^n y / dx^n) + a_1 (d^{n-1} y / dx^{n-1}) + a_2 (d^{n-2} y / dx^{n-2}) + \ldots$$

$$+ a_{n-1} (dy / dx) + a_n y = R (x) \qquad (1.4)$$

represents a *linear* differential equation of *first degree* but of nth order.

§ 2. ODEs of first order and first degree (simple cases)

Let $y = f(x)$ be a differentiable function of a (real) variable x with its derivative $dy/dx \equiv f'(x)$. A most general differential equation of the *first order* and *first degree* involving y and its derivative is given by Eq. (1.2). Precisely enough, it is also called a *linear differential equation* (of the first order and first degree) as it contains the powers of both y and its derivative as *one*. On the other hand, the following is a linear differential equation of *first degree* but of *second order*:

$$P (d^2 y / dx^2) + Q (dy / dx) + R y = S, \qquad (2.1)$$

where the coefficients P, Q, R and S are, in general, functions of x alone.

The class of differential equations is very large and there are many methods for their solution. In the present Section, we mainly deal with the equations of the type given by Eq. (1.2).

Theorem 2.1. If $f(x)$ is continuous on some interval I of a real line and b is a constant the differential equation

$$f'(x) = b f(x) \quad (2.2); \quad \text{has a solution} \quad f(x) = a\, e^{bx}, \qquad (2.3)$$

a being an arbitrary constant of integration.

Example 2.1. Rate of increase of the population of a city is proportional to the population. If it increases from 40,000 to 60,000 in 40 years what will be the population in 80 years?

Solution. Let $P(t)$ be the population after t years. The rate of its growth then will be $dP/dt = bP$ (as given). Its solution, by Eq. (2.3), is

$$P = a\, e^{bt}. \tag{2.4}$$

Applying the given conditions ($t = 0$, $P = 40,000$; and $t = 40$, $P = 60,000$) we get $a = 40,000$ and $e^{40b} = P/a = 60,000/40,000 = 3/2$. Therefore, when $t = 80$, Eq. (2.6) determines

$$P = 40,000 \times (3/2)^{t/40} = 40,000 \times (3/2)^2 = 90.000. \; //$$

§ 3. First order and first degree linear differential equations (general cases)

The ODEs of general type given by Eq. (1.2) are discussed in detail in this Section. The following methods for their solution are considered:

(i) Separable variables, (ii) Exact differential equations, (iii) Making the equation exact by multiplication of some additional factor called an *integrating factor*, (iv) Homogeneous equations, etc.

3.1. Separable variables form: Let the equation (1.1) be of the form

$$dy/dx = f_1(x)/f_2(y), \quad \text{or} \quad f_1(x).\,dx - f_2(y).\,dy = 0, \quad (3.1)$$

where the coefficients $f_1(x)$ and $f_2(y)$ are functions of the respective variables alone. Integrating each term in Eq. (3.1) separately w.r.t. their respective variables, we get

$$\int f_1(x).\,dx - \int f_2(y).\,dy = c, \tag{3.2}$$

c being an arbitrary constant of integration.

3.2. Homogeneous form: A differential equation of the form

$$dy/dx = f_1(x, y)/f_2(x, y), \tag{3.3}$$

where f_1, f_2 are some homogeneous functions of both variables x and y and have the same degree is said to be in a *homogeneous form*. Setting $y = v.x$ so that

$$dy/dx = v + x.\,dv/dx,$$

where v is a new variable depending on x, reduces the Eq. (3.3) to

$$v + x.\, dv/dx = f_1\,(x,\, v.x)\,/f_2\,(x,\, v.x) \equiv h\,(v) \;\Rightarrow\; dv\,/\{h\,(v) - v\} = dx/x.$$

This is a differential equation in v and x, where the variables are separated. Integrating it term wise w.r.t. the respective variables, we get

$$\int [1\,/\,\{h\,(v) - v\}.\, dv] \;=\; \ln x + c.$$

Finally, v is replaced by $y\,/\,x$ in above equation to yield the solution of Eq. (3.3).

3.3. Reducible to homogeneous form: A differential equation of the form
$$dy\,/\,dx \;=\; (a_1\,x + b_1\,y + c_1)\,/\,(a_2\,x + b_2\,y + c_2) \qquad\qquad (3.4)$$

can be reduced to a homogeneous form by introducing new set of variables ξ, η and some constants h and k :

$$x = \xi + h,\;\; y = \eta + k \;\;\Rightarrow\;\; dy\,/\,dx \;=\; \{d\,(\eta + k)\,/\,d\xi\}\,(d\xi/dx) \;=\; d\eta\,/\,d\xi.$$

Thus, the differential equation (3.4) reduces to

$$d\eta/d\xi = \{a_1\,\xi + b_1\eta + (a_1 h + b_1 k + c_1)\}/\{a_2\,\xi + b_2\eta + (a_2 h + b_2 k + c_2)\}.$$

Choosing the constants h and k so as to satisfy the simultaneous equations
$$a_1\,h + b_1\,k + c_1 = 0, \qquad \text{and} \qquad a_2\,h + b_2\,k + c_2 = 0,$$

the differential equation further simplifies to

$$d\eta\,/\,d\xi \;=\; (a_1\,\xi + b_1\,\eta)\,/\,(a_2\,\xi + b_2\,\eta). \qquad\qquad (3.5)$$

This is in a homogeneous form given by Eq. (3.3): both $a_1\,\xi + b_1\,\eta$ and $a_2\,\xi + b_2\,\eta$ being homogeneous functions of ξ and η of degree *one*. Setting $\eta = v.\xi$ and proceeding as in the preceding Sub-section we may find the solution of Eq. (3.5).

3.4. A linear form: The equation (1.2) represents a most general linear differential equation in variables x and y. Dividing it by $P\,(x)$ we may have its special form:

$$dy / dx + P_1(x) . y = R_1(x), \tag{3.6}$$

where $P_1 \equiv Q/P$ and $R_1 \equiv R/P$. It can be integrated by multiplying by a factor $e^{\int P_1 \, dx}$ (called the *integrating factor* of the differential equation):

$$e^{\int P_1 \, dx} (dy/dx) + (P_1 . e^{\int P_1 \, dx}) y = R_1 . e^{\int P_1 \, dx}.$$

The LHS of above equation is the derivative of the product function $y . e^{\int P_1 \, dx}$. Hence, the equation is directly integrable and its solution is:

$$e^{\int P_1 \, dx} . y = \int R_1 . e^{\int P_1 \, dx} . dx + c$$

$\Rightarrow$

$$y = e^{-\int P_1 \, dx} . \{ \int R_1 . e^{\int P_1 \, dx} . dx + c \}. \tag{3.7}$$

3.5. Reducible to a linear form: *Bernoulli*'s equation

$$dy / dx + P_1(x) . y = R_1(x) . y^n \tag{3.8}$$

can be reduced to a linear form as follows. Dividing it by y^n:

$$y^{-n} . (dy / dx) + P_1(x) . y^{1-n} = R_1(x),$$

and putting $y^{1-n} = v$ so that $(1 - n) y^{-n} (dy / dx) = dv / dx$ above equation reduces to

$$dv / dx + (1 - n) . P_1 . v = (1 - n) R_1,$$

which is of the form as in Eq. (3.6).

Also, a differential equation of the form given by

$$f'(y) . (dy / dx) + P_1(x) . f(y) = R_1(x), \tag{3.9}$$

where $f'(y) \equiv df / dy$, can be reduced to a linear form. Putting

$$f(y) = v \text{ so that } f'(y) . (dy / dx) = dv / dx,$$

above equation reduces to

$$dv / dx + P_1 . v = R_1,$$

which is of the form as in Eq. (3.6).

3.6. Exact form: A differential equation obtainable by differentiating its solution directly (without performing any other mathematical operations) is called *exact*. A necessary and sufficient condition for the differential equation (3.3) modified as:

$$f_1(x, y).\, dx + f_2(x, y).\, dy = 0, \tag{3.10}$$

to be exact is

$$\partial f_1 / \partial y = \partial f_2 / \partial x. \tag{3.11}$$

The following rules to have the integrating factors of a differential equation are noteworthy:

3.6.1. When the equation (3.10) is homogeneous in its variables x and y, and

$$F(x, y) \equiv f_1.\, x + f_2.\, y \neq 0,$$

then the integrating factor is

$$1 / F(x, y) \equiv 1 / (f_1.\, x + f_2.\, y). \tag{3.12}$$

3.6.2. When the equation (3.10) is of the form

$$f_1(x.\, y).\, y.\, dx + f_2(x.\, y).\, x.\, dy = 0, \tag{3.13}$$

and

$$F(x, y) \equiv (f_1.\, y).\, x - (f_2.\, x).\, y = (f_1 - f_2).\, x\, y \neq 0,$$

then the integrating factor is

$$1/F(x, y) \equiv 1/(f_1 - f_2).\, x. \tag{3.14}$$

3.6.3. When

$$(\partial f_1 / \partial y - \partial f_2 / \partial x) / f_2 \equiv F(x) \tag{3.15}$$

is a function of x alone the integrating factor is $e^{\int F(x)\, dx}$.

3.6.4. When

$$(\partial f_1 / \partial y - \partial f_2 / \partial x) / f_1 \equiv -G(y) \tag{3.16}$$

is a function of y alone the integrating factor is $e^{\int G(y)\, dy}$.

3.6.5. When the equation (3.10) is of the form

$$x^a y^b.(a_1.\, y.\, dx + b_1.\, x.\, dy) + x^h y^k.(h_1.\, y.\, dx + k_1.\, x.\, dy) = 0, \tag{3.17}$$

where $a, b, a_1, b_1, h, k, h_1, k_1$ are all constants, then the integrating factor is $x^l y^m$. The values of l, m are found by application of the condition given by Eq. (3.11) for exactness.

3.7. Change of variables: When the variables (both dependent and independent) are transformed, some of the differential equations reduce to any of the forms discussed above. As such, they are observed to become integrable.

§ 4. Differential equations of first order and of any degree

As introduced in the § 1, the equation (1.3) represents a first order differential equation of some degree. The following three forms are suggested for achieving solution of some of the differential equations of this type.

4.1. Equations solvable for p: Let the equation (1.3) be of nth degree in p and have the form

$$\{p - f_1(x, y)\}.\{p - f_2(x, y)\} \dots \{p - f_n(x, y)\} = 0. \qquad (4.1)$$

Thus, there result n differential equations each of the form as in Eq. (1.1):

$$p \equiv dy / dx = f_i(x, y), \quad i = 1, 2, \dots, n. \qquad (4.2)$$

Solutions of these equations are discussed in the preceding Section. Let Eq. (4.2) have solutions

$$F_i(x, y, c_i) = 0, \quad i = 1, 2, \dots, n \qquad (4.3)$$

giving rise to a general solution

$$F_1(x, y, c_1). F_2(x, y, c_2) \dots F_n(x, y, c_n) = 0.$$

Since the original differential equation (4.1) is of first order only its complete integral should contain only one arbitrary constant. So, without loss of generality, above n constants can be replaced by a single constant, say c. Thus, above general solution reduces to

$$F_1(x, y, c). F_2(x, y, c) \dots F_n(x, y, c) = 0. \qquad (4.4)$$

4.2. Equations solvable for *y*: When the equation (1.3) is solvable for *y*, we have

$$y = f(x, p). \tag{4.5}$$

Its differentiation w.r.t. *x* yields a relation

$$p = df/dx = \{(\partial f/\partial x) + (\partial f/\partial p).(dp/dx)\} \equiv \varphi(x, p, dp/dx),$$

which is a differential equation in the (independent) variable *x* and (dependent) variable *p*. It can be integrated by some method discussed in the preceding Section. Let its solution be

$$F(x, p, c) = 0. \tag{4.6}$$

Elimination of *p* between Eqs. (4.5) and (4.6) yields the desired solution. In case, this elimination is not feasible, we solve Eq. (4.6) for *x*:

$$x = g(y, p, c), \tag{4.7}$$

and leave the solution in the form given by Eqs. (4.5) and (4.7) treating *p* as a parameter. This method is especially useful when the differential equation does not contain the independent variable *x*.

4.3. Equations solvable for *x* : Let the equation (1.3) be solvable for *x*:

$$x = f(y, p). \tag{4.8}$$

Differentiating it w.r.t. *y*, we get

$$1/p \equiv dx/dy = df/dy = \{(\partial f/\partial y) + (\partial f/\partial p).(dp/dy)\} \equiv \varphi(y, p, dp/dy),$$

which is a differential equation in the (independent) variable *y* and (dependent) variable *p*. It can be integrated by some method discussed in the preceding Section. Let its solution be

$$F(y, p, c) = 0. \tag{4.9}$$

Elimination of *p* between Eqs. (4.8) and (4.9) yields the desired solution. In case, this elimination is not feasible, we solve Eq. (4.9) for *y*:

$$y = g(x, p, c), \tag{4.10}$$

and leave the solution in the form given by Eqs. (4.8) and (4.10), where p acts as a parameter.

4.4. Clairaut's form: The equation of the form

$$y = p.x + f(p) \qquad (4.11)$$

is known after F.C. Clairaut. Its differentiation w.r.t. x yields

$$p = p + \{x + (df/dp)\}.(dp/dx)$$

$$\Rightarrow \quad dp/dx = 0 \quad (4.12a); \qquad \text{or,} \qquad df/dp + x = 0. \quad (4.12b)$$

The first alternative has integral $p = c$; which, in conjunction with Eq. (4.11), yields the desired general solution of Eq. (4.11):

$$y = c.x + f(c). \qquad (4.13)$$

On the other hand, Eq. (4.12b) represents a differential equation in the variables x (independent) and p (dependent), where the variables are separated. So, its integral is

$$f(p) + x^2/2 = 0. \qquad (4.14)$$

Elimination of p between Eqs. (4.11) and (4.14) yields a solution which does not contain any arbitrary constant and is neither a particular case of the solution given by Eq. (4.13). Such a solution is termed as a *singular solution* of the differential equation (4.11).

4.5. A more general form: A more general equation of the form

$$y = x.f(p) + F(p) \qquad (4.15)$$

can also be solved similarly. Its differentiation w.r.t. x yields

$$p = f(p) + \{x.(df/dp) + dF/dp\}.(dp/dx),$$

or

$$\{p - f(p)\}.(dx/dp) - x.(df/dp) = dF/dp. \qquad (4.16)$$

A comparison of this equation with Eq. (1.2) shows that it is linear in the variables p (treated as an independent variable) and the dependent variable x. The solution of such equations has been discussed before.

§ 5. Linear differential equations of any order with constant coefficients

A linear differential equation of nth *order* and of *degree one* in the variables x (independent) and y (dependent) is represented by the equation (1.4). Let the coefficients $a_0, a_1, a_2, \ldots, a_n$ therein be constant. The solution of such differential equations is discussed in the following.

First, we consider a special form of such equation with vanishing function $R(x)$:

$$a_0 (d^n y / dx^n) + a_1 (d^{n-1} y / dx^{n-1}) + a_2 (d^{n-2} y / dx^{n-2}) + \ldots$$

$$+ a_{n-1} (dy / dx) + a_n y = 0. \tag{5.1}$$

Let $y = e^{mx}$ be a solution of this equation, then there holds

$$e^{mx} . (a_0 m^n + a_1 m^{n-1} + a_2 m^{n-2} + \ldots + a_{n-1} m + a_n) = 0,$$

implying that m is a root of the nth degree (algebraic) equation

$$a_0 m^n + a_1 m^{n-1} + a_2 m^{n-2} + \ldots + a_{n-1} m + a_n = 0. \tag{5.2}$$

This equation is called an *auxiliary equation* or *indicial equation* of the differential equation. There arise different situations for the roots of this auxiliary equation.

5.1. All real and distinct roots: Let $m_1, m_2, \ldots, m_n$ be all real and distinct roots of the auxiliary equation giving rise to n independent solutions $e^{m_1 x}, e^{m_2 x}, \ldots, e^{m_n x}$ or a complete solution (called the *complementary function*)

$$y = A_1 e^{m_1 x} + A_2 e^{m_2 x} + \ldots + A_n e^{m_n x}. \tag{5.3}$$

5.2. Some equal roots: Let the auxiliary equation have *two* equal real roots, say $m_1 = m_2$ and $n - 2$ distinct (real) roots $m_3, \ldots, m_n$. Correspondingly, the solution given by Eq. (5.3) becomes

$$y = (A_1 + A_2) e^{m_1 x} + A_3 e^{m_3 x} + \ldots + A_n e^{m_n x}.$$

Since the sum $A_1 + A_2$ of two arbitrary constants can be replaced by a single constant, say B, the number of arbitrary constants of integration reduces from n to $n-1$. As such, above solution cannot represent a general solution of the nth order differential equation (5.1). To overcome this difficulty, we proceed as follows for a general solution of such a differential equation. Beginning with a simple (*second order*) differential equation

$$a_0 (d^2 y / dx^2) + a_1 (dy / dx) + a_2 y = 0, \qquad (5.4)$$

or, setting $a_1 = -2a_0 m_1$ and $a_2 = a_0 m_1^2$,

$$a_0 \{d^2 y / dx^2 - 2m_1 (dy/dx) + m_1^2 y\} \equiv a_0 (d / dx - m_1)^2 y = 0,$$

so that the auxiliary equation has two equal roots. Putting

$$(d / dx - m_1) y = u, \qquad (5.5)$$

above differential equation becomes

$$a_0 (d / dx - m_1) u = 0, \qquad \text{i.e.} \qquad du / dx - m_1 u = 0,$$

which has a solution $u = A_2. e^{m_1 x}$. Accordingly, Eq. (5.5) can be rewritten as

$$dy / dx - m_1 y = A_2. e^{m_1 x}.$$

Being a linear differential equation, by Eq. (3.7), it has a solution

$$y. e^{-m_1 x} = A_1 + A_2.\int e^{m_1 x}.e^{-m_1 x}. dx = A_1 + A_2. x$$

$$\Rightarrow$$

$$y = (A_1 + A_2. x). e^{m_1 x}$$

giving a complete integral of the differential equation (5.4) whose auxiliary equation has equal roots. Similarly, the nth order differential equation (5.1), whose auxiliary equation has *two* equal roots, has a complete integral

$$y = (A_1 + A_2. x). e^{m_1 x} + A_3 \, e^{m_3 x} + ... + A_n \, e^{m_n x}. \qquad (5.6)$$

Generalizing this concept to the case when the auxiliary equation (5.2) has r equal roots a complete integral of Eq. (5.1) is given by

$$y = (A_1 + A_2. x + A_3.x^2 + ... + A_r. x^r). e^{m_1 x} + A_{r+1} \, e^{m_{r+1} x} + ... + A_n \, e^{m_n x}$$

$$(5.7)$$

5.3. Complex roots: Let the auxiliary equation have a complex root, say $\alpha + i\beta$. The equation then also has its complex conjugate $\alpha - i\beta$ as another root and $n - 2$ distinct (real) roots $m_3, \ldots, m_n$. Correspondingly, the solution given by Eq. (5.3) becomes

$$y = A_1 e^{(\alpha + i\beta)x} + A_2 e^{(\alpha - i\beta)x} + A_3 e^{m_3 x} + \ldots + A_n e^{m_n x}. \quad (5.8a)$$

The sum of first two terms can be simplified as

$$e^{\alpha x}.\{A_1 e^{i\beta x} + A_2 e^{-i\beta x}\} = e^{\alpha x}.\{A_1 (\cos \beta x + i \sin \beta x)$$

$$+ A_2 (\cos \beta x - i \sin \beta x)\} = e^{\alpha x}.\{(A_1 + A_2). \cos \beta x + i (A_1 - A_2). \sin \beta x\}$$

$$= e^{\alpha x}.(B_1. \cos \beta x + B_2. \sin \beta x) = C_1. e^{\alpha x}. \cos (\beta x + C_2),$$

where we have set

$$A_1 + A_2 = B_1 = C_1. \cos C_2 \qquad \text{and} \qquad i (A_1 - A_2) = B_2 = - C_1.\sin C_2.$$

Hence, Eq. (5.8a) can also be written as

$$y = (B_1. \cos \beta x + B_2. \sin \beta x). e^{\alpha x} + A_3 e^{m_3 x} + \ldots + A_n e^{m_n x} \qquad (5.8b)$$

$$= C_1. e^{\alpha x}. \cos (\beta x + C_2) + A_3 e^{m_3 x} + \ldots + A_n e^{m_n x}. \qquad (5.8c)$$

5.3.1. Repeated complex roots: If the complex roots $\alpha \pm i\beta$ are repeated the solution in Eq. (5.8a) becomes

$$y = (A_1 + A_2.x). e^{(\alpha + i\beta)x} + (A_3 + A_4. x). e^{(\alpha - i\beta)x} + A_5 e^{m_5 x} + \ldots + A_n e^{m_n x}$$

$$= \{(B_1 + B_2 .x). \cos \beta x + (B_3 + B_4.x). \sin \beta x\}. e^{\alpha x}$$

$$+ A_5 e^{m_5 x} + \ldots + A_n e^{m_n x}. \qquad (5.9)$$

5.4. Pair of roots $\alpha \pm \sqrt{\beta}$: If the auxiliary equation has two roots $\alpha + \sqrt{\beta}$ and $\alpha - \sqrt{\beta}$, the solution in Eq. (5.3) becomes

$$y = A_1 e^{(\alpha + \sqrt{\beta})x} + A_2 e^{(\alpha - \sqrt{\beta})x} + A_3 e^{m_3 x} + \ldots + A_n e^{m_n x}. \qquad (5.10a)$$

Analogous to above, the sum of first two terms can be simplified as

$$e^{ax}.\{A_1 e^{x\sqrt{\beta}} + A_2 e^{-x\sqrt{\beta}}\} = e^{ax}[A_1\{\cosh(x\sqrt{\beta}) + \sinh(x\sqrt{\beta})\}$$

$$+ A_2\{\cosh(x\sqrt{\beta}) - \sinh(x\sqrt{\beta})\}]$$

$$= e^{ax}.\{(A_1 + A_2).\cosh(x\sqrt{\beta}) + (A_1 - A_2).\sinh(x\sqrt{\beta})\}$$

$$= e^{ax}.(B_1.\cosh(x\sqrt{\beta}) + B_2.\sinh(x\sqrt{\beta})) = C_1.e^{ax}.\cosh(x\sqrt{\beta} + C_2),$$

where we have put

$$A_1 + A_2 = B_1 = C_1.\cosh C_2 \qquad \text{and} \qquad (A_1 - A_2) = B_2 = C_1.\sinh C_2.$$

Hence, the soln. in Eq. (5.10a) can also be written as

$$y = \{B_1.\cosh(x\sqrt{\beta}) + B_2.\sinh(x\sqrt{\beta})\}.e^{ax} + A_3 e^{m_3 x} + \ldots + A_n e^{m_n x}$$
$$\text{(5.10b)}$$

$$= C_1.e^{ax}.\cosh(x\sqrt{\beta} + C_2) + A_3 e^{m_3 x} + \ldots + A_n e^{m_n x}. \qquad \text{(5.10c)}$$

§ 6. The Particular Integral

We return to the discussion of integration of the differential equation of the type given by Eq. (1.4), where the RHS does not vanish any more. Treating the LHS of the equation as some function, say $f(D)$ of the differential operator $D \equiv d/dx$ operated upon y, a more general form of the differential equation can be considered as:

$$f(D)y = R(x). \qquad (6.1)$$

The *particular integral* of the differential equation is defined as

$$\{1/f(D)\} R(x) \qquad \text{also written as} \qquad \{f(D)\}^{-1} R(x).$$

The operators $f(D)$ and $1/f(D) \equiv \{f(D)\}^{-1}$ are inverse to each other, i.e.

$$f(D)[\{f(D)\}^{-1} R(x)] = R(x). \qquad (6.2)$$

In particular,

$$D\{D^{-1}R(x)\} = R(x) \quad \Rightarrow \quad D^{-1}R(x) = \int R(x)\,dx. \quad (6.3)$$

Similarly, we may evaluate $(D-a)^{-1}R(x)$, where a is a constant.

Theorem 6.1. $(D - a)^{-1} R(x) = e^{ax} . \int R(x) . e^{-ax} . dx.$ (6.4)

Theorem 6.2. We have

$$\{f(D)\}^{-1} R = A_1 . e^{a_1 x} . \int R. e^{-a_1 x} . dx + A_2 . e^{a_2 x} . \int R. e^{-a_2 x} . dx$$

$$+ ... + A_n . e^{a_n x} . \int R. e^{-a_n x} . dx. \qquad (6.5)$$

Proof. Breaking $1/f(D)$ into partial fractions:

$$1/f(D) = A_1 / (D - a_1) + A_2 / (D - a_2) + ... + A_n / (D - a_n),$$

and applying Eq. (6.4), we derive Eq. (6.5). //

Note 6.1. In case $R(x) = x^m$, $\{f(D)\}^{-1} x^m$ can be evaluated by expanding $\{f(D)\}^{-1}$ in ascending powers of D and the expansion so obtained operates on the function x^m.

Theorem 6.3. For any constant a satisfying $f(a) \neq 0$, there holds

$$\{1/f(D)\} e^{ax} = \{1/f(a)\} e^{ax} \qquad (6.6)$$

Corollary 6.1. For any constant a satisfying $f(a) \neq 0$, there hold

$$\left.\begin{array}{l} \{1/f(D^2)\} \cos ax = \{1/f(-a^2)\} \cos ax, \\[2mm] \{1/f(D^2)\} \sin ax = \{1/f(-a^2)\} \sin ax. \end{array}\right\} \qquad (6.7)$$

Note 6.2. If $f(D)$ also contains the odd powers of D, to operate cos ax, and sin ax by $1/f(D)$, we proceed as follows:

Separating the terms containing odd and even powers of D in $f(D)$:

$$f(D) = f_1(D^2) + D.f_2(D^2),$$

we obtain,

$$\{1/f(D)\} \cos ax = [1 / \{f_1(D^2) + D.f_2(D^2)\}] \cos ax$$

$$= [1 / \{f_1(-a^2) + D.f_2(-a^2)\}] \cos ax, \text{ by Eqs. (6.7)}.$$

Putting $f_1(-a^2) = a_1$, $f_2(-a^2) = a_2$ and introducing the operator $a_1 - a_2 D$ in numerator as well as in the denominator the RHS of above relation reduces to

$$(a_1 - a_2 D) [\{1 / (a_1 - a_2 D) (a_1 + a_2 D)\} \cos ax]$$

$$= (a_1 - a_2 D) [\{1 / (a_1^2 - a_2^2 D^2)\} \cos ax]$$

$$= (a_1 - a_2 D) [\{1 / (a_1^2 + a_2^2 . a^2)\} \cos ax], \text{ again by Eqs. (6.7)}$$

$$= \{a_1 . \cos ax - a_2 . D (\cos ax)\} / (a_1^2 + a^2 a_2^2)$$

$$= \{a_1 . \cos ax + a.a_2 . \sin ax)\} / (a_1^2 + a^2 a_2^2). \tag{6.8}$$

Similarly, we may derive

$$\{1 / f(D)\} \sin ax = \{a_1 . \sin ax - a.a_2 . \cos ax)\} / (a_1^2 + a^2 . a_2^2). \tag{6.9}$$

Theorem 6.4. For a non-zero constant a, there hold

$$\left.\begin{array}{l}\{1 / (D^2 + a^2)\} \cos ax = (x / 2a) \sin ax, \\[2mm] \{1 / (D^2 + a^2)\} \sin ax = - (x / 2a) \cos ax.\end{array}\right\} \tag{6.10}$$

Theorem 6.5. $\{1/ f(D)\} \{e^{ax} . u(x)\} = e^{ax} . \{1/ f(D + a)\} u(x). \tag{6.11}$

Note 6.3. When $f(a) = 0$, $\{1/ f(D)\} e^{ax}$ may be evaluated by Theorem 5.5 by taking $u(x) = 1$.

Theorem 6.6. Denoting derivative of $f(D)$ w.r.t. D by $f'(D)$, we get

$$\{1 / f(D)\} \{x. u(x)\} = x.\{1 / f(D)\}u(x) + [d\{f(D)\}^{-1}/ dD] u(x)$$

$$= x.\{1/ f(D)\}u(x) - [f'(D) /\{ f(D)\}^2] u(x). \tag{6.12}$$

§ 7. Homogeneous Linear Differential Equations of Any Order

A differential equation of the form

$$(x^n . D^n + a_1 x^{n-1} . D^{n-1} + a_2 . x^{n-2}. D^{n-2} + \ldots$$

$$+ a_{n-1} x. D + a_n)y = R(x) \tag{7.1}$$

is called a *homogeneous linear differential equation*. It is also called *Cauchy-Euler* or simply *Euler differential equation*. Applying the change of independent variable:

$$x = e^t \quad \Rightarrow \quad dx/dt = e^t \Rightarrow dt/dx = 1/e^t = 1/x, \qquad (7.2)$$

and therefore

$$D \equiv d/dx = (dt/dx)(d/dt) = (1/x).\, d/dt \Rightarrow x.\, D = d/dt \equiv D'. \qquad (7.3)$$

Operating the equation (7.3) by itself, there also results

$$x.D\,(x.\,D) = (d/dt)(d/dt),\ \text{i.e.}\quad x^2.D^2 + x.\,D = (d/dt)(d/dt)$$

$$\Rightarrow$$

$$x^2.\,D^2 = (d/dt)(d/dt) - x.\,D = (d/dt)(d/dt - 1) = D'(D' - 1), \quad (7.4)$$

by Eq. (7.3). Operating it again by $x.\,D \equiv D'$, we derive

$$x^3.D^3 + 2x^2.\,D^2 = D'\{D'(D' - 1)\}$$

$$\Rightarrow$$

$$x^3.\,D^3 = D'\{D'(D' - 1)\} - 2D'(D' - 1)\} = (D' - 2)\{D'(D' - 1)\}$$

$$= D'(D' - 1)(D' - 2), \qquad (7.5)$$

for Eq. (7.4) and the commutative properties of the operators D', $D' - 1$ and $D' - 2$.

Similarly, by the method of finite induction, we may also derive

$$x^n.\,D^n = D'(D' - 1)(D' - 2) \ldots \{D' - (n - 1)\}. \qquad (7.6)$$

Putting from Eqs. (7.3) – (7.6), the equation (7.1), reduces to a linear form with constant coefficients discussed in the preceding Section:

$$f(D')y = R(x). \qquad (7.7)$$

7.1. Complementary function: Setting $y = x^m$ so that

$$Dy = m.\,x^{m-1}, \qquad D^2y = m.\,(m - 1).\,x^{m-2}, \ldots,$$

$$D^n y = m.\,(m - 1)(m - 2) \ldots (m - n + 1)\, x^{m-n}$$

$$\Rightarrow$$

$$x.\,Dy = m\,x^m, \qquad x^2D^2y = m.\,(m - 1)\,x^m, \ldots,$$

$$x^n.\,D^n y = m.\,(m - 1).\,(m - 2) \ldots (m - n + 1)\, x^m,$$

and substituting for all above in the LHS of Eq. (7.1), we derive the auxiliary equation

$$\{ m (m-1) (m-2) \dots (m-n+1) + m (m-1) (m-2) \dots (m-n+2) a_1$$

$$+ \dots + m\, a_{n-1} + a_n \}\, x^m = 0, \qquad (7.8)$$

which is of nth degree in m. For n distinct roots $m_1, m_2, \dots, m_n$ the complementary function of the solution of Eq. (7.1) is

$$\text{C.F.} = A_1 x^{m_1} + A_2 x^{m_2} + \dots + A_n x^{m_n}. \qquad (7.9)$$

7.2. Particular integral: In analogy with the discussion in the preceding section, the particular integral of Eq. (7.7) can be defined by

$$\{ f(D') \}^{-1} R \equiv \{ 1 / f(D') \}\, R.$$

Theorem 7.1. $\qquad (D' - a)^{-1} R (x) = x^a . \int R (x) . x^{-(a+1)}. \, dx. \quad (7.10)$

Note 7.1. The result can also be derived directly from Eq. (6.4). With the change of variables as per Eq. (7.2) and $D' \equiv d/dt$, the result in Eq. (6.4) is rewritten as

$$(D' - a)^{-1} R (e^t) = e^{at} . \int R (e^t). e^{-at}. \, dt = x^a . \int R (x). x^{-a}. (1/x). \, dx$$

assuming the form given by Eq. (7.10).

The other cases discussed in the preceding Section can be similarly established for the differential equation (7.7).

§ 8. Differential Equations Reducible to Homogeneous Linear Form

A differential equation

$$\{ (ax + b)^n . D^n + a_1. (ax + b)^{n-1}. D^{n-1} + \dots$$

$$+ a_{n-1} .(ax + b). D + a_n \}\, y = R (x), \qquad (8.1)$$

with $a, b, a_1, \dots, a_n$ as constants, can be reduced to a homogeneous linear form with constant coefficients. Applying the change of variable: $ax + b = z$ so that

$$dz/dx = a, \ Dy \equiv dy/dx = (dy/dz).(dz/dx) = a \,(dy/dz) \equiv a.D'\, y,$$

$$D^2 y \equiv (d/dx)\,(dy/dx) = (d/dx)\,\{a\,(dy/dz)\}$$

$$= a\,(d^2y/dz^2)\,(dz/dx) = a^2\,(d^2y/dz^2) = a^2.\,D'^{\,2}\, y.$$

Similarly, $D^n y = a^n.\, D'^{\,n} y$. Substitution for these derivatives reduces the equation (8.1) to

$$\{\,(a.z)^n.\,D'^{\,n} + a_1.\,(az)^{n-1}.\,D'^{\,n-1} + \ldots + a_{n-1}.\,(az).\,D' + a_n\,\} y$$

$$= R\,\{(z-b)/a\},$$

or, on division by a^n,

$$\{\,z^n.\,D'^{\,n} + (a_1/a).\,z^{n-1}.\,D'^{\,n-1} + \ldots + (a_{n-1}/a^{n-1}).\,z.\,D'$$

$$+ (a_n/a^n)\,\}y = (1/a^n).\,R\,\{(z-b)/a\},$$

which is a homogeneous linear differential equation in the independent variable z and dependent variable y.

§ 9. Simultaneous Differential Equations

The preceding sections dealt with the solutions of a single differential equation involving one independent variable and one dependent variable. In dynamics, sometimes we encounter with a system of equations involving some variables and the derivatives of independent variables. The simplest class of such equations is the one containing one independent variable and other variable(s) as its functions.

9.1. Simultaneous linear differential equations with constant coefficients: Let t be an independent variable and $x\,(t)$ and $y\,(t)$ be two dependent variables. We consider the simultaneous equations

$$f_1\,(D)\,x\,(t) + g_1\,(D)\,y\,(t) = R_1\,(t), \quad f_2\,(D)\,x\,(t) + g_2\,(D)\,y\,(t) = R_2\,(t), \ \ (9.1)$$

where $D \equiv d/dt, f_1, g_1, f_2, g_2$ being rational functions of the differential operator D with constant coefficients and R_1, R_2 are functions of the independent variable t. We seek a solution of above pair of differential equations satisfying both equations. Operating these equations by $g_2\,(D)$ and $g_1\,(D)$ respectively and taking their difference, the terms containing variable y get eliminated:

$$g_2(D)\{f_1(D)x(t)\} - g_1(D)\{f_2(D)x(t)\}$$

$$\equiv \{g_2(D)f_1(D) - g_1(D)f_2(D)\}x(t)\} = g_2(D)R_1(t) - g_1(D)R_2(t), \quad (9.2)$$

where we have used the commutative property of the differential operators:

$$g_1(D)g_2(D)y(t) = g_2(D)g_1(D)y(t). \quad (9.3)$$

The resulting equation (9.2) is a linear differential equation of some order depending upon the resultant power of D in the operator $(g_2 f_1 - g_1 f_2)$ with constant coefficients and can be solved by methods given in Section 5. Let the solution be $x = \psi(t)$. Substitution for this value of x in any of the two equations in (9.1) yields a differential equation in $y(t)$ and t:

$$g_1(D)y(t) = R_1(t) - f_1(D)\psi(t), \quad \text{or} \quad g_2(D)y(t) = R_2(t) - f_2(D)\psi(t),$$

which are also linear of the order depending on that of g_1, g_2 and can be solved by some method similarly.

9.2. Simultaneous differential equations in three variables: Let x, y and z be the rectangular Cartesian coordinates of a moving point in three-dimensional Euclidean space. The locus of such points is a surface represented by some functional relation

$$F(x, y, z) = 0. \quad (9.4)$$

The number of independent variables involved in this equation is at most two and the remaining one as a function of the two independent variables. The first two coordinates: x and y are usually taken as the independent variables while z as a function of x and y. Let us consider the simultaneous equations

$$dx / P = dy / Q = dz / R, \quad (9.5)$$

where P, Q, R are some functions of the variables x, y, z. The equations (9.5) determine the direction ratios of a tangent line to a curve. Taking any two ratios in Eq. (9.5) and integrating the resulting differential equations there result the surfaces represented by

$$u_1(x, y, z) = c_1 \qquad \text{and} \qquad u_2(x, y, z) = c_2, \quad (9.6)$$

where c_1, c_2 are arbitrary constants of integration. The equations (8.6) holding simultaneously determine the curve of intersection of the (solution) surfaces. Since the arbitrary constants may be chosen in an infinite number of ways there exist doubly infinite number of such curves. There are different methods of integration of the equations (9.5) illustrated in the following examples.

§ 10. Solution of ODE by variation of parameters method

10.1. Solution of a linear ODE of the first order

Consider a differential equation

$$dy / dx + P(x).y = Q(x). \tag{10.1}$$

Its associated *homogenous* form

$$dy / dx + P(x).y = 0, \quad \text{or} \quad dy / y + P(x) dx = 0,$$

has integral

$$\ln y + \int P(x)\, dx = \ln c, \quad \text{or} \quad y.\exp\{\int P(x)\, dx\} = c$$

$$\Rightarrow$$

$$y_c = c.\exp\{-\int P(x)\, dx\}, \tag{10.2}$$

where c is an arbitrary constant of integration. Presuming that replacement of the constant c by some suitable function $u(x)$ in Eq. (10.2) provides a solution of the original differential equation (10.1):

$$y_P = u(x).\exp\{-\int P(x)\, dx\}, \tag{10.3}$$

so as to the solution y_P with its derivative

$$dy_P / dx = u'(x).\exp\{-\int P(x)\, dx\} - u.P.\exp\{-\int P(x)\, dx\} \tag{10.4}$$

satisfy the original diff. Eq. (10.1). Thus, putting from Eqs. (10.3) and (10.4), the Eq. (10.1) turns

$$u'(x).\exp\{-\int P(x)\, dx\} = Q(x) \quad \Rightarrow \quad u'(x) = Q(x).\exp\{\int P(x)\, dx\}.$$

Integrating it, we thus evaluate the function $u(x)$:

$$u(x) = \int [Q(x).\exp\{\int P(x)\, dx\}].\, dx. \tag{10.5}$$

The sum of functions y_c and y_P :

$$y = y_c + y_P = [c + \int Q(x). \exp \{\int P(x) \, dx\}]. \exp \{-\int P(x) \, dx\},$$

i.e.

$$y. \exp \{\int P(x) \, dx\} = \int Q(x). \exp \{\int P(x) \, dx\} + c \qquad (10.6)$$

provides the complete soln. of Eq. (10.1).

Note 10.1. The method (demonstrated above) is not preferable for the solution of 1st order linear ODE as there is a much shorter method by introduction of an integrating factor

$$\text{I.F.} = \exp \{\int P(x) \, dx\}.$$

Multiplying the ODE (10.1) by this I.F.:

$$\{dy/dx + P(x). y\}. \exp \{\int P(x) \, dx\} \equiv (d/dx) \{y. \exp \{\int P(x) \, dx\}$$

$$= Q(x). \exp \{\int P(x) \, dx\},$$

and integrating the resulting equation, one immediately gets the solution given by Eq. (10.6).

10.2. Solution of linear ODE of II order with constant coefficients

Consider an ODE

$$a_0 (d^2 y / dx^2) + a_1 (dy / dx) + a_2. y = Q_1 (x).$$

Dividing it by $a_0 \neq 0$, and putting $P \equiv a_1/a_0$, $Q \equiv a_2/a_0$, $R \equiv Q_1/a_0$, it may be re-written as

$$y'' + P. y' + Q. y = R(x) \qquad (10.7)$$

where primes denote derivation w.r.t. x. Its associated homogenous eqn. (with vanishing RHS)

$$y'' + P. y' + Q. y = 0 \qquad (10.8)$$

gives rise to its auxiliary equation $m^2 + Pm + Q = 0$ for some (algebraic) number m. Its roots m_1, m_2 determine the complimentary function C.F. [cf. § 5]:

$$y_c = A_1. \exp (m_1 x) + A_2. \exp (m_2 x), \qquad (10.9)$$

where A_1, A_2 are arbitrary constants of integration. It is also presumed that the roots m_1, m_2 are distinct and real (numbers). In general, let the

ODE (10.8) possesses two distinct solutions, say y_1 and y_2, so that above C.F. assumes the form

$$y_c = A_1. y_1 + A_2. y_2. \tag{10.10}$$

Like Case 10.1, let replacement of the constants A_1, A_2 by some functions $u_1(x)$ and $u_2(x)$ in Eq. (10.10) provides a solution of the original differential equation (10.7):

$$y_P = u_1. y_1 + u_2. y_2, \tag{10.11}$$

so that the soln. y_P and its derivatives:

$$y'_P = u'_1. y_1 + u_1. y_1' + u'_2. y_2 + u_2. y_2',$$

$$y''_P = (u''_1. y_1 + 2u'_1. y_1' + u_1. y_1'') + (u''_2. y_2 + 2u'_2. y_2' + u_2. y_2'')$$

should satisfy the original diff. Eq. (10.7). Thus, putting these values, the Eq. (10.7) yields

$$u_1.\{y_1'' + P. y_1' + Q. y_1\} + u_2.\{y_2'' + P. y_2' + Q. y_2\} + (u''_1. y_1 + u'_1. y_1' +$$

$$u''_2. y_2 + u'_2. y_2') + P (u'_1. y_1 + u'_2. y_2) + (u'_1. y_1' + u'_2. y'_2) = R(x).$$

As per hypothesis, both y_1, y_2 being solutions of homogeneous ODE (10.8), make the coefficients of u_1, u_2 in above equation zero reducing above equation:

$$(d/dx) (u'_1. y_1 + u'_2. y_2) + P (u'_1. y_1 + u'_2. y_2) + (u'_1. y_1' + u'_2. y'_2) = R(x).$$

Choosing the functions u_1, u_2 so as to satisfy

$$u'_1. y_1 + u'_2. y_2 = 0, \tag{10.12}$$

above equation further simplifies to

$$u'_1. y_1' + u'_2. y'_2 = R(x). \tag{10.13}$$

Solving these simultaneous linear eqns. for u'_1, u'_2:

$$u'_1 / \begin{vmatrix} y_2 & 0 \\ y'_2 & -R \end{vmatrix} - u'_2 / \begin{vmatrix} y_1 & 0 \\ y'_1 & -R \end{vmatrix} = 1 / \begin{vmatrix} y_1 & y_2 \\ y'_1 & y'_2 \end{vmatrix}$$

$\Rightarrow$

$$u'_1 = -R y_2 / (y_1 y'_2 - y'_1 y_2) \quad \text{and} \quad u'_2 = R y_1 / (y_1 y'_2 - y'_1 y_2),$$

and integrating the last two equations, we evaluate the functions u_1, u_2. Thus, the general soln. of the original ODE (10.7) is obtained:

$$y = y_c + y_P = (A_1 y_1 + A_2 y_2) + (u_1 y_1 + u_2 y_2). \qquad (10.14)$$

§ 11. Normal form of an ODE

Linear ODEs with constant coefficients have been studied in § 5 while § 7 deals with homogenous linear ODEs having variable coeffici‒ ents. Currently, we consider linear ODEs with any choice of variable coefficients and transform it under some suitable change of dependent variable(s) to reduce it into a normal form. For simplicity, let us consid‒ er a second order linear ODE

$$y'' + P(x). y' + Q(x). y = 0, \qquad (11.1)$$

where P, Q are any arbitrary functions of the independent variable x. Above form of ODE has been termed as a standard form (cf. Example 10.2). We aim to transform this form of ODE into one free from the term containing the first derivative term. Choosing two arbitrary func‒ tions $u(x)$ and $v(x)$, let us subject the variable y to the change:

$$y = u(x). v(x), \qquad (11.2)$$

$\Rightarrow$

$$y' = u'(x). v + u(x). v'(x), \qquad \text{and} \qquad y'' = u''. v + 2u' v' + u v''.$$

Plugging in these values in Eq. (11.1), the ODE is transformed to

$$(u''. v + u v'' + P. u v' + Q. u v) + (2v' + P. v) u' = 0. \qquad (11.3)$$

Choosing the functions u, v so as to make the coefficient of u' zero in above equation, there results

$$2v' + P. v = 0; \qquad (11.4)$$

reducing Eq. (11.3) to

$$u''. v + u v'' + P. u v' + Q. u v = 0. \qquad (11.5)$$

Differentiating Eq. (11.4) w.r.t. x, there also result

$$v' = -(P/2).v \quad \text{and} \quad v'' = -(1/2). (P' v + P v') = -(P' v /2) + (P^2 v) /4;$$

which further reduce Eq. (11.5) to the desired form:

$$u'' + R(x).\, u = 0, \qquad\qquad (11.6)$$

where

$$R(x) \equiv -P'/2 - P^2/4 + Q.$$

Definition 11.1. The form of ODE (11.6) is called a *normal form* of the given ODE (11.1).

CHAPTER 10

DIFFERENTIAL EQUATIONS (PARTIAL)

§ 1. Introduction

An equation involving *one* or *more* partial derivatives of an unknown function, say *u*, of *two* or *more* (independent) variables, say $x, y, z, \ldots$, is called a *partial differential equation* (PDE). Time parameter *t* is usually taken as one of these (independent) variables while remaining ones are the spatial coordinates. The order of the highest derivative of *u* in the equation is called the *order* of the PDE.

1.1. First order PDEs: A general PDE of *first* order and of any degree in (unknown) variable *z* depending on independent variables *x*, *y* may be represented by

$$f(x, y, z, p, q) = 0, \qquad (1.1)$$

where *p*, *q* as per Eqs. (1.1.6) are the first order partial derivatives of *z* w.r.t. *x* and *y* respectively. An integral of Eq. (1.1) determining *z* as a function of *x* and *y* provides a solution of the PDE. If the solution contains as many arbitrary constants of integration as the number of independent variables it is called a *complete* integral. Thus, with two arbitrary constant *a* and *b*, a functional relation

$$g(x, y, z, a, b) = 0, \qquad (1.2)$$

satisfying Eq. (1.1) may be taken as a complete integral of the PDE.

Note 1.1. If *x*, *y*, *z* denote the rectangular Cartesian coordinates of a point in 3-dimensional Euclidean space E_3, Eq. (1.2) then represents a family of surfaces in the space.

Example 1.1. With arbitrary constant *a* and *b*, the equation

$$z = (a + x)(b + y) \qquad (1.3)$$

provides a complete integral of the first order PDE $z = pq$.

Solution. Differentiating Eq. (1.3) partially w.r.t. *x*, *y*:

$$p \equiv \partial z / \partial x = b + y, \qquad q \equiv \partial z / \partial y = a + x,$$

and eliminating a, b from Eq. (1.3), one immediately gets the PDE. //

Example 1.2. With three arbitrary constant a, b, c, the equation

$$x^2/a^2 + y^2/b^2 + z^2/c^2 = 1 \qquad (1.4)$$

provides a complete integral of the second order simultaneous PDEs

$$p\,z = (p^2 + r\,z)\,x \quad (1.5a); \qquad q\,z = (q^2 + t\,z)\,y, \qquad (1.5b)$$

where

$$r \equiv \partial p / \partial x = \partial^2 z / \partial x^2, \quad t \equiv \partial q / \partial y = \partial^2 z / \partial y^2, \qquad (1.6)$$

are the second order partial derivatives of z w.r.t. x, y repectively.

Solution. Differentiating Eq. (1.4) successively w.r.t. x, y partially :

$$x/a^2 + p\,z/c^2 = 0 \qquad (1.7a); \quad y/b^2 + q\,z/c^2 = 0, \qquad (1.7b)$$

$$1/a^2 + p^2/c^2 + rz/c^2 = 0 \qquad (1.8a); \quad 1/b^2 + q^2/c^2 + t\,z/c^2 = 0. \quad (1.8b)$$

Subtraction of Eq. (1.7a) from x^{th} multiple of Eq. (1.8a) yields Eq. (1.5a). Also, subtraction of Eq. (1.7b) from y^{th} multiple of Eq. (1.8b) yields Eq. (1.5b). //

Example 1.3. The equation

$$(x - y)^2\,(p^2 + q^2 + 1) = 16\,(p - q)^2 \qquad (1.9)$$

represents the first order PDE of the familyof spheres of radius 4 and centre on the plane $x = y$.

Solution. Taking centres of the spheres at points (a, a, z_1) a family of spheres is represented by
$$(x - a)^2 + (y - a)^2 + (z - z_1)^2 = 16 \qquad (1.10)$$

Differentiating it partially w.r.t. x, y :

$$x - a + (z - z_1)\,p = 0 \qquad (1.11a); \quad y - a + (z - z_1)\,q = 0. \qquad (1.11b)$$

Substitution for $x - a$ and $y - a$ in Eq. (1.10), thus, determines

$$(z - z_1)^2 \cdot (p^2 + q^2 + 1) = 16. \tag{1.12}$$

On the other hand, difference of Eqs. (1.11) gives

$$x - y + (z - z_1)(p - q) = 0 \quad \Rightarrow \quad (z - z_1)^2 \cdot (p - q)^2 = (x - y)^2. \tag{1.13}$$

Elimination of $(z - z_1)^2$ from Eqs. (1.12) and (1.13) finally yields PDE. //

The following two examples deal with the PDEs satisfied by family of surfaces involving arbitrary set(s) of functions.

Example 1.4. With an arbitrary set of functions $f(x, y)$, a family of surfaces

$$z = y^2 + 2 f(1/x + \ln y), \tag{1.14}$$

satisfies the PDE

$$p x^2 + y q = 2y^2.$$

Solution. Differentiating Eq. (1.14) partially w.r.t. x, y :

$$p = 2 f' \cdot (- 1/x^2) \quad \Rightarrow \quad p x^2 + 2 f' = 0,$$

and

$$q = 2 y + 2 f' \cdot (1/y) \quad \Rightarrow \quad q y - 2 f' = 2 y^2.$$

Elimination of f' from the last resulting relations yields the PDE. //

Example 1.5. With arbitrary sets of functions $f(x)$ and $g(y)$ a family of surfaces

$$z = y f(x) + x g(y), \tag{1.15}$$

satisfies the PDE

$$s x y = px + qy - z, \tag{1.16}$$

where s is the second order partial derivative of z as per Eqs. (1.1.7).

Solution. Differentiating Eq. (1.15) partially w.r.t. x, y :

$$\left. \begin{array}{l} p = y. f'(x) + g(y) \Rightarrow f'(x) = \{p - g(y)\}/ y, \\[2mm] q = f(x) + x. g'(y) \Rightarrow g'(y) = \{q - f(x)\}/ x. \end{array} \right\} \tag{1.17}$$

Further partial derivation of p w.r.t. y gives

$$s = f'(x) + g'(y);$$

which, for Eqs. (1.17), reduces to

$$s = \{p - g(y)\}/y + \{q - f(x)\}/x = \{px + qy - x.g(y) - y f(x)\}/xy.$$

Its multiplication by $x\,y$ and use of Eq. (1.16), finally gives the PDE. //

1.2. First order PDEs of general form

Let a PDE involve n independent variables $x_1, x_2, \ldots , x_n$, an unknown) variable z (depending upon x_i 's, $i = 1, 2, \ldots, n$), and first order partial derivatives

$$p_1 = \partial z / \partial x_1, \qquad p_2 = \partial z / \partial x_2, \ldots , \qquad p_n = \partial z / \partial x_n, \qquad (1.18)$$

of z w.r.t. $x_1, x_2, \ldots , x_n$ respectively. A first order PDEs of general form is then expressed by

$$f(x_1, x_2, \ldots , x_n, z, p_1, p_2, \ldots , p_n) = 0. \qquad (1.19)$$

A solution to the first order PDE is a function $z(x_1, x_2, \ldots , x_n)$ that satisfies the Eq. (1.19).

§ 2. Classification of first order PDEs

First order PDEs can be classified as *linear, semi-linear, quasi-linear* and *non-linear* PDEs. As before, let a PDE be expressed in terms of an unknown variable z dependening on two independent variables x, y and its partial derivatives p, q given by Eqs. (1.1.6). With known coefficients P, Q, R and T the first three types of above PDEs are distinguished by the following equations:

(i) **Linear:** $P(x, y)\, p + Q(x, y)\, q + R(x, y)\, z = T(x, y);$ (2.1)

(ii) **Semi-linear:** $P(x, y)\, p + Q(x, y)\, q = T(x, y, z);$ (2.2)

(iii) **Quasi-linear:** $P(x, y, z)\, p + Q(x, y, z)\, q = R(x, y, z).$ (2.3)

Note 2.1. Particularly, when the coefficients P, Q, R in quasi-linear Eq. (2.3) are independent of z, it becomes *linear*. The first order PDEs of type other than aove are called *non-linear*. For instance, the PDE

$pq = z\,x$ is non-linear. The most general non-linear PDE of first order may be expressed by Eq. (1.1).

Such equations are encountered in the fields of continuum mechanics, gas dynamics, hydrodynamics, heat and mass transfer, wave theory, acoustics, multiphase flows, chemical engineering, etc.

Note 2.2. The system of ODEs

$$dx / P(x, y, z) = dy / Q(x, y, z) = dz / R(x, y, z) \qquad (2.4)$$

is known as the *characteristic system* of equations.

2.1. Solution of quasi-linear / linear equations

As seen above, a first order quasi-linear PDE contains only the *first* degree terms of the first order partial derivatives p, q of an unknown (*dependent*) variable z w.r.t. independent variables x, y. It is represented by Eq. (2.3), where (all the known) coefficients P, Q, R are functions of variables x, y, z and z is a function of x, y.

Theorem 2.1. Every integral
$$f(x, y, z) = \text{const.} \qquad (2.5)$$

of Eq. (2.3) is also an integral of the PDE

$$P(\partial / \partial x) + Q(\partial / \partial y) + R(\partial / \partial z) f = 0, \qquad (2.6)$$
and conversely.

Proof. Differentiating Eq. (2.5) partially w.r.t. x, y :

$$\partial f / \partial x + (\partial f / \partial z)\, p = 0 \quad \Rightarrow \quad p = -f_x / f_z,$$
and
$$\partial f / \partial y + (\partial f / \partial z)\, q = 0 \quad \Rightarrow \quad q = -f_y / f_z, \qquad (2.7)$$
where
$$f_x = \partial f / \partial x, \qquad f_y = \partial f / \partial y \quad \text{and} \quad f_z = \partial f / \partial z. \qquad (2.8)$$

Substitution for p, q in Eq. (2.3) transforms it to Eq. (2.6) establishing the first part of the theorem.

Conversely, when integral given by Eq. (2.5) satisfies PDE (2.6), the same substitutions for p, q reduce Eq. (2.6) to Eq. (2.3). //

Theorem 2.2. Every integral of the system of equations (2.4) satisfies Eq. (2.6) and conversely.

Proof. Let Eq. (2.5) be an integral of Eqs. (2.4). Therefore, there hold Eqs. (2.7) and

$$df = (\partial f / \partial x)\, dx + (\partial f / \partial y)\, dy + (\partial f / \partial z)\, dz = 0, \qquad (2.9)$$

by Taylor's theorem. Also, Eqs. (2.4) imply

$$dx = Pk, \qquad dy = Qk, \qquad dz = Rk\,;$$

for which Eq. (2.9) transforms to Eq. (2.6) for any $k \neq 0$. This establishes the first part of the theorem and the converse part can also be proved similarly. //

Theorem 2.3. If Eq. (2.5) and $g\,(x, y, z) = $ const. be two independent integrals of the Eqs. (2.4), then the functional relation

$$\phi\,(f, g) = 0, \qquad (2.10)$$

satisfies Eq. (2.6) and conversely.

Proof. By previous theorem, solutions of Eq. (2.4) satisfy Eqs. (2.6). Thus, together with Eq. (2.6), there also hold a similar equation satisfied by the function $g\,(x, y, z)$:

$$\{P\,(\partial / \partial x) + Q\,(\partial / \partial y) + R\,(\partial / \partial z)\}\, g = 0. \qquad (2.11)$$

On the other hand, partial derivations of Eq. (2.10) w.r.t. x, y, z yield

$$\partial \phi / \partial x \equiv (\partial \phi / \partial f)\,(\partial f / \partial x) + (\partial \phi / \partial g)\,(\partial g / \partial x) = 0,$$

$$\partial \phi / \partial y \equiv (\partial \phi / \partial f)\,(\partial f / \partial y) + (\partial \phi / \partial g)\,(\partial g / \partial y) = 0,$$

$$\partial \phi / \partial z \equiv (\partial \phi / \partial f)\,(\partial f / \partial z) + (\partial \phi / \partial g)\,(\partial g / \partial z) = 0.$$

Multiplying these relations by P, Q, R respectively, adding the results and collecting the coefficients of $\partial \phi / \partial f$ and $\partial \phi / \partial g$, we get

$$\{P\,(\partial / \partial x) + Q\,(\partial / \partial y) + R\,(\partial / \partial z)\}\, \phi = 0,$$

for Eqs. (2.6) and (2.11). Thus, the function ϕ satisfies Eq. (2.6). This establishes the first part of the theorem and the converse part can also be proved similarly. //

Note 2.3. Above theorems, indeed, conclude a method called afer Lagrange for solution of PDE (2.3). Equations (2.4) are also called the *subsidiary equations* of the PDE (2.3).

2.2. Particular cases of linear ODEs

If one of the coefficients P or Q are zero, the general form of PDE (2.3) reduces to

$$\partial z / \partial y = R/Q \qquad \text{(alternately)} \qquad \partial z / \partial x = R/P;$$

whose solutions may be found easily. For instance, the solution of linear PDE

$$\partial z / \partial y = ax + by,$$

may be obtained on its (partial) integration w.r.t. y :

$$z = axy + by^2/2 + f(x),$$

for any set of arbitrary functions $f(x)$.

Example. 2.1. Obtain two independent solutions of the PDE

$$p.\cos(x+y) + q.\sin(x+y) = z. \qquad (2.12)$$

Solution. Its subsidiary equations are given by Eqs. (2.4):

$$dx / \cos(x+y) = dy / \sin(x+y) = dz / z. \qquad (2.13)$$

As per Eq. (2.3.1), each of above ratios are also equal to

$$(dx \pm dy) / \{\cos(x+y) \pm \sin(x+y)\}.$$

Thus, two independent solutions of PDE are determined by

$$dz / z = (dx \pm dy) / \{\cos(x+y) \pm \sin(x+y)\}. \qquad (2.14)$$

Writing $x + y = u$ so that $dx + dy = du$, one of above ODEs reduces to

$$dz \,/\, z \;=\; du \,/\, (\cos u + \sin u) = du \,/\, \sqrt{2}.\,\cos (\pi/4 - u), \qquad (2.15)$$

which is a linear ODE in separated variables z and u. Thus, integrating it term-wise w.r.t. resoective variables, there follows a solution

$$\sqrt{2}.\,\ln z = -\ln \tan (\pi/4 + \pi/8 - u/2) + \text{const.}$$

$\Rightarrow$

$$f(x, y, z) \equiv z^{\sqrt{2}}.\,\tan (3\pi/8 - u/2) = \text{const.}$$

On the other hand, accounting the negative sign in Eq. (2.14), the alternate solution is

$$dz \,/\, z = (dx - dy) \,/\, (\cos u - \sin u) = d\,(x - y) \,/\, (\cos u - \sin u),$$

or, by Eqs. (2.15)

$$du \,/\, (\cos u + \sin u) = d\,(x - y) \,/\, (\cos u - \sin u).$$

Separating the variables u and $x - y$ and integrating above ODE w.r.t. respective variables, there follws

$$x - y = \int \left\{ \frac{\cos u - \sin u}{\cos u + \sin u} \right\} du \; + \text{const.} = \ln (\cos u + \sin u) + \text{const.}$$

or,

$$y - x + \ln (\cos u + \sin u) = \text{const.}$$

$\Rightarrow$

$$g(x, y, z) \equiv \exp (y - x).\{\cos (x + y) + \sin (x + y)\} = \text{const.}$$

Hence, the general solution of PDE (2.12) is constituted by above two solutions:

$$g(x, y, z) = \phi (f). \,/\!/$$

Example. 2.2. Solve the PDE $(x^2 - y^2 - z^2)\,p + 2\,x\,y\,q = 2\,z\,x.$

Solution. Its subsidiary equations are:

$$dx \,/\, (x^2 - y^2 - z^2) \;=\; dy \,/2xy \;=\; dz \,/2zx. \qquad (2.16)$$

The last two of above ratios yield $dy \,/\, y = dz \,/\, z$ having integral

$$\ln y = \ln z + \ln a \;\Rightarrow\; y = a\,z, \quad a \text{ being a const.} \quad (2.17)$$

On the other hand, the first and the third ratios in Eqs. (2.16) yield

$$dx / \{x^2 - (a^2 + 1) z^2\} = dz / 2zx \Rightarrow 2x (dx / dz) - x^2 / z = -(a^2 + 1) z.$$

Putting $x^2 = v$ so that $2x (dx/dz) = dv/dz$, above ODE transforms as

$$dv / dz - v / z = -(a^2 + 1) z,$$

which is linear in variables v and z. Its integrating factor is

$$\text{I.F.} = \exp \{-\int (1/z) \, dz\} = 1/z.$$

Hence, solution of above ODE is

$$v / z + (a^2 + 1) \int dz = \text{const.},$$

or,

$$x^2 / z + (y^2 / z^2 + 1) z = (x^2 + y^2 + z^2) / z = \text{const.,} \qquad (2.18)$$

by Eq. (2.17). The two solutions given by Eqs. (2.17) and (2.18) constitute the general solution

$$\phi \{ y / z, (x^2 + y^2 + z^2) / z \} = 0. //$$

2.3. (Quasi-) linear PDEs involving *n* independent variables

Let a quasi-linear PDE be expressed in terms of an unknown (dependent) variable z which is a function of n independent variables $x_1, x_2, \ldots, x_n$. Denoting the partial derivatives of z w.r.t. variables x_i's, $i = 1, 2, \ldots, n$, by p_i's and employing $n + 1$ coefficients P_i's and R as functions of above $n + 1$ variables, a quasi-linear PDE of general form may be taken as

$$P_1 p_1 + P_2 p_2 + \ldots + P_n p_n = R. \qquad (2.19)$$

To get its general solution we consider its subsidiary equations

$$dx_1 / P_1 = dx_2 / P_2 = \ldots = dx_n / P_n = dz / R,$$

and evaluate n independent solutions $f_i = \text{const.}$ These solutions form a general solution of Eq. (2.19):

$$\phi (f_1, f_2, \ldots, f_n) = 0.$$

§ 3. Non-linear PDEs of first order

A general form of such PDEs is represengted by Eq. (1.1). The following method called after *Charpit* deals with the solution of above PDEs. We consider an additional PDE of general form involving an arbitrary constant a:

$$g(x, y, z, p, q, a) = 0, \tag{3.1}$$

and solve these equations for p, q:

$$p = p(x, y, z, a) \qquad \text{and} \qquad q = q(x, y, z, a). \tag{3.2}$$

For Taylor's theorem given by Eq. (5.22.1), there holds a linear PDE:

$$dz = (\partial z / \partial x) \, dx + (\partial z / \partial y) \, dy = p \, dx + q \, dy = 0. \tag{3.3}$$

Also, differentiating Eqs. (1.1) and (3.1) partially w.r.t. x and y:

$$f_x + f_z \cdot p + f_p \cdot p_x + f_q \cdot q_x = 0, \quad g_x + g_z \cdot p + g_p \cdot p_x + g_q \cdot q_x = 0, \tag{3.4}$$

and

$$f_y + f_z \cdot q + f_p \cdot p_y + f_q \cdot q_y = 0, \quad g_y + g_z \cdot q + g_p \cdot p_y + g_q \cdot q_y = 0, \tag{3.5}$$

where the suffixes denote partial derivatives w.r.t. respective suffixes in agreement with Eq. (1.1.6). Eliminating p_x from Eqs. (3.4) and q_y from Eqs. (3.5), there result

$$(f_x \cdot g_p - f_p \cdot g_x) + p \cdot (f_z \cdot g_p - f_p \cdot g_z) + q_x (f_q \cdot g_p - f_p \cdot g_q) = 0,$$

and

$$(f_y \cdot g_q - f_q \cdot g_y) + q \cdot (f_z \cdot g_q - f_q \cdot g_z) + p_y (f_p \cdot g_q - f_q \cdot g_p) = 0.$$

For Eq. (1.1.7) and rearrangement of the terms , their addition simplifies to

$$-f_p \cdot g_x - f_q \cdot g_y - (p.f_p + q.f_q) \, g_z + (f_x + p.f_z) \, g_p + (f_y + q.f_z) \, g_q = 0, \tag{3.6}$$

which is linear PDE in unknown variable g depending upon x, y, z, p, q. The latter five variables may be treated independent. Thus, in analogy with Theorems 2.1 and 2.2, every solution of linear PDE (3.6) shall also satisfy its subsidiary equations

$$\frac{dx}{-f_p} = \frac{dy}{-f_q} = \frac{dz}{-p\,f_p - q\,f_q} = \frac{dp}{f_x + p\,f_z} = \frac{dq}{f_y + q\,f_z} = \frac{dg}{0}. \quad (3.7)$$

The integrals of these subsidiary equations determine the values of p, q which, in association with Eqs. (3.2) and (3.3), will ultimately lead to the desired solution of PDE (1.1). The procedure is better explained by means of the following examples.

Example 3.1. Find a complete integral of the PDE

$$f(x, y, p, q) \equiv 2\,(qx + py + pq) + x^2 + y^2 = 0. \quad (3.8)$$

Solution. Evaluating the partial derivaives of F:

$$f_x = 2\,(q + x), \quad f_y = 2\,(p + y), \quad f_z = 0, f_p = 2\,(y + q), \quad f_q = 2\,(x + p),$$

the subsidiary equations are written as per Eqs. (3.7):

$$\frac{dx}{-(2y+q)} = \frac{dy}{-(2x+p)} = \frac{dz}{x^2 + y^2} = \frac{dp}{2(q+x)} = \frac{dq}{2(p+y)} = k \text{ (say)}.$$

As per Eqs. (2.3.1), each of above ratios is also equal to

$$(dx + dy + dp + dq) / 0 = k \implies dx + dy + dp + dq = 0,$$

having integral an integral

$$p + q + x + y \equiv (p + x) + (q + y) = a \text{ (const.)}. \quad (3.9)$$

On the otherhand, terms in Eq. (3.8) may be re-arranged as

$$2(p + x)\,(q + y) + (x - y)^2 = 0 \implies (p + x)\,(q + y) = -(x - y)^2 / 2. \quad (3.10)$$

From Eqs. (3.9) and (3.10), there follows

$$(p + x) - (q + y) = \sqrt{\{(p + x + q + y)^2 - 4\,(p + x)\,(q + y)\}}$$

$$= \sqrt{\{a^2 + 2\,(x - y)^2\}}. \quad (3.11)$$

Addition and subtraction of Eqs. (3.9) and (3.11), thus, determine

$$p = -x + a/2 + \sqrt{\{a^2 + 2(x-y)^2\}}/2,$$

and

$$q = -y + a/2 - \sqrt{\{a^2 + 2(x-y)^2\}}/2.$$

Hence, Eq. (3.3) bedomes

$$dz = -(x\,dx + y\,dy) + a\,(dx + dy)/2 + \sqrt{\{a^2 + 2(x-y)^2\}}.(dx - dy)/2.$$

Integrating it term-wise w.r.t. respective variables, we get

$$2z + b = -(x^2 + y^2) + a\,(x+y) + \int\sqrt{\{a^2 + 2(x-y)^2\}}.d\,(x-y).$$

Applying Eq. (6.2.34), integral in above equation can be evaluated as

$$\{(x-y)/\sqrt{2}\}.\sqrt{\{a^2/2 + (x-y)^2\}} + (a^2/2\sqrt{2}).\ln\,\{x-y+\sqrt{\{a^2/2 + (x-y)^2\}}.//$$

Example 3.2. Find a complete integral of the PDE

$$f(x, y, p, q) \equiv p^2 + q^2 - 2pq.\tanh 2y - \operatorname{sech}^2 2y = 0. \qquad (3.12)$$

Solution. Evaluating the partial derivaives of F:

$$f_x = 0, \qquad f_y = -4.\operatorname{sech}^2 2y\,(pq - \tanh 2y), \qquad f_z = 0,$$

$$F_p = 2\,(p - q.\tanh 2y), \quad F_q = 2\,(q - p.\tanh 2y),$$

the subsidiary equations are written as per Eqs. (3.7):

$$\frac{dx}{2(-p+q.\tanh 2y)} = \frac{dy}{2(-q+p.\tanh 2y)} = \frac{dp}{0} = \frac{dq}{4(-pq+\tanh 2y).\operatorname{sech}^2 2y}.$$

Hence, $dp = 0$ has integral $p = a$ (const.). For such value of p, Eq. (3.12) becomes

$$q^2 - 2(a.\tanh 2y)\,q + (a^2 - \operatorname{sech}^2 2y) = 0,$$

determining

$$q = a.\tanh 2y + \sqrt{\{a^2\,(\tanh^2 2y - 1) + \operatorname{sech}^2 2y\}}$$

$$= a.\tanh 2y + \sqrt{(1 - a^2)}.\operatorname{sech} 2y, \qquad\qquad \text{by Eq. (5.9.25).}$$

For these values of p, q, the PDE (3.3) becomes

$$dz = a\,dx + \{\,a.\tanh 2y + \sqrt{(1 - a^2)}.\operatorname{sech} 2y\,\}.dy.$$

Integrating it termwise w.r.t. respective variables, we get

$$z + b = a \{x + (1/2) \ln \cosh 2y\} + \sqrt{(1 - a^2)}. \int \operatorname{sech} 2y. \, dy.$$

For Eq. (5.9.22), the integral in above equation can be evaluated as

$$\int \operatorname{sech} 2y. \, dy = \int \frac{2dy}{e^{2y} + e^{-2y}} = \int \frac{2e^{2y}.dy}{e^{4y} + 1} = \tan^{-1} e^{2y}. \; //$$

§ 4. Second order PDEs with variable coefficients

Let a PDE involve one (unknown) dependent variable z and two independent variables x, y. Secod order partial derivatives of z w.r.t. x, y are given by Eqs. (1.1.7). Analogus to Eq. (1.1), a general PDE of second order involving the second order partial derivatives r, s, t of z may be represented by

$$F(x, y, z, p, q, r, s, t) = 0. \tag{4.1}$$

Analogous to PDEs of first order discussed in § 2, the following equation

$$R\,r + S\,s + T\,t = H, \tag{4.2}$$

where the coefficients R, S, T, H are functions of x, y, z, p and q, represents a *quasi-linear* PDE of second order in a general form. Particularly, when R, S, T are functions of x, y only and H is linear in p, q, z :

$$H = P\,p + Q\,q + Z\,z + G, \tag{4.3}$$

together with all the coefficients P, Q, Z, G as functions of x, y only, the PDE (4.2) becomes *linear*. Thus, a general linear PDE of second order may be expressed as

$$R\,r + S\,s + T\,t + P\,p + Q\,q + Z\,z = G, \tag{4.4}$$

where all the coefficients R, S, T, P, Q, Z, G are functions of independent variables x, y only and R, S, T do not vanish simultaneously. A complete integral of PDE (4.4) given by equation

$$\phi(x, y, z) = 0,$$

represents a surface called an *integral surface* in the space E_3 determined by the rectangular Cartesian coordinate axes Ox, Oy and Oz. If on the integral surface there exist curves for which the second order derivatives r, s, t are discontinuous or indeterminate, the curves are called the *characteristic* curves of PDE.

When $G = 0$, the PDE (4.4) is called *homogenous*. Three different types of PDEs (4.4) are discussed below.

4.1. First type: The equations

$$r = G_1 (x, y), \qquad s = G_2 (x, y), \qquad t = G_3 (x, y), \qquad (4.5)$$

form the *first type* of above PDEs where the coefficients of derivatives are constants.

Example 4.1. Solve the PDE $r = \sin (x\, y)$.

Solution. Integrating it partially w.r.t. x, there follows the first order PDE

$$p = (- 1/y) \cos (x\, y) + \phi_1 (y).$$

Its further partial integration w.r.t. x yields the complete solution

$$z = (- 1/y^2) \sin (x\, y) + x.\ \phi_1 (y) + \phi_2 (y),$$

ϕ_1, ϕ_2 being arbitrary sets of functions of y only. //

Example 4.2. Solve the PDE $s = 1/x\, y^2 - 2x/y$.

Solution. Integrating it partially w.r.t. x, there follows the first order PDE

$$q = (1/y^2) \ln x - x^2/y + \phi_1 (y).$$

Its further partial integration w.r.t. y yields the complete solution

$$z = (- 1/y) \ln x - x^2.\ \ln y + \int \phi_1 (y)\, dy + \phi_2 (x),$$

$\phi_1 (y)$, $\phi_2 (x)$ being arbitrary sets of functions of y and x respectively. //

Example 4.3. Solve the PDE $t = x^2 . \cos (x\, y)$.

Solution. Integrating it partially w.r.t. y, there follows the first order PDE

$$q = x. \sin(x\,y) + \phi_1(x).$$

Its further partial integration w.r.t. y yields the complete solution

$$z = -\cos(x\,y) + y\,\phi_1(x) + \phi_2(x),$$

ϕ_1, ϕ_2 being arbitrary sets of functions of x only. //

4.2. Second type: The equations

$$R\,r + P\,p = G_1, \quad S\,s + P\,p = G_2, \quad S\,s + Q\,q = G_3, \quad T\,t + Q\,q = G_4, \quad (4.6)$$

form the *second type* of above PDEs, where all the coefficients of derivatives and G's are constants. These equations can be alternately written as

$$R\,(\partial p / \partial x) + P\,p = G_1, \quad S\,(\partial p / \partial y) + P\,p = G_2,$$

$$S\,(\partial q / \partial x) + Q\,q = G_3, \quad T\,(\partial p / \partial y) + Q\,q = G_4, \quad (4.6a)$$

Interestingly, these are linear ODEs of first order where p (or q) are the dependent variables.

Example 4.4. Solve the PDE $x\,r + p = 9x^2\,y^2$.

Solution. The equation can be viewed as a linear ODE in p (dependent) and x (independent) variables:

$$\partial p / \partial x + p / x = 9xy^2. \quad (4.7)$$

Its integrating factor is

$$\text{I.F.} = \int \exp(1/x)\,dx = \exp(\ln x) = x.$$

Thus, integrating Eq. (4.7) partially w.r.t. x (treating y as a constant), its solution is

$$px = 9y^2.\int x^2.dx + \phi_1(y) = 3x^3\,y^2 + \phi_1(y),$$

$\Rightarrow$

$$p = \partial z / \partial x = 3x^2\,y^2 + (1/x).\,\phi_1(y).$$

Integrating it further partially w.r.t. x, there follows the complete integralof (4.7) with two arbitrary sets of fucntions $\phi_1(y) + \phi_2(y)$:

$$z = x^3\,y^2 + (\ln x).\,\phi_1(y) + \phi_2(y). //$$

Example 4.5. Solve the PDE $ys + p = \cos(x+y) - y.\sin(x+y)$.

Solution. Re-writing the equation as a linear ODE in p (dependent) and y (independent) variables:

$$\partial p/\partial y + p/y = (1/y).\cos(x+y) - \sin(x+y), \qquad (4.8)$$

its integrating factor is

$$\text{I.F.} = \int \exp(1/y)\, dy = \exp(\ln y) = y.$$

Thus, integrating Eq. (4.8) partially w.r.t. y (treating x as a constant), its solution is

$$py = \int\{\cos(x+y) - y.\sin(x+y)\}.dy + \phi_1(x) = y.\cos(x+y) + \phi_1(x),$$

$\Rightarrow$

$$p = \partial z/\partial x = \cos(x+y) + (1/y).\phi_1(x).$$

Integrating it further partially w.r.t. x, there follows the complete integral of Eq. (4.8) with two arbitrary sets of fucntions $\phi_1(x) + \phi_2(y)$:

$$z = \sin(x+y) + (1/y).\int \phi_1(x)\, dx + \phi_2(y). \; //$$

Example 4.6. Solve the PDE $y\,t + 2q = (9y+6).\exp(2x+3y)$.

Solution. Re-writing the equation as a linear ODE in q (dependent) and y (independent) variables:

$$\partial q/\partial y + 2q/y = (9 + 6/y).\exp(2x+3y), \qquad (4.9)$$

its integrating factor is

$$\text{I.F.} = \int \exp(2/y)\, dy = \exp(2\ln y) = y^2.$$

Thus, integrating Eq. (4.9) partially w.r.t. y (treating x as a constant), its solution is

$$qy^2 = e^{2x}.\int \{ (3y^2 + 2y).\, 3e^{3y} \}.\, dy + \phi_1(x). \qquad (4.10)$$

Applying method of integration by parts repeatedly taking $3e^{3y}$ as the second function, above integral is evaluated:

$$(3y^2 + 2y).\, e^{3y} - \int\{ (6y+2).\, e^{3y} \}.dy$$

$$= (3y^2 + 2y).e^{3y} - (6y+2).e^{3y}/3 + \int (2e^{3y}).dy$$

$$= (3y^2 + 2y).e^{3y} - (6y + 2).e^{3y}/3 + (2/3).e^{3y} = 3y^2.e^{3y}.$$

Thus, on division by y^2, Eq. (4.10) reduces to

$$q = e^{2x}. (3e^{3y}) + (1/y^2). \phi_1 (x),$$

which, on further partial integration w.r.t. y gives the complete integral of the given PDE:

$$z = \exp(2x + 3y) - (1/y). \phi_1 (x) + \phi_2 (x). //$$

4.3. Third type: The equations

$$R r + S s + P p = G_1, \qquad S s + T t + Q q = G_2, \tag{4.11}$$

form the *third type* of above PDEs, where all the coefficients of derivatives and G's are constants. These equations can be alternately written as

$$\left. \begin{array}{l} R (\partial p/\partial x) + S (\partial p/\partial y) = G_1 - Pp, \\[2mm] S (\partial q/\partial x) + T (\partial q/\partial y) = G_2 - Qq, \end{array} \right\} \tag{4.12}$$

which can be viewed as linear PDEs in p (alternately q) as dependent and x, y as independent variables.

Example 4.7. Solve the PDE $x r + y s = 15x^2 y^2$.

Solution. We rewrite the equation as

$$\partial p/\partial x + (y/x) \partial p/\partial y = 15xy^2, \tag{4.13}$$

and view it as a linear PDE in derivatives of p and two independent variables x, y. Writing Lagrange subsidiary equations (2.4) for it:

$$\frac{dx}{1} = \frac{dy}{y/x} = \frac{dp}{15xy^2} \qquad \Rightarrow \qquad dy / y = dx / x.$$

Integrating it termwise we get one independent solution of PDE (4.13):

$$\ln y = \ln x + \ln a, \qquad \text{i.e.} \qquad y = ax, \tag{4.14}$$

involving an arbitrary constant a. Taking the first and third ratios in subsidary equations, we also derive

$$15a^2 x^3 = dp, \qquad\qquad \text{by Eq. (4.14).}$$

Integrating it termwise w.r.t. respective variables, the other solution is

$$p = 15a^2 x^4 / 4 + \text{const.} = 15x^2 y^2 / 4 + \text{const.}$$

A general solution of PDE (4.13) is constituted by aove two solutions:

$$p = \partial z/\partial x = 15x^2 y^2 + \phi_1 (y/x), \qquad\qquad \text{by Eq. (4.14).}$$

Integrating it partially w.r.t. x, we derive the complete integral of the given PDE:

$$z = 5x^3 y^2 + \int \phi_1 (y/x) \, dx + \phi_2 (y).$$

Putting $y/x = t$, so that $x = y/t$ and $dx = -(y/t^2) \, dt$, above integral reduces to

$$\int \phi_1 (y/x) \, dx = -y. \int \{ \phi_1 (t) / t^2 \} \, dt = y. \, \phi_3 (y/x), \qquad \text{say.}$$

Therefore,

$$z = 5x^3 y^2 + y. \, \phi_3 (y/x) + \phi_2 (y). \; //$$

Example 4.8. Solve the PDE $xy \, r + x^2 s - y p = x^3 e^y$.

Solution. We rewrite the equation as

$$x y \, (\partial p/\partial x) + x^2 (\partial p/\partial y) = y p + x^3 e^y, \qquad\qquad (4.15)$$

and view it as a linear PDE in derivatives of p and two independent variables x, y. Writing Lagrange subsidiary equations (2.4) for it:

$$\frac{dx}{xy} = \frac{dy}{x^2} = \frac{dp}{yp + x^3.e^y} \qquad\Rightarrow\qquad x \, dx - y \, dy = 0,$$

we find one independent solution of PDE (4.15):

$$x^2 - y^2 = a^2 \text{ (const.).} \qquad\qquad (4.16)$$

Taking the first and third ratios in subsidiary equations, a linear ODE in p (dependent) and x (independent) variable is obtained:

$$dp/dx - p/x = x^2. \, e^y / y. \qquad\qquad (4.17)$$

Its integrating factor is

$$\text{I.F.} = \int \exp(-1/x)\, dx = \exp(-\ln x) = 1/x.$$

Hence, Eq. (4.17) has solution

$$p/x = \int \{x/\sqrt{(x^2 - a^2)}\}.\exp\sqrt{(x^2 - a^2)}.\, dx + \text{const.}$$

$$= \exp\sqrt{(x^2 - a^2)} + \text{const.} \qquad \Rightarrow \qquad p = x.\, e^y + x.\, \phi_1\,(x^2 - y^2),$$

by Eq. (4.16). Integrating it partially w.r.t. x, we derive

$$z = x^2\, e^y /2 + \int x.\, \phi_1\,(x^2 - y^2)\, dx + \phi_2\,(y).$$

Setting $x^2 - y^2 = t$ so that $x\, dx = dt/2$, above integral reduces to

$$(1/2).\int \phi_1\,(t)\, dt = \phi_3\,(t) = \phi_3\,(x^2 - y^2),\ \text{say.}$$

Therefore, the complete integral of the PDE (4.15) is

$$z = x^2\, e^y /2 + \phi_3\,(x^2 - y^2) + \phi_2\,(y).\ //$$

Example 4.9. Solve the PDE $sy - 2xr - 2p = 6xy$.

Solution. We rewrite the equation as

$$y\,(\partial p/\partial y) - 2x\,(\partial p/\partial x) = 2p + 6x\,y, \qquad (4.18)$$

and view it as a linear PDE in derivatives of p and two independent variables x, y. Writing Lagrange subsidiary equations (2.4) for it:

$$\frac{dx}{-2x} = \frac{dy}{y} = \frac{dp}{2p + 6x.y} \qquad \Rightarrow \qquad dx/x + 2dy/y = 0, \qquad (4.19)$$

We derive the first independent solution of Eq. (4.18):

$$\ln x + 2\ln y = \ln a \qquad \Rightarrow \qquad x\,y^2 = a = \text{const.} \qquad (4.20)$$

Each of the ratios in the subsidiary equations also equals

$$\{-2y^3\, dx + 2y\,(p + xy)\, dy - y^2 dp\} / 0,$$

so there holds

$$-2y^3\, dx + 2y\,(p + xy)\, dy - y^2 dp = 0.$$

Introducing an additional zero term $-2xy^2\,dy + 2xy^2\,dy$ and arranging the terms above equation takes the form

$$2y\,(p + 2xy)\,dy - y^2\,(dp + 2x\,dy + 2y\,dx) = 0,$$

or,

$$2y\,(p + 2xy)\,dy - y^2\,.\,d\,(p + 2xy) = 0.$$

Putting $p + 2xy = t$ so that $d\,(p + 2xy) = dt$, and dividing by $-y^2\,(p + 2xy)$, above linear ODE in t and y takes the separated variables form:

$$dt\,/\,t - 2dy\,/\,y = 0.$$

Integrating it termwise w.r.t. respecyive variables, we get another solution

$$\ln t - 2\ln y = \text{const.}$$

$\Rightarrow$

$$t\,/\,y^2 = (p + 2xy)\,/\,y^2 = \text{const.},$$

or,

$$p + 2xy = y^2.\,\text{const.} = y^2.\,\phi_1\,(xy^2),\qquad \text{by Eq. (4.20).}$$

Inrtegrating it partially w.r.t. x, there follows

$$z = -x^2\,y + \int y^2.\,\phi_1\,(xy^2)\,dx + \phi_2\,(y).$$

Taking $x\,y^2 = t$ so that $y^2\,dx = dt$, above integral reduces to

$$\int \phi_1\,(t)\,dt = \phi_3\,(t) = \phi_3\,(xy^2),\qquad \text{say.}$$

Thus, the complete solution of given PDE with two sets of arbitrary functions is

$$z = -x^2\,y + \phi_3\,(xy^2) + \phi_2\,(y).\ //$$

§ 5. Second order linear PDEs in mechanics

In mechanics, a linear distance and the time parameter are denoted by variables x and t (especially for 1-dimensional motions). For higher dimensional motions the spatial coordinate x is augmented by y, z etc. Thus, taking u as the unknown (dependent) variable and x, y, z (special coordinates) as independent variables the following are some important *linear* PDEs of the *second* order in mechanics (wherein c is a constant and t is the time parameter):

Equation	1-dimensional	2-dimensional	3-dimensional
Heat	$c^2 \, \partial^2 u / \partial x^2$ $= \partial u / \partial t$		
Wave	$c^2 \, \partial^2 u / \partial x^2$ $= \partial^2 u / \partial t^2$	$c^2 \, (\partial^2 u / \partial x^2 +$ $\partial^2 u / \partial y^2)$ $= \partial^2 u / \partial t^2$	
Laplace		$\partial^2 u / \partial x^2 +$ $\partial^2 u / \partial y^2 = 0$	$\nabla^2 u \equiv \partial^2 u / \partial x^2$ $+ \partial^2 u / \partial y^2 +$ $\partial^2 u / \partial z^2 = 0$
Poisson		$\partial^2 u / \partial x^2 +$ $\partial^2 u / \partial y^2$ $= f(x, y)$	
Schrö-dinger's	$\partial^2 u / \partial x^2$ $+ u \, \partial u / \partial t = 0$		
Transport	$\partial u / \partial x +$ $\partial u / \partial t = 0$		
Burger's	$u \, (\partial u / \partial x) +$ $\partial u / \partial t = 0$		

A *solution* of a partial differential equation in some region $\Re$ of the space of the independent variable is a function which has all the partial derivatives appearing in the equation in some domain containing $\Re$ and satisfies the equation everywhere in $\Re$.

Note 5.1. We note that all $u = x^2 - y^2$, $u = e^x. \cos y$, $u = \ln (x^2 + y^2)$ which are entirely different from each other are solutions of two-dimensional Laplace equation.

The *unique solution* can be determined from the additional infor-mation given concerning the variables of the equation. We call the additional information as *initial conditions* or more generally the *boundary conditions* because they do not always refer to *zero* values of the independent variables.

Example 5.1. Solve the differential equation

$$\partial^2 u / \partial x^2 = 12 x^2 (t + 1) \tag{5.1}$$

given that

$$u(0) = \cos 2t, \quad \text{and} \quad (\partial u/\partial x)_{x=0} = \sin t. \tag{5.2}$$

Solution. We note that x and t are independent variables and u is a function of both x and t. Integrating the PDE (5.1) partially w.r.t. x, we get

$$\partial u / \partial x = 4 x^3 . (t + 1) + f(t), \tag{5.3}$$

where $f(t)$ is a function of t alone and is independent of x. Applying the second initial condition, we evaluate $f(t) = \sin t$. Integrating Eq. (5.3) again partially w.r.t. x, we obtain

$$u = x^4 . (t + 1) + x . \sin t + g(t),$$

where $g(t)$ is also independent of x. Now applying the first initial condition, we also evaluate $g(t) = \cos 2t$. Hence, the desired solution is

$$u = x^4 . (t + 1) + x . \sin t + \cos 2t. //$$

Example 5.2. Solve the differential equation

$$\partial^2 u / \partial x \, \partial y = \sin(x + y), \tag{5.4}$$

given that

$$u(0) = (y - 1)^2 \quad \text{and} \quad (\partial u/\partial x)_{y=0} = 1. \tag{5.5}$$

Solution. Here, x and y are independent variables and u is a function of both x and y. Integrating Eq. (5.4) partially w.r.t. y, we get

$$\partial u / \partial x = - \cos(x + y) + f(x), \tag{5.6a}$$

where $f(x)$ is independent of y. Applying the second initial condition, we evaluate $f(x) = 1 + \cos x$. Hence, Eq. (5.6a) reduces to

$$\partial u / \partial x = - \cos(x + y) + 1 + \cos x. \tag{5.6b}$$

Integrating it again partially w.r.t. x, we obtain

$$u = -\sin(x+y) + x + \sin x + g(y), \tag{5.7}$$

where $g(y)$ is independent of x. Now applying the first initial condition, we evaluate $g(y) = (y-1)^2 + \sin y$. Hence, Eq. (5.7) gives the desired solution

$$u = -\sin(x+y) + x + \sin x + (y-1)^2 + \sin y. \; //$$

§ 6. Homogeneous linear PDE

A *linear* PDE is *homogeneous* if each of its terms contains either u or one of its partial derivatives; if not, it is *non-homogeneous*. It may be noted that the 2-dimensional Poisson eqn. (with non-zero f) § 5 is non-homogeneous, whereas the other equations therein are homogeneous.

6.1. Separation of variables

For a homogeneous linear PDE, it is sometimes possible to find par-ticular solution in the form of a product of two independent functions. Let us consider a function $u(x, y)$ of two independent variables x and y in a product form:

$$u = X(x). \, Y(y), \tag{6.1}$$

where X is a function of x only and Y a function of y only. As imme-diate consequences of this we have

$$\left. \begin{array}{ll} \partial u / \partial x = X'(x).Y(y), & \partial u / \partial y = X(x).Y'(y), \\[2mm] \partial^2 u / \partial x^2 = X''(x).Y(y), & \partial^2 u / \partial y^2 = X(x).Y''(y), \end{array} \right\} \tag{6.2}$$

where the primes denote ordinary differentiation w.r.t. the respective variables. This method is applicable for the solution of partial differen-tial equations where variables are separable.

Example 6.1. Find the product solutions of the partial differential equation

$$\partial^2 u / \partial x^2 = 4 (\partial u / \partial y). \tag{6.3}$$

Solution. Let Eq. (6.1) provide a product solution then, in view of Eq. (6.2), Eq. (6.3) reduces to

$$X''.Y = 4\,X.Y' \quad \Rightarrow \quad X'' / 4X = Y' / Y. \tag{6.4}$$

Since the LHS of Eq. (6.4) is independent of y whereas the RHS is independent of x. But, the two sides being identically equal, we conclude that expressions on either sides must be constant, say k. Often, it is convenient to write this real constant as either λ^2 or $-\lambda^2$.

Case 2.1. Taking $k = \lambda^2 > 0$, Eq. (6.4) leads to

$$X'' - 4\lambda^2. X = 0 \qquad (6.5); \qquad Y' - \lambda^2. Y = 0. \qquad (6.6)$$

These equations have solutions (cf. Eq. (9.5.10b))

$$X = A_1. \cosh(2\lambda x) + B_1. \sinh(2\lambda x), \qquad (6.7)$$

(respectively)

$$Y = C_1. \exp(\lambda^2 y). \qquad (6.8)$$

Thus, a particular solution of Eq. (6.3), by Eq. (6.1), is

$$u = (A_1. \cosh 2\lambda x + B_1. \sinh 2\lambda x) \{C_1. \exp(\lambda^2 y)\}$$

$$= (A_2. \cosh 2\lambda x + B_2. \sinh 2\lambda x). \exp(\lambda^2 y), \qquad (6.9)$$

where we have put $A_2 = A_1. C_1$ and $B_2 = B_1.C_1$.

Case 6.2. Taking $k = -\lambda^2 < 0$, Eq. (6.4) leads to

$$X'' + 4\lambda^2. X = 0 \qquad (6.10); \qquad \text{and} \qquad Y' + \lambda^2. Y = 0. \quad (6.11)$$

These equations have solutions (cf. Eq. (9.5.8b))

$$X = A_3. \cos 2\lambda x + B_3. \sin 2\lambda x, \qquad (6.12)$$

(respectively)

$$Y = C_3. \exp(-\lambda^2 y). \qquad (6.13)$$

Thus, a particular solution of Eq. (6.3) is

$$u = (A_3. \cos 2\lambda x + B_3. \sin 2\lambda x) \{C_3.\exp(-\lambda^2 y)\}$$

$$= (A_4. \cos 2\lambda x + B_4. \sin 2\lambda x). \exp(-\lambda^2 y) \qquad (6.14)$$

where we have put $A_4 = A_3.C_3$ and $B_4 = B_3.C_3$.

Case 6.3. Taking $k = 0$, Eq. (6.4) leads to $X'' = 0$ and $Y' = 0$ having their solutions $X = A_5.x + B_5$ and $Y = C_5$ respectively. Thus, a particular solution of Eq. (6.3) is

$$u = (A_5.x + B_5). C_5 = A_6.x + B_6, \qquad (6.15)$$

where $A_6 = A_5.C_5$ and $B_6 = B_5.C_5$. //

Definition 6.1. (Superposition principle) If u_1, u_2, u_3, ... , u_n, are some solutions of a linear partial differential equation, then their linear combination

$$U = c_1. u_1 + c_2. u_2 + c_3. u_3 + ... + c_n. u_n , \qquad (6.16)$$

where c_i's, $i = 1, 2, 3, ... , n$, are constants, is also a solution.

Example 6.2. Solve the following partial differential equations:

(i) $\partial u / \partial x + y = 0,$ **(ii)** $\partial u / \partial x + u = e^y,$ **(iii)** $\partial^2 u / \partial x \, \partial y = 1.$

Solution. (i) Integrating the differential equation partially w.r.t. x, we get

$$u + x y = f(y),$$

where $f(y)$ is an arbitrary constant of integration with respect to x and is a function of y alone.

(ii) This is a linear (in the dependent variable u and its partial derivative $\partial u/\partial x$) partial differential equation of first order. The integrating factor is $e^{\int 1 \, dx} = e^x$. Therefore, its partial integration w.r.t. x (treating the other independent variable y constant) gives

$$u. e^x = e^y. \int e^x. \, dx + f(y) = e^y. e^x + f(y) \Rightarrow u = e^y + e^{-x}. f(y),$$

where $f(y)$ is independent of x.

(iii) Integrating the differential equation partially w.r.t. y (treating the other independent variable x constant), we get

$$\partial u / \partial x = y + f(x),$$

where $f(x)$ is an arbitrary constant of integration (with respect to y) and is a function of x alone. Integrating it further partially w.r.t. x (now treating y constant), we get

$$u = x\,y + \int f(x).dx + g\,(y),$$

where $g\,(y)$ is an arbitrary constant of integration w.r.t. x and is a function of y alone. //

§ 7. One-dimensional wave equation

Consider a perfectly flexible elastic string of (natural) length l stretched between two points O $(x = 0)$ and A $(x = l)$ with uniform tension T. Let the string be displaced slightly from its initial position of rest and released while the end points remain fixed. The string will then vibrate and the vibrations are governed by the one dimensional wave equation written as

$$c^2\,(\partial^2\,u\,/\,\partial x^2) = \partial^2\,u\,/\,\partial t^2. \qquad (7.1)$$

Assuming the initial conditions

$$x = 0, \qquad u\,(0,\,t) = 0, \qquad (7.2a)$$

$$x = l, \qquad u\,(l,\,t) = 0, \qquad (7.2b)$$

for all values of t and taking the initial deflection and initial velocity as the following

$$u\,(x,\,0) = f(x) \qquad (7.3); \qquad (\partial u\,/\,\partial t)_{t\,=\,0} = g\,(x), \qquad (7.4)$$

we need to find a solution of Eq. (7.1). This is carried out in the following three steps.

Step 1. Analogous to Eq. (6.1), let

$$u = X\,(x).\,T\,(t) \qquad (7.5)$$

be a trial solution of Eq. (7.1) so that, in analogy with Eq. (6.2), there hold the relations

$$\left.\begin{array}{l} u_x \equiv \partial u\,/\,\partial x = X'\,(x).\,T\,(t), \quad u_t \equiv \partial u\,/\,\partial t = X\,(x).\,T'\,(t) \\[2mm] u_{xx} \equiv \partial^2\,u/\,\partial x^2 = X''\,(x).T\,(t), \quad u_{tt} \equiv \partial^2\,u/\,\partial t^2 = X\,(x).T''\,(t) \end{array}\right\} \quad (7.6)$$

reducing Eq. (7.1) to the form

$$c^2 X''.T = X.T'' \qquad \Rightarrow \qquad X''/X = T''/c^2 T. \quad (7.7)$$

The expressions on the two sides of Eq. (7.7) are functions of independ-ent) variables x and t respectively so either of them must be constant, say k. Thus, the equations (7.7) lead to

$$X'' - k. X = 0 \quad (7.8); \quad \text{and} \qquad T'' - c^2.k.T = 0. \qquad (7.9)$$

Step 2. We now seek the solutions of these differential equations so that Eq. (7.5) satisfies the initial conditions given by Eq. (7.2) for every value of t :

$$u(0, t) = X(0).T(t) = 0, \qquad (7.10a)$$

$$u(l, t) = X(l). T(t) = 0. \qquad (7.10b)$$

We discuss both vanishing and non-vanishing choices of the function T: $T = 0$ gives a trivial solution of Eq. (7.1): $u \equiv X.T = 0$, which is of no interest to us. Also, when $T \neq 0$, from Eq. (7.10), there result

$$X(0) = 0, \text{ as well as } X(l) = 0. \qquad (7.11)$$

In the following, we will discuss various possibilities for the constant k.

(i) If $k = 0$, the ODE (7.8) has a general solution $X = a.x + b$; which, for the initial conditions given by Eq. (7.11), determines

$$0 = a.0 + b \Rightarrow b = 0; \text{ and } 0 = X(l) = a.l + 0 \Rightarrow a = 0, \text{ as } l \neq 0.$$

Hence, the solution of Eq. (7.8) becomes $X = 0 \Rightarrow u \equiv X. T = 0$ (as before for $T = 0$).

(ii) If $k > 0$, say ρ^2, then Eq. (7.8) assumes the form $X'' - \rho^2 X = 0$ having a solution

$$X = A e^{\rho x} + B e^{-\rho x}. \qquad (7.12)$$

Applying the initial conditions given by Eq. (7.11), we evaluate the con-stants A and B:

$$X(0) = 0 = A + B \qquad \text{and} \qquad X(l) = 0 = A e^{\rho l} + B e^{-\rho l}.$$

Eliminating B from these we get

$$A.(e^{\rho l} - e^{-\rho l}) = 0 \quad \Rightarrow \quad A = 0, \text{ for } e^{\rho l} - e^{-\rho l} \neq 0.$$

Hence, $B = -A = 0$. Thus, Eq. (7.12) again yields $X = 0 \Rightarrow u = 0$ as before.

(iii) When $k < 0$, say $-\rho^2$, then Eq. (7.8) reduces to $X'' + \rho^2 X = 0$ having a general solution

$$X = A \cos \rho x + B \sin \rho x. \tag{7.13}$$

Applying the initial conditions given by Eq. (7.11), we derive $X(0) = 0 = A$, and $X(l) = B \sin \rho l = 0$. The latter relation gives two alternatives: either $B = 0$, or

$$\sin \rho l = 0. \tag{7.14}$$

The choice $B = 0$ (together with $A = 0$ as seen above) again reduces Eq. (7.13) to $X = 0$ implying $u = 0$, which is trivial. But, the latter alternative Eq. (7.14) yields $\rho l = 0$, or $n\pi$ (n being an integer). Since $l \neq 0$ and ρ (the mass per unit length of the string) too cannot be zero leaving the possibility

$$\rho = n\pi / l. \tag{7.15}$$

Setting $B = 1$ (for simplicity), we obtain infinitely large number of solutions of Eq. (7.8):

$$X \equiv X_n(x) = \sin(n\pi x / l), \, n = 1, 2, 3, \ldots \tag{7.16}$$

For Eq. (7.15), $k \equiv -\rho^2 = -(n\pi / l)^2$, the equation (7.9) takes the form

$$T'' + \lambda_n^2 .T = 0 \qquad (7.17); \quad \text{where} \qquad \lambda_n \equiv c\rho = cn\pi / l. \tag{7.18}$$

A general solution of Eq. (7.17) is

$$T_n(t) = A_n . \cos(\lambda_n t) + B_n . \sin(\lambda_n t). \tag{7.19}$$

Thus, Eqs. (7.16) and (7.19) determine solutions of Eq. (7.1):

$$u_n(x, t) \equiv X_n(x).T_n(t) = \{A_n.\cos(\lambda_n t) + B_n.\sin(\lambda_n t)\} \sin(n\pi x / l). \tag{7.20}$$

Definition 7.1. Above solutions of the wave equation (7.1) are called the *eigen functions* and the values λ_n given by Eq. (7.18) as the *eigen values* of the vibrating string.

Step 3. From Eq. (7.20), we derive

$$u_n(x, 0) = A_n. \sin(n\pi x / l) = f(x), \qquad (7.21)$$

for Eq. (7.3); and

$$\partial u_n / \partial t = \{-A_n. \sin(\lambda_n t) + B_n. \cos(\lambda_n t)\}.\lambda_n. \sin(n\pi x / l)$$

$$\Rightarrow \qquad (\partial u_n / \partial t)_{t=0} = B_n \lambda_n \sin(n\pi x / l) = g(x), \qquad (7.22)$$

for Eq. (7.4). Comparing Eqs. (7.21) and (7.22), we note that the functions $f(x)$ and $g(x)$ are not independent and are connected by a linear relation

$$(B_n \lambda_n / A_n) f(x) = g(x).$$

As Eq. (7.1) is linear and homogeneous we cannot accept Eq.(7.20) as a solution. Instead, we consider the infinite series

$$u(x, t) \equiv \sum_{n=1}^{\infty} u_n(x, t)$$

$$= \sum_{n=1}^{\infty} \{A_n. \cos(\lambda_n t) + B_n. \sin(\lambda_n t)\} \sin(n\pi x / l). \qquad (7.23)$$

At $t = 0$, we, therefore, have

$$u(x, 0) \equiv f(x) = \sum_{n=1}^{\infty} \{A_n. \sin(n\pi x / l). \qquad (7.24)$$

This gives an expansion of $f(x)$ in a *sine* series. Hence, in view of Eq. (13.5.16), we get

$$A_n = (2/l) \int_{x=0}^{l} f(x) \sin(n\pi x / l), dx, n = 1, 2, 3, \ldots \qquad (7.25)$$

Also, in view of Eq. (7.22), we have

$$(\partial u / \partial t)_{t=0} \equiv \sum_{n=1}^{\infty} (\partial u_n / \partial t)_{t=0} = g(x) = \sum_{n=1}^{\infty} B_n \lambda_n \sin(n\pi x / l). \qquad (7.26)$$

This, in analogy with Eq. (7.25), determines

$$B_n \lambda_n = (2/l) \int_{x=0}^{l} g(x). \sin(n\pi x / l). dx,$$

or, for Eq. (7.18),

$$B_n = (2 / cn\pi) \int_{x=0}^{l} g(x) \sin(n\pi x / l). \, dx, \quad n = 1, 2, 3, \ldots \quad (7.27)$$

Thus, Eq. (7.23) gives a solution of the wave equation (7.1) where the constants A_n, B_n are determined by Eqs. (7.25) and (7.27).

7.1. Particular case: When the string is released from rest $g(x) = 0$ for every x in the interval $0 \le x \le l$, it follows from Eq. (7.27) that $B_n = 0$ reducing the solution (7.23) to

$$u(x, t) = \sum_{n=1}^{\infty} A_n. \cos(\lambda_n t). \sin(n\pi x / l). \quad (7.28)$$

Example 7.1. Let a vibrating string of length 30 cms. satisfies the wave equation

$$4 (\partial^2 u / \partial x^2) = \partial^2 u / \partial t^2, \quad 0 < x < 30, \ 0 < t. \quad (7.29)$$

Let the ends of the string be fixed and it is set in motion with zero initial velocity from the initial position:

$$u(x, 0) = f(x) = \begin{cases} x/10, & 0 \le x \le 10, \\ (30 - x)/20, & 10 \le x \le 30. \end{cases} \quad (7.30)$$

Find the displacement $u(x, t)$ of the string.

Solution. A solution of the wave equation is given by equation (7.28). For $c = 2$, $l = 30$ determining $\lambda_n = cn\pi/l = 2n\pi/30 = n\pi/15$, the equation (7.28) becomes

$$u(x, t) = \sum_{n=1}^{\infty} A_n \cos(n\pi t /15). \sin(n\pi x /30). \quad (7.31)$$

Also, Eq. (7.25) determines A_n:

$$A_n = (2/30). \int_{x=0}^{30} f(x). \sin(n\pi x / 30). \, dx$$

$$= (1/15) \{ \int_{x=0}^{10} (x/10). \sin(n\pi x/30). \, dx$$

$$+ \int_{x=10}^{30} \{(30 - x) / 20\}. \sin(n\pi x/30). \, dx \}$$

$$= (1/150) \int_{x=0}^{10} x. \sin (n\pi x/30). \, dx + (1/10) \int_{x=10}^{30} \sin (n\pi x/30). dx$$

$$- (1/300) \int_{x=10}^{30} x. \sin (n\pi x/30). dx$$

$$= (1/5n\pi) \left\{ \left[-x.\cos (n\pi x/30) \right]_0^{10} + \int_{x=0}^{10} \cos (n\pi x / 30). \, dx \right\}$$

$$+ (3/n\pi) \left[-\cos (n\pi x/30) \right]_{10}^{30}$$

$$+ (1/10n\pi) \left\{ \left[x.\cos (n\pi x/30) \right]_{10}^{30} - \int_{x=10}^{30} \cos (n\pi x / 30). \, dx \right\}$$

$$= (- 2/n\pi) \cos (n\pi / 3) + (6/n^2\pi^2) \left[\sin (n\pi x/30) \right]_0^{10}$$

$$+ (3/n\pi)\{ - \cos n\pi + \cos (n\pi/3)\} + (1/n\pi)\{3 \cos n\pi - \cos (n\pi/3)\}$$

$$- (3/n^2\pi^2) \left[\sin (n\pi x/30) \right]_{10}^{30}$$

$$= (6/n^2 \pi^2) \sin (n\pi/3) - (3/n^2\pi^2) \{\sin n\pi - \sin (n\pi/3)\}$$

$$= (9/n^2 \pi^2) \sin (n\pi/3).//$$

§ 8. Second order PDEs in n independent variables

Analogous to Sub-section 1.2, we now consider a PDE involving n independent variables $x_1, x_2, \ldots, x_n$, an unknown variable z (depending upon x_i's, $i = 1, 2, \ldots, n$), and partial derivatives of z up to second order. The first order derivatives of z w.r.t. x_i's are defined by Eqs. (1.20). In continuation to the same notation, let the second order partial derivatives of z w.r.t. x_i's be denoted by

$$p_{ij} = \partial p_i / \partial x_j = \partial^2 z / \partial x_j \partial x_j, \qquad i, j = 1, 2, \ldots, n. \tag{8.1}$$

A second order PDEs of general form is then expressed by

$$F (x_1, x_2, \ldots, x_n, z; p_1, p_2, \ldots, p_n; p_{11}, p_{12}, \ldots, p_{1n};$$

$$p_{21}, p_{22}, \ldots, p_{2n}; \ldots; p_{n1}, p_{n2}, \ldots, p_{nn}) = 0. \tag{8.2}$$

A solution to the first order PDE is a function $z (x_1, x_2, \ldots, x_n)$ satisfying the Eq. (8.2). If the function F is a linear sum of z and its derivatives, the PDE will becalled *linear*. In that case it may be written as

$$\sum_{i,j=1}^{n} R_{ij}\, p_{ij} + \sum_{i=1}^{n} P_i\, p_i + Zz = G, \qquad (8.3)$$

where all the coefficients R_{ij}, P_i, Z and G are functions of x_i's only.

8.1. Classification of linear PDEs of second order

Let $\lambda_1, \lambda_2, \ldots, \lambda_n$ be the eigen values of the coefficient matrix $((R_{ij}))$.

(i) The PDE is called *elliptic* if all the λ_i's are either positive or all negative.

(ii) The PDE is called *parabolic* if all the λ_i's are either positive or negative or zero.

(iii) The PDE is called *hyperbolic* if only one of λ_i's is negative leaving the rest all positive; or only one of λ_i's is positive and all the rest negative.

(iv) The PDE is called *ultrahyperbolic* if more than one of λ_i's is positive and more than one of them negative, and also none of them are zero.

8.2. Particularly, a second order linear PDE involving only two independent variables x, y and dependent (unknown) variable z, so that its second order partial derivatives are equal, is given by Eq. (4.4). Following conclusion of Sub-section 8.1, the PDE is classified as

(i)   If $S^2 - 4RT < 0$ the PDE is *elliptic*,

(ii) If $S^2 - 4RT = 0$ the PDE is *parabolic*,

(iii) If $S^2 - 4RT > 0$ the PDE is *hyperbolic*.

CHAPTER 11

DIFFUSION EQUATION

§ 1. The diffusion equation

It is a partial differential equation (PDE). In physics, it describes the behavior of the collective motion of micro–particles in a material resulting from the random movement of each micro–particle. In mathematics, it is applicable in common to a subject relevant to the Markov process as well as in various other fields, such as the material sciences, information science, life science, social science, and so on. These subjects described by the diffusion equation are generally called *Brown problems*. The equation is usually written as:

$$\partial \varphi(\mathbf{r}, t) / \partial t = \nabla \cdot \{ D(\phi, \mathbf{r}) \nabla \varphi(\mathbf{r}, t) \},$$

where $\phi(\mathbf{r}, t)$ is the density of the diffusing material at location $\mathbf{r}$ and time t, $D(\phi, \mathbf{r})$ is the collective diffusion coefficient for density ϕ at location $\mathbf{r}$, and ∇ represents the vector differential operator. If the diffusion coefficient depends on the density then the equation is nonlinear, otherwise it is linear.

More generally, when D (not to be confused with derivation) is a symmetric positive definite matrix, the equation describes aniso-tropic diffusion, which is written (for 3-dimensional case) as:

$$\partial \varphi(\mathbf{r}, t) / \partial t = \sum_{i,j=1}^{n} \frac{\partial}{\partial x^i} \left\{ D_{ij}(\varphi, \mathbf{r}) \frac{\partial \varphi(\mathbf{r}, t)}{\partial x^j} \right\}.$$

If D is constant, then the equation reduces to the following linear differential equation:

$$\partial \varphi(\mathbf{r}, t) / \partial t = D \nabla^2 \varphi(\mathbf{r}, t)$$

also called the *heat equation*.

§ 2. Heat equation

It is also an important PDE governing the temperature u in a body in space. We obtain this model of temperature distribution under the following assumptions.

Physical assumptions

(i)   The *specific heat* σ and the *density* ρ of the material of the body are constant. No heat is produced or it disappears in the body.

(ii)   Experiments show that, in a body, heat flows in the direction of decreasing temperature, and the rate of flow is proportional to the gradient of the temperature; that is, the velocity **v** of the heat flow in the body is of the form

$$\mathbf{v} = -K \operatorname{grad} u, \tag{2.1}$$

where $u(x, y, z, t)$ is the temperature at a point and time t.

(iii) The *thermal conductivity K* is constant, as is the case for homogeneous material and non-extreme temperatures.

Under these assumptions we can model heat flow as follows.

Let R be a region in the body bounded by a surface S with outer unit normal vector **n** such that the (Gauss) divergence theorem applies. Then **v . n** is the component of **v** in the direction of **n**. Hence, $|\mathbf{v} . \mathbf{n}\, \Delta A|$ is the amount of heat *leaving R* (if **v . n** > 0 at some point P) or *entering R* (if **v . n** < 0 at P) per unit time at some point P of S through a small portion ΔS of S of area ΔA. Hence the total amount of heat that flows across S from R is given by the surface integral $\iint\limits_{S} \mathbf{v} . \mathbf{n}\, dA$. Using Gauss divergence theorem, we now convert our surface integral into a volume integral over the region R. Because of (2.1) this gives

$$\iint\limits_{S} \mathbf{v.n}\, dA = -K \iint\limits_{S} (\operatorname{grad} u) . \mathbf{n}\, dA$$

$$= -K \iiint\limits_{R} \operatorname{div} (\operatorname{grad} u)\, dx\, dy\, dz = -K \iiint\limits_{R} (\nabla^2 u)\, dx\, dy\, dz. \tag{2.2}$$

On the other hand, the total amount of heat in R can be written as

$$H = \iiint\limits_{R} (\sigma\rho\, u)\, dx\, dy\, dz$$

with σ and ρ as before. Hence the time rate of decrease of H is

$$-\partial H / \partial t = -\iiint\limits_{R} (\sigma\rho)\,(\partial u / \partial t)\,dx\,dy\,dz\,.$$

This must be equal to the amount of heat leaving R because no heat is produced or disappears in the body. From Eq. (2.2), we thus obtain

$$-\partial H / \partial t = -\iiint\limits_{R} (\sigma\rho)\,(\partial u / \partial t)\,dx\,dy\,dz = -K\iiint\limits_{R} (\nabla^2 u)\,dx\,dy\,dz\,,$$

or (dividing by $-\sigma\rho$ and putting $c^2 = K/\sigma\rho$)

$$\iiint\limits_{R} (\partial u / \partial t - c^2\,(\nabla^2 u)\,dx\,dy\,dz = 0.$$

Since this holds for any region R in the body, the integrand (if continuous) must be zero everywhere. That is,

$$\partial u / \partial t = c^2\,(\nabla^2 u).\qquad(2.3)$$

This is the **heat equation** (the fundamental PDE) modeling heat flow. It gives the temperature $u\,(x, y, z, t)$ in a body of homogeneous material in space. The constant c^2 is the *thermal diffusivity*. K is the *thermal conductivity*, σ the *specific heat*, and ρ the *density* of the material of the body. $\nabla^2 u$ is the Laplacian of u and, with respect to the Cartesian coordinates x, y, z. The heat equation is also called the **diffusion equation** because it also models chemical diffusion processes of one substance or gas into another.

CHAPTER 12

DYNAMICS

§ 1. Motion of a particle in a straight line

A particle, by definition, is a body of negligible size. So, it can be regarded as located at a point. The mass of the particle is the quantity of matter that it contains and it is a scalar quantity.

There are three *laws of motion* in Newtonian mechanics:

Law 1: Every particle continues in a state of rest or of motion with constant speed in a straight line unless its is acted upon by some external force(s).

Indeed, this law defines a force:

Definition 1.1. The quantity changing (or tending to change) the state of rest or the uniform motion of a particle is called a *force*.

The second law of motion introduces the concept of the *momentum* of a particle (defined as the product of mass and velocity of the particle). The mass being scalar and velocity a vector, the moment is also a vector.

Law 2: The rate of change of momentum of a particle is proportional to the force acting on a particle and it has the same direction as the force:

$$\mathbf{F} = d\,(m\mathbf{v})\,/\,dt\ =\ m\,(d\mathbf{v}\,/\,dt)\ =\ m\mathbf{f}. \qquad (1.1)$$

Note 1.1. If $m = 1$ kg. and $\mathbf{f}$ has magnitude of 1 metre/sec.^{2} then the magnitude of force $\mathbf{F}$ is $1 \times 1 = 1$ Newton (abbreviated as 1 N).

Note 1.2. In the present context, only the particles of constant mass are considered. (Notably, a rocket is an example of a *varying mass*.)

§ 2. Uniform acceleration

Let a particle start moving from a fixed point O along the straight line OA. Let P and Q be two positions described by the particle in time intervals (measured from O) t and $t + \delta t$ such that OP $= s$ and OQ $= s +$

δs. Thus, the particle is displaced through the distance δs in time δt. The velocity of the particle at P is defined as the rate of displacement per unit time. So, it is given by

$$v \equiv \lim_{\delta t \to 0} (\delta s / \delta t) \equiv ds / dt = \dot{s}. \qquad (2.1)$$

Further, let it attain velocity $v + \delta v$ at Q so that the change in velocity from P to Q is δv. The acceleration of the particle at P is defined as the rate of change of velocity per unit time. Thus, it is given by

$$f(t) \equiv \lim_{\delta t \to 0} (\delta v / \delta t) \equiv dv / dt = \dot{v} \equiv \ddot{s}. \qquad (2.2)$$

Applying the chain rule of differentiation dv / dt can be written as

$$dv / dt = (dv / ds)(ds / dt) = (dv / ds) v = (1/2)(dv^2 / ds).$$

Accordingly, Eq. (2.2) may be rewritten as

$$f(t) = v(dv / ds) = (1/2)(dv^2 / ds). \qquad (2.3)$$

When f is constant, Eq. (2.2) on integration with respect to t, yields

$$ft + c_1 = v(t), \qquad (2.4)$$

where c_1 is an arbitrary constant of integration. If the particle started from O (where $t = 0$) with initial velocity u, Eq. (2.4) determines $c_1 = u$. Accordingly, Eq. (2.4) assumes the form

$$ds / dt = v = u + ft. \qquad (2.5)$$

Integrating it further with respect to t, one obtains

$$s = ut + (1/2)ft^2 + c_2,$$

c_2 being also a constant of integration. Applying initial conditions: $t = 0$, $s = 0$ at O above relation determines $c_2 = 0$. Hence, we have

$$s = ut + (1/2)ft^2. \qquad (2.6)$$

Finally, integrating Eq. (2.3) with respect to s, we also derive

$$\int f \, ds + c_3 = \int v(dv / ds) \, ds = \int v \, dv,$$

i.e.

$$f s + c_3 = (1/2) v^2. \tag{2.7}$$

For $s = 0$ and $v = u$ at O, above relation determines $c_3 = u^2 / 2$. Therefore, Eq. (2.7) reduces to

$$v^2 = u^2 + 2 f s. \tag{2.8}$$

The relations (2.5) and (2.8) also yield another useful relation

$$v + u \equiv (v^2 - u^2) / (v - u) = 2 f s / f t = 2 s / t$$

$$\Rightarrow$$

$$s = (u + v) t / 2. \tag{2.9}$$

Note 2.1. The Eqs. (2.5), (2.6), (2.8) and (2.9) are called *equations of motion* of a particle moving along a straight line with constant (or uniform) acceleration. The study of such motions is called *kinematics*.

Theorem 2.1. The distance described by the particle in t^{th} second is

$$s_t = u + f(t - 1/2). \tag{2.10}$$

Proof. Applying the formula (2.6), we have

$$s_t \equiv s_t - s_{t-1} = u t + f t^2 / 2 - \{u (t - 1) + f (t - 1)^2 / 2\} = u + f .(2t - 1)/2,$$

which is same as in Eq. (2.10). //

Example 2.1. A car travelling with constant acceleration along a straight road moves through a distance of 160 metres as the velocity increases from 10 m./sec. to 30 m./sec. Find the acceleration and time taken.

Solution. Given $u = 10$ m./sec., $v = 30$ m./sec. and $s = 160$ m., the formula (2.8) determines

$$(30)^2 - (10)^2 = 2 f \times 160 \quad \Rightarrow \quad f = 20 \times 40 / 2 \times 160 = 2 \cdot 5 \text{ m./sec.}^2$$

Hence, Eq. (2.5) yields

$$30 - 10 = (2 \cdot 5) t \quad \Rightarrow \quad t = 20/2 \cdot 5 = 8 \text{ sec. //}$$

Example 2.2. A particle moving in a straight line with constant acceleration moves 26 metres in the 4th second and 30 m. in the 5th second of its motion. Find (i) its acceleration, (ii) initial velocity, and (iii) the distance travelled in the 6th second.

Solution. Applying Eq. (2.10) we get

$$s_4 = u + f(4 - 1/2) = u + (7/2)f = 26,$$

and

$$s_5 = u + f(5 - 1/2) = u + (9/2)f = 30.$$

Solving these simultaneous linear equations for u and f we obtain

$$f = 4 \text{ m./sec.}^2 \quad \text{and} \quad u = 12 \text{ m./sec.}$$

Applying Eq. (2.10) again we, therefore, have

$$s_6 = u + f(6 - 1/2) = 12 + 4(11/2) = 34 \text{ m.} \; //$$

§ 3. Vertical motion under gravity

3.1. Downward motion: Let a particle be dropped
from a point O with velocity zero falling vertically
downward under gravity $g = 9.8$ m./sec.^{2}. If it reaches

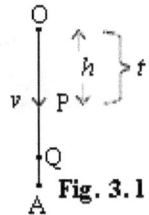

Fig. 3.1

some position P (OP $= h$ metres) in time t secs. and acquires the velocity
v m./sec. there, the equations of motion at P are

$$v = 0 + gt = gt, \quad h = 0.t + (1/2) gt^2 = (1/2) gt^2, \left.\vphantom{\begin{array}{c} a \\ b \end{array}}\right\} \quad (3.1)$$

$$v^2 = 0^2 + 2gh = 2gh.$$

The first and the third of these relations also yield

$$v^2 / v = 2gh / gt \implies v = 2h / t \implies h = vt / 2. \quad (3.2)$$

The distance fallen in t^{th} second is given by Eq. (2.10):

$$h_t = g(t - 1/2). \quad (3.3)$$

Example 3.1. A stone falls freely from the top of a tower of height
45 metres. Find the time taken to reach the ground and the speed with
which the stone hits the ground if it is:

(i) released from rest at the top of the tower;

(ii) thrown vertically downwards from the top of the tower with speed 20 m./ sec. (Take $g = 10$ m./sec.^{2})

Solution. (i) Given $u = 0$ and $h = 45$ m., the relations (3.1) determine

$$v^2 = 2 \times 10 \times 45 \quad \Rightarrow \quad v = 30 \text{ m./sec. and} \quad t = v/g = 30/10 = 3 \text{ secs.}$$

(ii) The relations (3.1) get modified as

$$v = u + gt, \qquad h = ut + (1/2) gt^2, \quad v^2 = u^2 + 2gh; \tag{3.4}$$

which, for the present data: $u = 20$ m./sec and $h = 45$ m., determine

$$v^2 = (20)^2 + 2 \times 10 \times 45 = 1300 \quad \Rightarrow \quad v = 10\sqrt{(13)} \text{ m./sec.}$$

and

$$t = (v - u) / g = \{10 \sqrt{(13)} - 20\}/10 = \sqrt{(13)} - 2 \approx 1\cdot 6 \text{ sec. } //$$

3.2. Upward motion (*against gravity*): Let a particle be projected vertically upwards from a point O with initial velocity u m./sec. and attains a height h m. (above O) in t secs. and acquires the velocity v m./sec. there. The gravity g always acting vertically downwards opposes the particle's (upward) motion. As such the gravity retards the vertical motion. Hence, the equations of motion in this case are

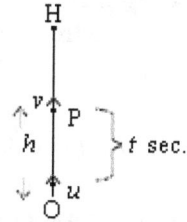

Fig. 3.2

$$v = u - gt, \qquad h = ut - (1/2) gt^2, \qquad v^2 = u^2 - 2gh. \tag{3.5}$$

The first and the last of above relations also yield

$$(u^2 - v^2) / (u - v) = 2gh / gt \quad \Rightarrow \quad u + v = 2h / t;$$

i.e.

$$h = (u + v) t / 2, \tag{3.6}$$

in analogy with Eq. (2.9).

Retardation caused by the gravity ultimately diminishes the velocity to zero. Let H be the highest point reached by the particle (where velocity becomes zero) in time t_H (measured from O). The equations of motion at H are

$$0 = u - g t_H, \qquad h_H = u t_H - (1/2) g (t_H)^2, \qquad 0^2 = u^2 - 2g h_H.$$

$\Rightarrow$

$$t_H = u/g \quad \text{and} \quad h_H = u^2/2g. \tag{3.7}$$

After reaching the highest point the particle will start falling vertically downwards (with initial velocity zero). It will return to the point of projection O after falling through the vertical height h_H given by Eq. (3.7). The time of falling from H to O can be calculated from the second of Eqs. (3.1):

$$(t_O)^2 = 2h_H/g = u^2/g^2, \qquad \text{by Eq.(3.7).}$$

This determines

$$t_O = u/g = t_H, \tag{3.8}$$

by Eq. (3.7). Thus, we have:

Note 3.1. The time of ascent from O to the highest point H equals the time of descent from H to O. Hence, the time of flight from the point of projection O to the highest point H and subsequently coming back to O is $2u/g$ secs.

Further, the velocity with which the particle returns back to O can be measured from the first of Eqs. (3.1) and (3.8):

$$v_O = g\, t_O = u, \tag{3.9}$$

equalling the velocity of projection.

§ 4. Kinematics in two dimensions

Let a particle move along a given curve C in a plane. Let it start from a fixed point A (on C) at time $t = 0$ and describe two neighbouring positions P and Q in time t and $t + \delta t$ respectively. Taking OA as the initial line, O as origin the polar coordinates of P, Q be (r, θ) and $(r + \delta r, \theta + \delta\theta)$ respectively. The length OP = r is called the *radius vector* of P and the angle $\angle AOP = \theta$ as the *vectorial angle* of P. The particle describes the infinitesimal arc-length PQ = δs in time δt. In limiting case, when Q $\rightarrow$ P, the length of arc PQ equals to that of chord PQ and the direction of chord PQ becomes tangential to the curve. Thus, the velocity of the particle at P is given by Eq. (2.1), is which is directed towards the tangent to the curve at P, hence called the *tangential velocity*. Its resolved parts

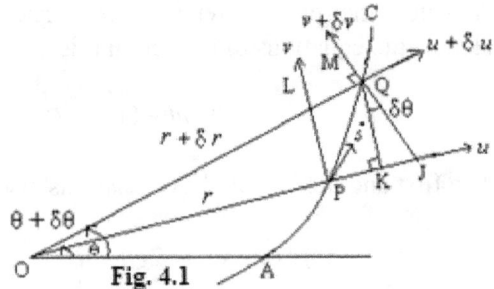

Fig. 4.1

along and perpendicular to the radius vector are called the *radial* and *transverse* velocities respectively.

Theorem 4.1. The radial and transverse velocities of a particle at a point P (r, θ) are:

$$\text{Rad. Vel.} = \dot{r} \quad (4.1); \quad \text{Trans.Vel.} = r\dot{\theta} \quad (4.2)$$

Corollary 4.1. The tangential velocity and its components along and perpendicular to the radius vector are related by

$$\dot{s}^2 = \dot{r}^2 + (r\dot{\theta})^2, \tag{4.3a}$$

or, equivalently

$$(\delta s)^2 = (\delta r)^2 + r^2 (\delta\theta)^2. \tag{4.3b}$$

§ 5. Radial and transverse accelerations

Let the two components of velocity: along and perpendicular to the radius vectors OP and OQ be denoted as:

At P: Rad.Vel. $= u$, Trans.Vel. $= v$;

At Q: Rad.Vel. $= u + \delta u$, Trans.Vel. $= v + \delta v$.

While the particle moves from P to Q, the change in radial velocity (cf. Fig. 4.1) is

$$(u + \delta u) \cos \delta\theta - (v + \delta v) \sin \delta\theta - u.$$

$\delta\theta$ being small, $\cos \delta\theta \rightarrow 1$, $\sin \delta\theta \rightarrow \delta\theta$, and dropping the smaller quantity $\delta u \, \delta\theta$, the above change tends to

$$u + \delta u - v \, \delta\theta - u = \delta u - v \, \delta\theta.$$

Therefore, the radial acceleration, by Eq. (2.2), is

$$\lim_{\delta t \to 0} (\delta u - v \, \delta\theta) / \delta t = du / dt - v \, d\theta / dt = d\dot{r} / dt - (r\dot{\theta})\dot{\theta},$$

by Eqs. (4.1) and (4.2). Thus, we have

$$\text{Rad. Acc.} = \ddot{r} - r\dot{\theta}^2. \tag{5.1}$$

On the other hand, the change in transverse velocity (cf. Fig. 4.1) is

$$(v + \delta v) \cos \delta\theta + (u + \delta u) \sin \delta\theta - v,$$

which, for reasons stated above, tends to $v + \delta v + u\,\delta\theta - v = \delta v + u\,\delta\theta$. Therefore, the transverse acceleration, by Eq. (2.2), is

$$\lim_{\delta t \to 0} (\delta v + u\,\delta\theta)\,/\,\delta t \;=\; dv\,/\,dt + u\,d\theta\,/\,dt = \; d\,(r\dot\theta)\,/\,dt + \dot r\,\dot\theta,$$

by Eqs. (4.1) and (4.2). The same also simplifies to

$$\text{Trans. Acc.} \;=\; 2\dot r\,\dot\theta + r\ddot\theta \;=\; (1/r)\,d\,(r^2\,\dot\theta)\,/\,dt. \tag{5.2}$$

§ 6. Tangential and normal components of velocity and acceleration

It is seen in § 4, that the velocity of the particle at any point P is wholly directed towards the tangent to the curve. Hence, there is no component of velocity in a direction (called the *normal* to the curve) perpendicular to the tangent. As in § 5, let $v\,(=\dot s)$ and $v + \delta v$ be the tangential velocities of the particle at two neighbouring positions P and Q and the time elapsing from P to Q be δt. If these velocities are inclined at an angle $\delta\psi$, change in the tangential velocity is

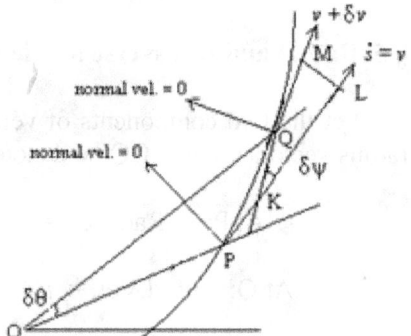

Fig. 6.1

$$(v + \delta v) \cos \delta\psi - v \to v + \delta v - v = \delta v.$$

Therefore, by Eq. (2.2), the tangential acceleration is

$$\text{Tang. Acc.} = \lim_{\delta t \to 0} (\delta v\,/\,\delta t) \;=\; dv\,/\,dt \;=\; d\,\dot s\,/\,dt \equiv \ddot s. \tag{6.1}$$

Similarly, the change in the normal velocity is

$$(v + \delta v) \sin \delta\psi - 0 \qquad \to \qquad v\,\delta\psi.$$

Therefore, by Eq. (2.2), the normal acceleration is

$$\text{Norm. Acc.} = \lim_{\delta t \to 0} (v\,\delta\psi\,/\,\delta t) \;=\; v\,d\psi\,/\,dt \;=\; (v\,d\psi\,/\,ds)\,(ds\,/\,dt)$$

$$= (v / \rho) v = v^2 / \rho = \dot{s}^2 / \rho, \tag{6.2}$$

where $\rho = ds/d\psi$ is the radius of curvature of the curve at P.

§ 7. Motion along a (horizontal) circle

In a circle the radius being perpendicular to the tangent the normal direction coincides with the radial direction. Also, the transverse direction coincides with the tangential direction. Further, the radius of a circle being constant (say *a*):

$$r = a \implies \dot{r} = \ddot{r} = 0. \tag{7.1}$$

Accordingly, Eqs. (4.1) and (4.2) reduce to

$$\text{Rad. Vel.} = \text{Norm.Vel.} = \dot{r} = 0, \tag{7.2}$$

$$\text{Tang. Vel.} = \text{Trans. Vel.} = \dot{s} = a\dot{\theta}. \tag{7.3}$$

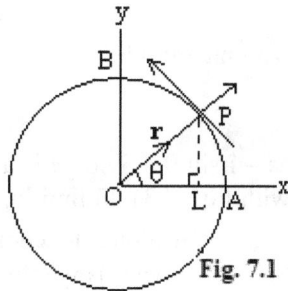

Fig. 7.1

Note 7.1. The arc-length *s* of a circle subtending an angle θ at its centre is given by $s = a\theta$. The radius *a* being constant this relation, on differentiation with respect to *t*, yields Eq. (7.3).

Definition 7.1. The quantity $\dot{\theta} \equiv d\theta / dt$ is called the angular velocity of the particle at P. When constant it is usually denoted by Greek letter ω.

Also, the relations (7.1) and (7.3) reduce Eqs. (5.1), (5.2), (6.1) and (6.2) as:

$$\text{Rad. Acc.} = -a\dot{\theta}^2, \quad \text{Norm. Acc.} = a^2\dot{\theta}^2 / a = a\dot{\theta}^2, \tag{7.4}$$

which are in opposite directions. Further,

$$\text{Trans. Acc.} = a\ddot{\theta}, \quad \text{Tang. Acc.} = \ddot{s} = a\ddot{\theta}, \tag{7.5}$$

which are equal in magnitude as well as in direction.

7.1. Above results in terms of vectors: Taking the centre O of the circle as origin and the rectangular coordinate axes O*x* and O*y* along two mutually perpendicular radii OA and OB (cf. Fig. 7.1) of the circle the position vector of P is described as

$$\mathbf{r} = \overline{OP} = \overline{OL} + \overline{LP} = (OP)\,(\hat{\mathbf{i}}\cos\theta + \hat{\mathbf{j}}\sin\theta)$$

$$= (a\cos\theta)\,\hat{\mathbf{i}} + (a\sin\theta)\,\hat{\mathbf{j}}, \qquad (7.6)$$

where $\hat{\mathbf{i}}$, $\hat{\mathbf{j}}$ are the unit vectors acting along the respective coordinate axes. Differentiating Eq. (7.6) with respect to the time parameter t we, thus, find the velocity vector

$$\dot{\mathbf{r}} \equiv d\mathbf{r}/dt = \{(-a\sin\theta)\,\hat{\mathbf{i}} + (a\cos\theta)\,\hat{\mathbf{j}}\}\,\dot{\theta}$$

$$= a\dot{\theta}\,(-\hat{\mathbf{i}}\sin\theta + \hat{\mathbf{j}}\cos\theta), \qquad (7.7)$$

with magnitude

$$|\dot{\mathbf{r}}| \equiv a\dot{\theta}, \qquad (7.8)$$

as $-\hat{\mathbf{i}}\sin\theta + \hat{\mathbf{j}}\cos\theta$ is a unit vector. The relation (7.8) is in conformity with Eq. (7.3). In limiting case, when Q → P the infinitesimal vector $\overline{OQ}$ $- \overline{OP} = \delta\mathbf{r}$ along the chord PQ describes the tangent vector to the curve. So, the velocity is wholly directed towards the tangent to the curve.

Further differentiation of Eq. (7.7) with respect to t yields the acceleration vector:

$$\ddot{\mathbf{r}} = a\ddot{\theta}\,(-\hat{\mathbf{i}}\sin\theta + \hat{\mathbf{j}}\cos\theta) + a\dot{\theta}^{2}\,(-\hat{\mathbf{i}}\cos\theta - \hat{\mathbf{j}}\sin\theta)$$

$$= (\ddot{\theta}/\dot{\theta})\,\dot{\mathbf{r}} - \dot{\theta}^{2}\,\mathbf{r}, \qquad (7.9)$$

by Eqs. (7.6) and (7.7). Thus, the components of acceleration vector along the tangent $\mathbf{r}$ and along the radius vector $\mathbf{r}$ are:

$$\text{Tang. Acc.} = (\ddot{\theta}/\dot{\theta})\,|\dot{\mathbf{r}}| = \ddot{\theta}\,|\dot{\mathbf{r}}|/\dot{\theta} = a\ddot{\theta}, \quad \text{by Eq. (7.8);}$$

and

$$\text{Rad. Acc.} = \dot{\theta}^{2}\,|-\mathbf{r}| = \dot{\theta}^{2}\,|\overline{PO}| = a\dot{\theta}^{2},$$

directed towards the centre. These values are also in conformity with Eqs. (7.5) and (7.4) respectively.

7.2. Circular motion with uniform angular velocity

In case the particle describes a circular motion with uniform (i.e. constant) angular velocity

$$\dot{\theta} = \omega \ (\text{constant}) \implies \ddot{\theta} = 0, \qquad (7.10)$$

various components of velocity and acceleration further reduce to

Rad.Vel. = Norm.Vel. = 0   (7.11); Tang.Vel. = Trans.Vel. = $a\omega$;   (7.12)

$$- \text{ Rad. Acc.} = \text{Norm. Acc.} = a\omega^2, \qquad (7.13)$$

$$\text{Tang. Acc.} = \text{Trans. Acc.} = 0. \qquad (7.14)$$

Conclusively, the velocity of the particle is directed towards the tangent to the (circular) path while the acceleration is wholly directed towards the centre (i.e. along the inward normal) of the circle. It is of magnitude $a\omega^2$.

The results in terms of vectors, as expressed by Eqs. (7.7) – (7.9), for a circular motion with uniform angular velocity, reduce to

$$\mathbf{r} = a\omega\,(-\hat{\mathbf{i}}\sin\theta + \hat{\mathbf{j}}\cos\theta), \qquad |\dot{\mathbf{r}}| = a\omega, \qquad (7.15)$$

and

$$\ddot{\mathbf{r}} = -\omega^2\,\mathbf{r}. \qquad (7.16)$$

These results are also in conformity with Eqs. (7.12) and (7.13) respe– ctively. We also note that the tangential component of acceleration in Eq. (7.9) vanishes for Eq. (7.10); which conforms with Eq. (7.14).

Note 7.2. Summarizing above results we have

	Radial	Transverse	Tangential	Normal
Velocity	$\dot{r}$	$r\dot{\theta}$	$\dot{s} = v$	0
Acceleration	$\ddot{r} - r\dot{\theta}^2$	$2\dot{r}\dot{\theta} + r\ddot{\theta}$	$\ddot{s}$	v^2/ρ
Velocity in a circle	0	$\dot{s} = a\dot{\theta}$	$\dot{s} = a\dot{\theta}$	0
Acceleration in a circle	$-a\dot{\theta}^2$	$A\ddot{\theta}$	$a\ddot{\theta}$	$a\dot{\theta}^2$
Velocity in uniform circular motion	0	$A\omega$	$a\omega$	0
Acc. in uniform circular motion	$-a\omega^2$	0	0	$a\omega^2$

Example 7.1. A point is moving in a circle of radius 4 metres with velocity t^2 m./sec. Find its acceleration if its direction is equally inclined

to the radius and the tangent.

Solution. The velocity at any point P is given by Eq. (7.3):

$$v = \dot{s} = a\dot{\theta} = t^2 \text{ (as per hypothesis)}$$

$$\Rightarrow$$

$$\dot{\theta} = t^2/a = t^2/4 \Rightarrow \ddot{\theta} = t/2.$$

Hence, the radial and transverse components of the acceleration, by Eqs. (7.4) and (7.5), are

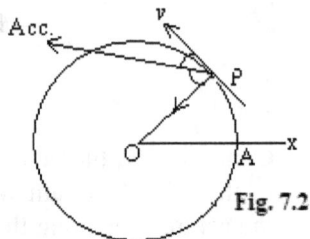

Fig. 7.2

$$\text{Rad. Acc. (along PO)} = a\dot{\theta}^2 = 4t^4/16 = t^4/4,$$

and

$$\text{Trans. Acc. (along the tangent at P)} = a\ddot{\theta} = 4t/2 = 2t.$$

The radial and tangential directions being perpendicular (for a circle) and their resultant equally inclined from them (as given) the component accelerations are equal to each other:

$$t^4/4 = 2t \Rightarrow t = 2.$$

Hence, either of the component accelerations are $2t = 4$ with their resultant $4\sqrt{2}$ m./sec.^{2}. //

Example 7.2. A point P describes a circle of radius a. At certain time t it attains a speed $2a\pi/(b+t)$, b being a constant. Show that when its speed is $a\pi/b$ the distance travelled by it is $(2a\pi \ln 2)$. Also, find the speed of P when it has described one round of the circle.

Solution. (i) The velocity at P is given by Eq. (4.3):

$$v = \dot{s} = 2a\pi/(b+t) \qquad\qquad (7.17)$$

$$\Rightarrow$$

$$ds = \{2a\pi/(b+t)\}\, dt.$$

Integrating it with respect to respective variables on either side we get

$$s = 2a\pi. \ln(b+t) + c, \qquad\qquad (7.18)$$

where c is a constant of integration. If the particle started from a point A (Fig. 7.2) where $s = 0$, $t = 0$, above equation determines $c = -2a\pi.\ln b$.

Accordingly, Eq. (7.18) reduces to

$$s = 2a\pi.\{\log_e (b + t) - 2a\pi.\log_e b\} = 2a\pi\{\log_e (1 + t / b). \qquad (7.19)$$

For a given speed the time is determined by Eq. (7.17). So, when $v = a\pi/b$, we get

$$1/ b = 2 / (b + t) \qquad \Rightarrow \qquad t = b.$$

The distance travelled at that instant is found by Eq. (7.19):

$$s = 2a\pi.\{\ln (1 + 1) = 2a\pi.\ln 2.$$

(ii) The point P makes one complete round of the circle when $s = 2a\pi$. Hence, Eq. (7.19) determines the time for that:

$$1 = \ln (1 + t / b) \qquad \Rightarrow \qquad 1 + t / b = e \qquad \Rightarrow \qquad b + t = b\,e.$$

Also, the relation (7.17) then determines its speed after one round $v = 2a\pi / be.$ //

Example 7.3. A point moves in a circle of radius 24 cms. After t secs. it acquires the speed $(t^2 + 4t)$ m./sec. Find the acceleration and its direction after 2 seconds.

Solution. (i) The relation (7.3) gives the speed of the point:

$$v = \dot{s} = t^2 + 4t \quad \Rightarrow \quad \ddot{s} = 2t + 4.$$

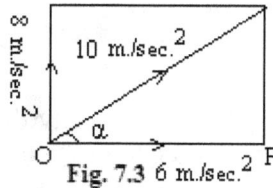
Fig. 7.3

Evaluating these quantities at $t = 2$ secs. we get $v = 12$ and $\ddot{s} = 8$. Hence, the radial and tangential components of the acceleration are obtained by Eqs. (7.4) and (7.5):

$$\text{Rad. Acc.} = v^2 / \rho = (12)^2 /24 = 6 \quad \text{and} \quad \text{Tang. Acc.} = \ddot{s} = 8.$$

So, their resultant acceleration is $\sqrt{(8^2 + 6^2)} = 10$ m./ sec.^{2}

(ii) The inclination α of the resultant acceleration to the radius vector is given by $\tan \alpha = 8/6 = 4/3.$ //

§ 8. Simple harmonic motion

Let a particle P move along a straight
line under an acceleration always directed
towards a fixed point (say O) in the line of motion and the magnitude of
the acceleration is proportional to the distance OP.

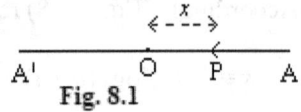

Fig. 8.1

If the particle starts from a point A (distant a from O) with velocity
zero and describes a position P (OP $= x$) after time t secs. (measured
from A). The motion of the particle is caused by an acceleration always
directed towards O and proportional to the distance OP $= x$. The equation
of motion is

$$\ddot{x} = -\mu x, \tag{8.1}$$

where μ is the constant of proportionality. The negative sign in above
equation is due to the fact that the variables x and t are measured in opposite directions: x from O to P while t from A to P. So, they are inverse-
ly proportional making $\dot{x}$ and $\ddot{x}$ negative. Multiplying (8.1) by $2\dot{x}$ and
integrating either side with respect to t we get

$$\int (2\dot{x}\,\ddot{x})\,dt = -\mu \int (2x\dot{x})\,dt + c, \quad \text{or,} \quad (\dot{x})^2 = -\mu x^2 + c. \tag{8.2}$$

Initially, there hold $x = a$ and $\dot{x} = 0$ at A; for which Eq. (8.2) determines
$c = \mu a^2$. Hence, Eq. (8.2) reduces to

$$v^2 \equiv (\dot{x})^2 = \mu (a^2 - x^2). \tag{8.3}$$

Thus, the velocity diminishes to zero when $x = \pm a$, i.e. when the
particle is at A (OA $= a$) and at A' (OA' $= -a$), where A, A' are the end
points of the motion. On the other hand, the particle acquires the maximum velocity at the centre O of its path (where $x = OO = 0$). This maximum velocity at O compels the particle to move beyond O (towards A').
During the motion along OA' the acceleration (as per hypothesis) being
directed towards the centre O retards the velocity causing the particle to
come to rest at A'. Thereafter, the acceleration (being of maximum value
μa) at A' and directed towards O pulls the particle to retrace its path
from A' to O. Again, at O (where $x = 0$) the velocity becomes maximum
that takes the particle beyond O (towards A). Thus, the particle oscillates
between the end points A, A' of the path. Such a motion is called the
simple harmonic motion.

8.1. Distance in terms of time: Considering the negative square root of Eq. (8.3)

$$v \equiv \dot{x} = - \sqrt{\{\mu (a^2 - x^2)\}}, \tag{8.4}$$

separating the variables:

$$- dx / \sqrt{(a^2 - x^2)} = \sqrt{(\mu)} \, dt.$$

On further integration with respect to t, it yields

$$\cos^{-1}(x/a) = t \sqrt{\mu} + c_1. \tag{8.5}$$

Initially, there hold $x = a$ and $t = 0$ at the starting point A; for which Eq. (8.5) gives $c_1 = \cos^{-1}(1) = 0$. Hence, Eq. (8.5) finally assumes the form

$$x = a \cos(t \sqrt{\mu}). \tag{8.6}$$

This relation describes a position P ($OP = x$) after time t (measured from A). On the other hand, the relation also determines t for a given value of x:

$$t = (1 / \sqrt{\mu}) \cos^{-1}(x/a). \tag{8.7}$$

Thus, the particle reaches the mid-point O (where $x = 0$) in time

$$t_1 = (1/\sqrt{\mu}) \cos^{-1}(0) = \pi / 2\sqrt{\mu};$$

and the other end A' (where $x = - OA' = - a$) in time

$$t_2 = (1/\sqrt{\mu}) \cos^{-1}(-1) = \pi / \sqrt{\mu}.$$

The time t_2 being twice of t_1, the particle takes equal time for the intervals AO and OA'. Also, the particle makes a complete oscillation (describing motion from A to A' and, thereafter, back to A) in time

$$2t_2 = 2\pi / \sqrt{\mu} = T \text{ (say)}. \tag{8.8}$$

Definition 8.1. The time T given by Eq. (8.8) is called the *periodic time* of the simple harmonic motion.

In other words, the particle makes

$$n \equiv 1/T = \sqrt{\mu} / 2\pi \tag{8.9}$$

number of oscillations per unit time. This number n is called the *frequency* of the S.H.M. Also, the distances OA $= a$ $=$ OA' are called the *amplitude* of the motion.

Note 8.1. It is seen above that the acceleration is maximum at the end points A $(x = a)$ and A' $(x = -a)$ and it is minimum (with value zero) at the mid-point O $(x = 0)$. On contrary, the velocity is minimum (with value zero) at the end points but maximum (with value $a\sqrt{\mu}$) at O.

8.2. Motion in a vertical circle: A circular motion has been discussed in the previous chapter. Consider a position $P(x, y)$ of the particle on the circle given by Eq. (1.1.2). If the particle moves along the circle with constant angular velocity ω, the acceleration at P is completely directed towards the radius vector PO and is $a\omega^2$ as given by Eq. (7.13). Its horizontal component along AO (cf. Fig. 7.1), i.e. x-axis, is

$$x = -a\omega^2 \cos\theta = -\omega^2 x, \qquad (8.10)$$

which is proportional to x. Integrating $\dot\theta = \omega$ with respect to t we get

$$\theta = \omega t + c.$$

Let the particle start from A (where $t = 0$) and the vectorial angle θ be measured from OA (positive in the anti-clockwise direction). Thus, initially at A both t and θ vanish identically. Hence, above equation determines $c = 0$. Accordingly, the equation reduces to

$$\theta = \omega t. \qquad (8.11)$$

Hence, the particle describes the whole circle in time $T = 2\pi/\omega$, which is the same as the periodic time of a S.H.M. given by Eq. (8.10) performed by the foot L of P on the horizontal diameter AOA'. Thus, we have the:

Theorem 8.1. When P describes a vertical circle its foot on x-axis (taken along a horizontal diameter AOA') describes a S.H.M. along the diameter.

Example 8.1. A particle describing a S.H.M. takes 4 secs. to travel from one end to another end of its path. It acquires the greatest speed 3π m./sec Find the amplitude of the motion and the distance of the particle from the mid-point of its path when its speed is π m./sec.

Solution. The periodic time of a S.H.M. is given by Eq. (8.8):

$$2\pi /\sqrt{\mu} = 8 \qquad \Rightarrow \qquad \sqrt{\mu} = \pi/4. \qquad (8.12)$$

The maximum velocity of the particle is obtainable from Eq. (8.3) when $x = 0$:

$$v^2 = \mu a^2 = (3\pi)^2, \qquad \text{by hypothesis;}$$

or, for Eq. (8.12)

$$a = 3\pi /\sqrt{\mu} = 12 \text{ m.}$$

Further, Eq. (8.3) also determines the distance x for a given velocity $v = \pi$:

$$\pi^2 = (\pi^2 /16)(12^2 - x^2) \qquad \Rightarrow \qquad x^2 = 12^2 - 16 = 128.$$

Therefore,

$$x = \sqrt{(128)} = 8\sqrt{2} \text{ m. //}$$

Example 8.2. A particle describing a S.H.M. makes 5 complete oscillations per second and has its path 4 cms. long. Find the greatest magnitude of its acceleration.

Solution. The frequency of a S.H.M. is given by Eq. (8.9):

$$5 = \sqrt{\mu} / 2\pi \qquad \Rightarrow \qquad \sqrt{\mu} = 10\pi.$$

Further, amplitude of the motion, being half of its path, is $a = 2$ cms. $= 2/100$ m. The acceleration is given by Eq. (8.1) and it is greatest at the end points of the path, i.e. when $x = a$:

$$\ddot{x} = -(10\pi)^2 \times 2/100 = -2\pi^2 \text{ m./sec.}^2. //$$

§ 9. Projectile on a horizontal plane

Let a particle be projected from a point O with velocity u in a vertical plane at an angle α to the horizontal line OB in the plane. If, after time t (measured from O), the particle attains some position P and move with velocity v at an angle θ to the horizontal line through P that lies in the vertical plane.

Taking O as origin, the line OB as x-axis and the vertical line OC in the plane of projection as y-axis, let P have coordinates (x, y). The horizontal component of velocity at O is $u \cos \alpha$, which remains the same throughout the motion as there is no any horizontal force affecting the motion. On the other hand, the vertical component of velocity $u \sin \alpha$ is continually diminished due the gravity force of the Earth always acting vertically downwards. Similarly, the horizontal and vertical components of velocity at P are $v \cos \theta$ and $v \sin \theta$ respectively. Following Eq. (2.5), there hold:

$$v \sin \theta = u \sin \alpha - gt, \qquad (9.1)$$

and

$$v \cos \theta = u \cos \alpha. \qquad (9.2)$$

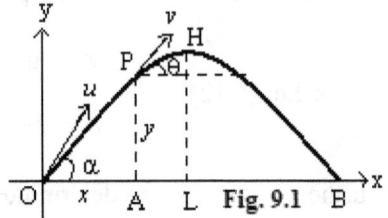

Fig. 9.1

These relations determine the velocity and its direction at P. Squaring the relations and adding them we get

$$v^2 = (u \sin \alpha - gt)^2 + (u \cos \alpha)^2 = u^2 - 2gut \sin \alpha + g^2 t^2, \qquad (9.3)$$

and

$$\tan \theta = (u \sin \alpha - gt) / u \cos \alpha = \tan \alpha - gt / u \cos \alpha. \qquad (9.4)$$

Following Eq. (2.6), the horizontal and vertical distances: x and y of P from O (described in time t) are obtained by

$$x = (u \cos \alpha) t, \qquad y = (u \sin \alpha) t - (1/2) gt^2. \qquad (9.5)$$

Eliminating t from these relations we derive the Cartesian equation of the path:

$$y = x \tan \alpha - gx^2 / 2 u^2 \cos^2 \alpha \qquad (9.6)$$

Differentiating Eq. (9.5) with respect to t the horizontal and vertical components of velocity may be obtained:

$$\dot{x} \equiv dx/dt = u \cos \alpha, \qquad \dot{y} \equiv dy/dt = u \sin \alpha - gt. \qquad (9.7)$$

Theorem 9.1. The particle thrown as above describes:

(i) the highest height: $\qquad$ LH $= (u^2 \sin^2 \alpha) / 2g,$ $\qquad$ (9.8)

(ii) the time of its flight from O to B $= (2u \sin \alpha) / g,$ $\qquad$ (9.9)

and

(iii) the range (on horizontal plane) OB $= (u^2 \sin 2\alpha) / g.$ $\qquad$ (9.10)

Note 9.1. Thus we note that the time of flight t_B (from O to B) is twice the time t_H to reach the highest point.

(iii) For above value of t_B the first of Eqs. (9.5) determines the range OB:

$$OB = (u \cos \alpha)(2u \sin \alpha) / g = (u^2 \sin 2\alpha) / g. \text{//}$$

Corollary 9.1. The maximum (horizontal) range is u^2/g and the angle of projection for the same is $\pi/4$.

Definition 9.1. The path so traced by the particle is called a *projectile*.

Theorem 9.2. The projectile has a parabolic path with vertex upwards.

Example 9.1. A particle is projected with velocity 20 m./sec. at an angle of elevation of 30° from a point on level ground. If $g = 10$ m./sec.^{2}, find the:

(i) greatest height reached, **(ii)** time of flight, **(iii)** horizontal range,

(iv) direction in which it is moving after: (a) 1/2 sec., (b) 3/2 secs.

Solution. For $u = 20$, $\alpha = 30°$ and $g = 10$, the Eqs. (9.8) – (9.10) determine:

(i) greatest height $= (20 \sin 30°)^2 / 2 \times 10 = (400/4) / 20 = 5$ metres;

(ii) time of flight $= 2 \times 20 \sin 30° / 10 = 2$ secs.;

(iii) range on horizontal plane $= 20 \times 20 \sin 60° / 10 = 20\sqrt{3}$ metres;

(iv) (a) Putting for u, α and $t = 1/2$ in Eq. (9.4), we obtain

$$\tan \theta = \tan 30° - (10 \times 1/2)/20.\cos 30° = 1/\sqrt{3} - 1/2\sqrt{3} = 1/2\sqrt{3}.$$

(b) When $t = 3/2$, Eq. (9.4) gives

$$\tan \theta = \tan 30° - (10 \times 3/2)/20.\cos 30° = 1/\sqrt{3} - \sqrt{3}/2 = -1/2\sqrt{3}. \text{//}$$

Example 1.2. A particle is projected with velocity 60 m./sec. at an angle $\tan^{-1}(4/3)$ to the horizontal from a point O on a horizontal plane. If $g = 10$ m./sec.^{2}, find:

(i) when it is at a height of 99 metres above the plane;

(ii) the horizontal distances from O of the particle at the times when it is at a height of 99 m?

Solution. (i) Given that $u = 60$, $g = 10$, $\alpha = \tan^{-1}(4/3)$ so that $\tan \alpha = 4/3$, $\sin \alpha = 4/5$ and $\cos \alpha = 3/5$. Putting for these values and $y = 99$ in Eq. (9.5) we obtain

$$99 = (60 \times 4/5)\,t - 10\,t^2/2, \quad \text{or} \quad 5t^2 - 48\,t + 99 = 0,$$

determining $t = 3$ and $33/5 = 6\cdot6$ secs. Thus, the particle is at above height at two times: 3 secs. and 6·6 secs. after its projection.

(ii) For these values of t, the first of Eq. (9.5) determines:

$$x_1 = (60.\cos\alpha).3 = (60 \times 3/5).3 = 108 \text{ m.};$$

and

$$x_2 = (60.\cos\alpha).6\cdot6 = (60 \times 3/5).6\cdot6 = 237\cdot6 \text{ m.} \; //$$

§ 10. Projectile on an inclined plane

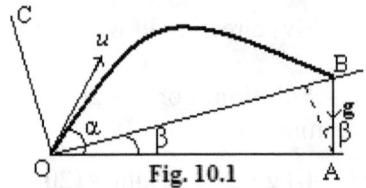

Fig. 10.1

Let OB be a plane inclined at an angle β to the horizontal plane OA and a particle be projected from the point O with velocity u in a vertical plane at an angle $\alpha \; (> \beta)$ to the horizontal line OA up the inclined plane. The path of the particle will, thus, lie in a vertical plane through a line of greatest slope of the inclined plane. Let the particle strikes the inclined plane again at a point B. In the following, we shall discuss the time of flight from O to B (up the inclined plane) and the range OB. Let OC be a line perpendicular to OB. It is convenient to discuss the motion along OB and OC. The acceleration due to gravity has components:

$$g \sin\beta \text{ (along BO)} \quad \text{and} \quad g \cos\beta \quad \text{(along CO)};$$

or, equivalently

$$-g \sin\beta \text{ (along OB)} \quad \text{and} \quad -g \cos\beta \quad \text{(along OC)}.$$

Also, the components of initial velocity at O along OB and OC are:

$$u \cos(\alpha - \beta) \quad \text{and} \quad u \sin(\alpha - \beta) \quad \text{respectively.}$$

Initially, there hold $s = 0$ and $t = 0$ at O. Also, at B the distance travelled along OC is zero. But, it is also given by Eq. (2.6):

$$s_B = 0 = t\,u \sin (\alpha - \beta) - (g \cos \beta)\, t^2 /2,$$

determining $t_O = 0$ (when the particle is at O), and the time of flight from O to B is given by

$$t_B = 2u \sin (\alpha - \beta) / g \cos \beta. \qquad (10.1)$$

During this time of flight the particle describes the horizontal distance OA with constant horizontal velocity $u \cos \alpha$:

$$OA = (u \cos \alpha)\, t_B = 2u^2 \cos \alpha \,.\, \sin (\alpha - \beta) / g \cos \beta. \qquad (10.2)$$

In the right triangle OAB, we have the range on the inclined plane:

$$OB = (OA) / \cos \beta = 2u^2 \cos \alpha .\, \sin (\alpha - \beta) / g \cos^2 \beta. \qquad (10.3)$$

10.1. Maximum range on the inclined plane

For any angles A and B, there holds

$$\sin (A + B) - \sin (A - B) = 2 \cos A .\, \sin B. \qquad (10.4)$$

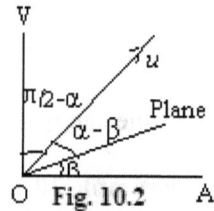

Fig. 10.2

Taking $\angle A = \alpha$ and $\angle B = \alpha - \beta$, we derive from Eq. (10.4):

$$\sin (2\alpha - \beta) - \sin \beta = 2 \cos \alpha .\, \sin (\alpha - \beta). \qquad (10.5)$$

Hence, Eq. (10.3) reduces to

$$OB = u^2 \{\sin (2\alpha - \beta) - \sin \beta\} / g \cos^2 \beta. \qquad (10.6)$$

For a given β, OB attains a maximum value when $\sin (2\alpha - \beta)$ attains so:

$$\sin (2\alpha - \beta) = 1 \;\Rightarrow\; 2\alpha - \beta = \pi/2 \;\Rightarrow\; \alpha - \beta = \pi/2 - \alpha. \qquad (10.7)$$

Thus, OB becomes maximum when the angle of projection is such that the direction of projection bisects the angle between the inclined plane and the vertical line OV through O. Consequently, Eq. (10.6) determines the maximum range on the inclined plane:

$$OB_{max.} = u^2 (1 - \sin \beta) / g \cos^2 \beta = u^2 / g (1 + \sin \beta). \qquad (10.8)$$

10.2. Particle projected down an inclined plane

Let an inclined plane OB be considered down the horizontal plane OA at some angle, say $-\beta$ and a particle be projected from O on the same. If it strikes the inclined plane at some position B the range OB (on the inclined plane) can now be calculated by replacing β by $-\beta$ in Eqs. (10.3) or (10.6):

$$OB = 2u^2 \cos \alpha . \sin (\alpha + \beta) / g \cos^2 \beta$$

$$= u^2 \{\sin (2\alpha + \beta) + \sin \beta\} / g \cos^2 \beta. \quad (10.9)$$

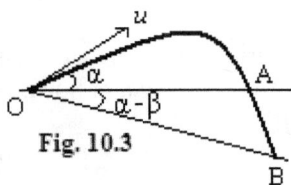

Fig. 10.3

It becomes maximum when

$$\sin (2\alpha + \beta) = 1 \quad \Rightarrow \quad 2\alpha + \beta = \pi/2 \quad \Rightarrow \quad \alpha = \pi/4 - \beta/2, \quad (10.10)$$

and the maximum range (on the inclined plane) is

$$OB_{max.} = u^2 (1 + \sin \beta) / g \cos^2 \beta = u^2/g (1 - \sin \beta). \quad (10.11)$$

The same can be directly obtained by replacing β by $-\beta$ in Eq. (10.8).

Example 10.1. A particle is projected with velocity 40 m./sec. at an angle 45° to the horizontal from a point on a plane inclined at an angle 30° to the horizontal . The path of the particle lies in a vertical plane containing a line of greatest slope of the plane. Taking $g = 10$ m./sec.^{2} , find the range:

(i) up the inclined plane, **(ii)** down the inclined plane, and **(iii)** maximum ranges up and down the inclined plane and the corresponding angles of projection.

Solution. Given $u = 40$, $\alpha = 45°$ and $\beta = 30°$, Eqs. (10.6) and (10.9) determine

(i) range up the plane $= 40 \times 40 \{\sin (90 - 30) - \sin 30\}/10. \cos^2 30$

$= 160 \{ (\sqrt{3} - 1) / 2\} \times 4/3 = 320 (\sqrt{3} - 1) / 3 = 78{\cdot}08$ m.

(ii) range down the plane

$= 40 \times 40 \{\sin (90 + 30) + \sin 30\} / 10. \cos^2 30$

$$= 160 \{(\sqrt{3} + 1)/2\} \times 4/3 = 320 (\sqrt{3} + 1) / 3 = 291{\cdot}413 \text{ m.}$$

(iii) Maximum range up the plane, by Eq. (10.8), is

$$40 \times 40 / 10.(1 + \sin 30) = 160 / (1 + 1/2) = 320/3 = 106{\cdot}667 \text{ m.}$$

The angle of projection for such a range, by Eq. (10.7), is

$$\alpha = (\pi/2 + \beta)/2 = 60°.$$

(iv) Maximum range down the plane, by Eq. (10.11), is

$$40 \times 40 / 10.(1 - \sin 30) = 160 / (1 - 1/2) = 320 \text{ m.}$$

The angle of projection for such a range is given by Eq. (10.10):

$$\alpha = \pi/4 - \beta/2 = 45° - 15° = 30°. //$$

Example 10.2. A particle is projected at an angle α to the horizontal from a point on a plane inclined at an angle β to the horizontal. The path of the particle lies in a vertical plane through a line of greatest slope of the plane. The particle strikes the plane when moving horizontally. Show that

$$\tan \alpha = 2 \tan \beta. \tag{10.12}$$

Solution. Let the particle attain some position P in time t where its velocity is v in the direction making angle θ to the horizontal. This direction is given by Eq. (9.4). As per hypothesis, it strikes the plane at B (cf. Fig. 10.1) after the time of flight t_B given by Eq. (10.1), where its direction becomes horizontal causing $\theta = 0$. Therefore, Eq. (9.4) assumes the form

$$0 = \tan \alpha - 2ug. \sin (\alpha - \beta) / ug.\cos \alpha .\cos \beta$$

$$= \tan \alpha - 2\sin (\alpha - \beta) / \cos \alpha..\cos \beta,$$

$$\Rightarrow$$

$$\sin \alpha . \cos \beta = 2 (\sin \alpha . \cos \beta - \cos \alpha . \sin \beta),$$

or,

$$2 \cos \alpha . \sin \beta = \sin \alpha . \cos \beta.$$

Division by $\cos \alpha . \cos \beta$ in above equation yields Eq. (10.12). //

Example 10.3. If the particle in the previous Example strikes the plane at right angles, show that

$$2 \tan (\alpha - \beta). \tan \beta = 1. \qquad (10.13)$$

Solution. The particle strikes the plane at B where the direction of its
velocity makes angle $\theta - \beta = \pi /2$ (as per hypothesis) from the inclined plane. This determines $\theta = \pi /2 + \beta$. Consequently, Eq. (9.4) reduces to

$$\tan (\pi /2 + \beta) = \tan \alpha - g\, t_B / u \cos \alpha = \tan \alpha - 2\sin (\alpha - \beta) / \cos \alpha. \cos \beta,$$

by Eq. (10.1). Multiplying throughout by $\cos \alpha. \sin \beta$ we get

$$2 \sin (\alpha - \beta). \tan \beta = (\sin \alpha + \cot \beta. \cos \alpha) \sin \beta$$

$$= \sin \alpha. \sin \beta + \cos \alpha. \cos \beta = \cos (\alpha - \beta).$$

Division by $\cos (\alpha - \beta)$ yields the result. //

§ 11. Energy

The capacity of the engine to do work is called *energy*. Like work, it is also a scalar quantity. The unit of the energy is also *joule*. The energy can be of different forms such as heat, light and chemical energy. Currently, we discuss the mechanical energy possessed by a particle. This (mechanical) form of energy is also of two types: *kinetic energy* and *potential energy*.

11.1. Kinetic energy: the capacity of a force to do work by virtue of its speed produces the *kinetic energy*.

If a particle of mass m moving with initial speed u is constantly retarded by a force **F** capable of bringing it to the rest position in distance s, then by Eq. (2.8), we have

$$0^2 = u^2 - 2fs = u^2 - 2 (F/ m) s \quad \Rightarrow \quad F s = (1/ 2) mu^2. \qquad (11.1)$$

But, Fs is the work done against the force in bringing the particle to rest. Thus, by virtue of its speed u the particle has the capacity to do the work of magnitude $(1/2) mu^2$. In other words, the kinetic energy of a particle of mass m moving at speed u is $(1/2) mu^2$.

11.2. Potential energy: the capacity of a force to do work by virtue of its position produces the *potential energy*.

Let a particle of mass m be released from rest at a height h above a table and it falls freely under gravity. By Eq. (3.1), we compute the speed of the particle when hitting the table:

$$v^2 = 0^2 + 2gh = 2gh.$$

Thus, the kinetic energy of the particle when reaching the table is

$$(1/2)\, mv^2 = mgh. \tag{11.2}$$

In other words, the particle possessed the capacity to do the work of magnitude mgh due to falling through the height h above the table. When the particle was at a height h above the table it possessed the potential energy mgh with reference to the level of the table.

The total mechanical energy of a particle of mass m is the sum of its kinetic and potential energies, i.e.

$$(1/2)\, mv^2 + mgh, \tag{11.3}$$

h being the height of the particle above the chosen level of reference

Definition 11.1. When the work done by a force or against a force in displacing a particle from a point A to another point B is independent of the path described from A to B, the force is said to be *conservative force*.

Definition 11.2. (*Principle of conservation of mechanical energy*) The sum of kinetic and potential energies of a system of particles under the action of conservative forces is constant.

Example 11.1. A particle is projected with velocity u at an angle of elevation α. Find its velocity at a height h above the level of the point of projection.

Solution. Let the horizontal plane OH through the point of projection O be taken as the level of reference for potential energy. Let m be the mass of the particle. At the point of projection the kinetic energy is $(1/2)\, mu^2$, while the potential energy is zero for the hypothesis. At height h, the kinetic energy is $(1/2)\, mv^2$ and the potential energy is mgh.

Therefore, by the principle of conservation of mechanical energy, we have

$$(1/2)\ mu^2 + 0 = (1/2)\ mv^2 + mgh$$

⇒

$$v^2 = u^2 - 2gh.$$

Therefore, the velocity at height h is $\sqrt{(u^2 - 2gh)}$. //

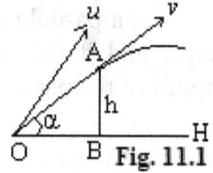

Fig. 11.1

§ 12. Momentum of a particle

Let a particle of mass m be moving with velocity **u**. The product mu has been called the *momentum* of the particle (cf. § 1). Precisely enough, it is the *linear* momentum as it is a vector in the same direction as **u** along the straight line through the particle.

If the mass is measured in *kg.* and the velocity in *metres / sec.* the unit of momentum is *kg. metres / sec.*

Example 12.1. Let a particle of mass m be moving in a straight line with velocity **u**. The particle is given a blow that trebles its speed. Find the change in momentum:

(i) if the particle continues to move in the same direction as before;

(ii) if the particle reverses its direction after the blow.

Solution. (i) Momentum before the blow = mu, and that after the blow = $3mu$. Therefore, the change in momentum = $3mu - mu = 2mu$. //

(ii) Momentum after the blow = $-3mu$. Hence, the change in momentum is $-3mu - mu = -4mu$. //

Example 12.2. A particle of mass 100 gms. moves with velocity 40 m./sec. along the unit vector **ĵ**. It is given a blow that reduces its velocity to 30 m./sec. in the direction of the unit vector **î**. Find the change in momentum.

Solution. Momentum before the blow = $(0\cdot1) \times 40$ **ĵ** $= 4$**ĵ** kg. m./sec. and that after the blow = $(0\cdot1) \times 30$ **î** $= 3$**î** kg. m./sec. Therefore, the change in momentum = 3**î** $- 4$**ĵ** kg. m./sec. The magnitude of this vector is $|3\hat{\mathbf{i}} - 4\hat{\mathbf{j}}| = \sqrt{(3^2 + 4^2)} = 5$. Hence, the change in momentum is 5 kg. m./sec. in the direction of the vector $3\,\hat{\imath} - 4\,\hat{\jmath}$. //

Definition 12.1. The change in momentum is called the *impulse* of the force acting for a given time.

§ 13. Conservation of linear momentum

Let two particles P_1, P_2 of masses m_1, m_2, moving with velocities u_1, u_2 respectively in the same straight line, make a direct collision. Let the respective particles acquire the velocities v_1, v_2 after the collision. Changes in their momentums are

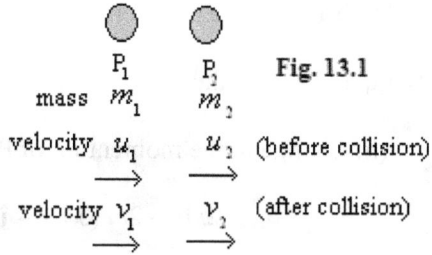

mass m_1 m_2

velocity u_1 u_2 (before collision)

velocity v_1 v_2 (after collision)

Fig. 13.1

$$P_1: m_1\, v_1 - m_1\, u_1 \equiv m_1\, (v_1 - u_1), \qquad P_2: m_2\, v_2 - m_2\, u_2 \equiv m_2\, (v_2 - u_2).$$

These changes are equal in magnitude but opposite in directions because the impulse given to P_1 by P_2 is equal in magnitude to the impulse given to P_2 by P_1 but in opposite direction. Thus, we have

$$m_1\, (v_1 - u_1) = -\, m_2\, (v_2 - u_2) \;\Rightarrow\; m_1\, u_1 + m_2\, u_2 = m_1\, v_1 + m_2\, v_2. \quad (13.1)$$

Thus, the sum of linear momentums of the particles before collision equals the sum of their linear momentums after the collision, i.e. this sum remains constant. This principle is called the *principle of conservation of momentum*.

Example 13.1. Two particles P and Q move in the same straight line approaching towards each other. Their masses are $3m$, $2m$ and their velocities are $2u$ and u respectively. If after their collision they move together, find their common velocity.

Solution. Momentums before and after collision are

$$3m\,(2u) + 2m\,(-u) = 4mu \quad \text{and} \quad (3m + 2m)\,v = 5mv,$$

where v is their common velocity. By Eq. (13.1) we, therefore, get

$$4mu = 5mv \qquad \Rightarrow \qquad v = 4u\,/\,5. \text{ //}$$

Example 13.2. A particle P of mass $2m$ moving with velocity $3\,\hat{\imath} + 4\hat{\jmath}$ collides with another particle Q of mass $3m$ moving with velocity $-2\,\hat{\imath} - \hat{\jmath}$. After their collision they move together with velocity **v**. Find **v** and the change in momentum of each particle.

Solution. (i) Momentums before and after collision are

$$2m\,(3\,\hat{\imath} + 4\,\hat{\jmath}) + 3m\,(-2\,\hat{\imath} - \hat{\jmath}) \;=\; 5m\,\hat{\jmath} \quad \text{and} \quad (2m + 3m)\,\mathbf{v} = 5m\mathbf{v}.$$

By Eq. (13.1), we then have $\;5m\hat{\jmath} \;=\; 5m\mathbf{v} \;\Rightarrow\; \mathbf{v} = \hat{\jmath}.$

(ii) Change in the momentum of P is

$$2m\mathbf{v} - 2m\,(3\,\hat{\imath} + 4\,\hat{\jmath}) = -\,6m\,(\hat{\imath} + \hat{\jmath}).$$

Similarly, the change in momentum of Q is

$$3m\mathbf{v} - 3m\,(-2\,\hat{\imath} - \hat{\jmath}) \;=\; 6m\,(\hat{\imath} + \hat{\jmath}).\;//$$

§ 14. Impact with a fixed surface

14.1. Perpendicular impact: When a ball is dropped from some height h to a horizontal floor, it rebounds to a height lesser than h. Its speed before striking the floor is greater than the one after leaving the floor when re-bouncing. During a short interval when the ball is in contact with the floor, it is first compressed and, thereafter, it regains its original shape. During compression the impulse of magnitude mu (m being the mass and u the velocity of the ball before striking the floor) makes the ball to come to rest. Afterwards, an impulse of magnitude mv (where v is the velocity with which it rebounds) expands the ball. The ratio of mv to mu, denoted by

$$e = v/u, \tag{14.1}$$

depends on the elasticity of the ball and the floor. This ratio remains constant for a given ball and the given floor. It is called the *coefficient of restitution*. Whenever, the ball strikes the floor perpendicularly, there holds above equation. The energy lost in striking the floor is in the form of heat and sound (making a bang). Consequently, $e < 1$. If, however, no energy is lost in impact, e becomes 1; that reduces Eq. (14.1) to $v = u$.

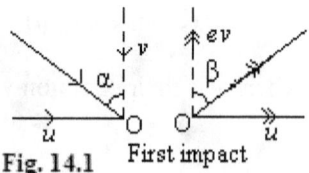

Fig. 14.1   First impact

14.2. Oblique impact

Let a ball strike a smooth horizontal plane at an angle α to the vertical and rebounds in a direction making an angle β to the vertical. Let u and v be the horizontal and vertical components of its velocity before striking the plane at O (Fig. 14.1). After impact, the horizontal component of velocity remains the same as before the impact (i.e. u) but the vertical component, by Eq. (14.1), becomes ev, where e is the coefficient of restitution between the ball and the plane. For clarity, the single arrows indicate the direction of velocities before the impact while the double arrows denote the direction of velocities after the impact.

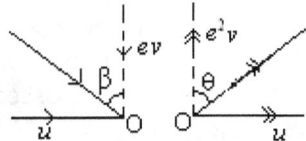

Fig. 14.2 Second impact

These components of velocities are connected by

$$\tan \alpha = u/v \qquad \text{and} \qquad \tan \beta = u/ev. \qquad (14.2)$$

Elimination of u/v from these relations determines

$$e = \tan \alpha / \tan \beta. \qquad (14.3)$$

Let the ball make another (second) impact with the plane (cf. Fig. 14.2). The horizontal and vertical components of its velocity before the second impact are u and ev respectively. Let the angle at which it hits the plane (second time) to the vertical be β. As before (during the first impact), the horizontal component of velocity after the (second) impact again remains the same, i.e. u while the vertical component of velocity after the second impact, by Eq. (14.1), becomes $e(ev) = e^2 v$. Let the ball leave the floor (after the second impact) at an angle θ to the vertical then, analogous to Eq. (14.2), we have

$$\tan \beta = u/ev \qquad \text{and} \qquad \tan \theta = u/e^2 v. \qquad (14.4)$$

Putting for u/ev and $1/e$ (obtainable from Eq. (14.3)) the second of relations (14.4) reduces to

$$\tan \theta = \tan^2 \beta / \tan \alpha. \qquad (14.5)$$

Note 14.1. If the ball strikes the horizontal plane with velocity u at an angle α to the vertical and after the (first) impact leaves the plane with velocity v in the direction making an angle β to the vertical, then the horizontal and vertical components of velocities before and after the impact are related by

$$u \sin \alpha = v \sin \beta, \qquad e\,(u \cos \alpha) = v \cos \beta. \qquad (14.6)$$

Elimination of u, v from these relations again yields Eq. (14.3). //

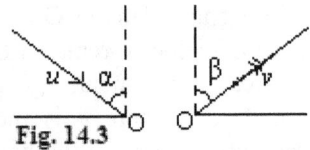

Fig. 14.3

Example 14.1. A golf ball is dropped from rest at a point above the floor. It takes 1 second to reach the floor. Show that the ball comes to rest 24 seconds after being released if the coefficient of restitution is 0·92.

Solution. Let the ball be dropped from H_o (with initial velocity $u = 0$) to the floor. Its velocity before striking the floor, by Eq. (3.1), is

$$v = 0 + g \times 1 = g.$$

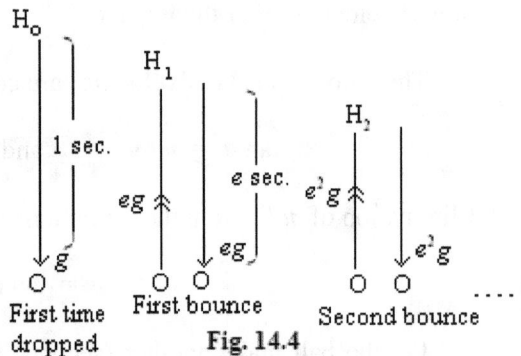

Fig. 14.4

After striking the floor, it rebounds with the velocity eg (cf. Eq. (14.1)). It reaches the highest point H_1 (by Eq. (3.7)) in time $t_{H_1} = eg\,/\,g = e$ sec. After attaining the greatest height, it again starts falling vertically downwards from H_1 with zero velocity and reaches the floor in same time (e sec.) with velocity (equal to the velocity of projection from the floor) eg. Thus, during the first bounce (rising up to the highest point H_1 and, thereafter, returning back to the point of pro-jecttion) the time elapsed is $2\,t_{H_1} = 2e$ secs.

Next, the ball rebounds second time with velocity (by Eq. (14.1)) $e.\,(eg) = e^2 g$. It takes the time

$$t_{H_2} = e^2 g\,/\,g = e^2 \text{ sec.}$$

to reach the highest point H_2 of its (second) bounce and again returns back to the point of projection with velocity $e^2 g$ in time e^2. Thus, the total time to complete the second bounce $= 2\,t_{H_2} = 2e^2$. Conclusively, the time elapsing in every bounce gets multiplied by e to that in the preceding bounce. Hence, the total time taken by the ball (measured from the point H_o to come to rest is

$$1 + 2e + 2e^2 + 2e^3 + \ldots + 2e^n + \ldots$$

$$= 1 + 2e \, (1 + e + e^2 + \ldots) = 1 + 2e \, / \, (1 - e)$$

$$= (1 + e) \, / \, (1 - e) = (1 + \cdot 92) \, / \, (1 - \cdot 92) = 1 \cdot 92 \, / \cdot 08 = 192 \, / \, 8 = 24 \text{ secs. } //$$

§ 15. Direct impact of two spheres

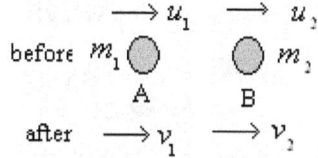

Let A and B be two spheres of masses m_1, m_2 moving in the same straight line with velocities u_1, u_2 before collision. Let their velocities, after collision, become v_1,

before m_1 $\xrightarrow{u_1}$ A $\xrightarrow{u_2}$ B m_2

after $\xrightarrow{v_1}$ $\xrightarrow{v_2}$

Fig. 15.1

v_2 respectively. By the principle of conservation of momentum applied along their line of centres, there holds the relation (13.1). Further, the relative velocity of A to B (before impact) is $u_1 - u_2$ and the same (after impact) is $v_1 - v_2$. If e is the coefficient of restitution between the two spheres there follows the relation

$$(-e) \, (u_1 - u_2) \; = \; v_1 - v_2, \tag{15.1}$$

from the Newton's experimental law. Thus, the relations (13.1) and (15.1) characterize the impact under consideration.

Example 15.1. A sphere A of mass 4 kg. moving at the speed of 5 m./sec. collides directly with another sphere B of mass 3 kg. approaching the first sphere at the speed of 4 m./sec. The coefficient of restitution between the spheres is 1/6. Find their velocities after the impact.

Solution. Since the spheres are approaching each other it can be in terpreted that the velocity u_2 (cf. Fig. 15.1) is -4 m./sec. Therefore, the equations (13.1) and (15.1) assume the forms:

$$4 \times 5 + 3(-4) = 8 = 4v_1 + 3v_2,$$

and

$$(-1/6) \, (5 + 4) = -3/2 = v_1 - v_2.$$

Solving these equations, we get $v_1 = 1/2 = 0 \cdot 5$ m./sec. and $v_2 = 2$ m./sec. //

§ 16. Relative motion

Let two objects A and B move with velocities v_A and v_B respectively. The velocity of A relative to B is $v_A - v_B$ and that of B relative to A is $v_B - v_A$.

Example 16.1. A particle A has velocity $3\,\hat{\imath} + 4\,\hat{\jmath}$ and that of B is $6\,\hat{\imath} - 2\,\hat{\jmath}$. Find their relative velocities to each other.

Solution. The velocity of A relative to B is

$$3\,\hat{\imath} + 4\,\hat{\jmath} - (6\hat{\imath} - 2\,\hat{\jmath}) = -3\,\hat{\imath} + 6\,\hat{\jmath}.$$

Similarly, the velocity of B relative to A is

$$6\hat{\imath} - 2\hat{\jmath} - (3\hat{\imath} + 4\,\hat{\jmath}) = 3\hat{\imath} - 6\hat{\jmath}. \; //$$

Note 16.1. If the particles move in the same direction along a line then velocity of A relative to B is the resultant velocity of v_A and $- v_B$. On the other hand, if they move in opposite directions along the same line then velocity of A relative to B is the resultant velocity of v_A and v_B.

Generalizing this concept of relative motion for the particles moving in different directions at an angle α to each other, the velocity of A relative to B is the resultant of forces v_A and $- v_B$ (inclined at angle $\pi - \alpha$). Thus, by Eq. (1.1.8), it is

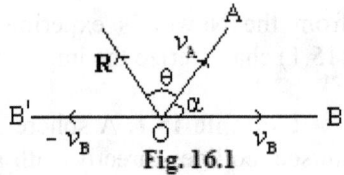

Fig. 16.1

$$R = \sqrt{\{(v_A)^2 + (v_B)^2 + 2\,v_A\,v_B \cos (\pi - \alpha)\}}, \qquad (1.1)$$

The direction of relative velocity to that of v_A can be calculated by Eq. (1.1.9)

$$\tan \theta = \{v_B \sin (\pi - \alpha)\} / \{v_A + v_B \cos (\pi - \alpha)\}$$

$$= (v_B \sin \alpha) / (v_A - v_B \cos \alpha). \qquad (1.2)$$

Example 16.2. A ship A is sailing North-East at a speed of 20 knots while another ship B is sailing on a bearing of 300° at a speed of 12 knots. Find the velocity of B relative to A.

Solution. *Knot* is the unit of velocity of a ship and is given in nautical miles per hour. (One nautical mile is approximately 2,000 metres). The direction of sailing of a ship from the true north (measured clockwise) is called the *"bearing"* in naval terms. Thus, the ship A is sailing at 45° angle (clockwise) to the true north while B is sailing at angle 60° (anti-clockwise) from the true north. The velocity of B relative to A, as per above discussion, is the resultant R of velocity of B and – 20 K. velocity of A (directed towards south–west). Thus, the angle between their directions is 30° + 45° = 75°. Hence, by Eq. (1.1.8), we have

$$R = \sqrt{\{(12)^2 + (-20)^2 + 2 \times 12(-20) \cos 75°\}} = \sqrt{(544 + 480.\cos 75°)}.$$

But,

$$\cos 75° = \cos (30 + 45) = \cos 30. \cos 45 - \sin 30. \sin 45 = (\sqrt{3} - 1) / 2\sqrt{2}$$

$$\Rightarrow$$

$$480. \cos 75 = 120.\sqrt{2}. (\sqrt{3} - 1)$$

$$\approx 120 \times 1·414 \times ·732 \approx 124·206.$$

Hence,

$$R = \sqrt{(544 + 124·206)}$$

$$= \sqrt{(668·206)} \approx 26 \text{ knots.}$$

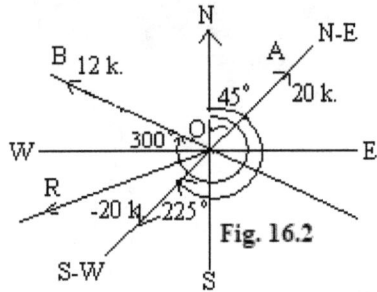

Fig. 16.2

The angle made by R to the South-West direction may be obtained by Eq. (1.1.9)

$$\tan \theta = (12.\sin 75) / (20 + 12. \cos 75)$$

$$= 12\{(\sqrt{3} + 1) / 2\sqrt{2}\} / \{20 + 12(\sqrt{3} - 1) / 2\sqrt{2}\}$$

$$= 3 \times 2·732 / (10\sqrt{2} + 3 \times ·732) = 8·196 / 16·336 \approx 1/2$$

$\Rightarrow \theta \approx 27°$ (measured clockwise). So, R has a bearing of 225 + 27 = 252° approximately. //

CHAPTER 13

FOURIER TRANSFORMS OF FUNCTIONS

§ 1. Introduction

The Fourier transform of the function f is traditionally denoted by adding a circumflex: $\hat{f}$. There are several common conventions for defining the Fourier transform of an integrable function $f: \mathrm{R} \rightarrow \mathrm{C}$.

Here we use the following definition:

$$\hat{f}(t) = \int_{-\infty}^{\infty} f(x) \cdot e^{-2i\pi xt} \cdot dt$$

for any real number t. When the independent variable x represents *time*, the transform variable t represents frequency (e.g. if time is measured in seconds, then the frequency is in *hertz*). Under suitable conditions, f is determined by $\hat{f}$ via the **inverse transform**:

$$f(x) = \int_{-\infty}^{\infty} \hat{f}(t) \cdot e^{2i\pi xt} \cdot dt$$

for any real number x. The reason for the negative sign convention in the definition of $\hat{f}(t)$ is that the integral produces the amplitude and phase of the function $f(x) \cdot e^{-2i\pi xt}$ at frequency zero (0), which is identical to the amplitude and phase of the function $f(x)$ at frequency t, which is what $\hat{f}$ (t) is supposed to represent.

Note 1.1. There is a close connection between the definition of Fourier series and the Fourier transform for functions f that are zero outside an interval. For such a function, we can calculate its Fourier series on any interval that includes the points where f is not identically zero. The Fourier transform is also defined for such a function. As we increase the length of the interval on which we calculate the Fourier series, then the Fourier series coefficients begin to look like the Fourier transform and the sum of the Fourier series of f begins to look like the inverse Fourier transform.

§ 2. Properties of the Fourier transform

Here we consider $f(x)$, $g(x)$ and $h(x)$ as *integrable functions* on the real line satisfying:

$$\int_{-\infty}^{\infty} |f(x)| \, dx < \infty.$$

We denote the Fourier transforms of these functions by $\hat{f}(t)$, $\hat{g}(t)$ and $\hat{h}(t)$ respectively.

The Fourier transform has the following basic properties:

(i) **Linearity:** For any complex numbers a and b, if

$$h(x) = a f(x) + b g(x), \qquad \text{then } \hat{h}(t) = a \cdot \hat{f}(t) + b \cdot \hat{g}(t).$$

(ii) **Translation / time shifting:** For any real number x_0, if

$$h(x) = f(x - x_0), \text{ then} \qquad \hat{h}(t) = \exp(-2\pi i x_0 t). \hat{f}(t).$$

(iii) **Modulation / frequency shifting:** For any real number t_0, if

$$h(x) = \exp(2\pi i x t_0). f(x), \text{ then} \qquad \hat{h}(t) = \hat{f}(t - t_0).$$

(iv) **Time scaling:** For a non-zero real number a, if

$$h(x) = f(ax), \qquad \text{then} \qquad \hat{h}(t) = (1/|a|)\hat{f}(t/a).$$

The case $a = -1$ leads to the *time-reversal* property, which states: if

$$h(x) = f(-x), \qquad \text{then} \qquad \hat{h}(t) = \hat{f}(-t).$$

§ 3. Fourier series

Let $f(x)$ be some function of x defined over some interval, say $[-\pi, \pi]$ of the real line and it satisfies certain conditions. We know, by Taylor's theorem of differential calculus, the function can be expanded in infinite series of terms in polynomial form about a given point x_0 in the interval:

$$f(x) = \sum_{n=0}^{\infty} a_n (x - x_0)^n. \tag{3.1}$$

Expansion of $f(x)$ in an infinite series whose terms need not be poly nomials is also possible. One of the most useful such expansion is the following trigonometric series:

$$f(x) = a_0 + (a_1 \cos x + a_2 \cos 2x + \ldots + a_n \cos nx + \ldots)$$

$$+ (b_1 \sin x + b_2 \sin 2x + \ldots + b_n \sin nx + \ldots)$$

$$= a_0 + \sum_{n=1}^{\infty} (a_n \cos nx + b_n \sin nx), \qquad (3.2)$$

so that each term in the series is term-wise integrable. Above series is called a *Fourier series* for the function $f(x)$ over the given interval $[-\pi, \pi]$. The coefficients $a_0, a_1, a_2, \ldots, a_n; b_1, b_2, \ldots, b_n$ are called the *Fourier coefficients*.

§ 4. Evaluation of Fourier coefficients

4.1. Integrating the series in Eq. (3.2) over the interval $[-\pi, \pi]$, and applying the properties of definite integrals:

$$\int_{-\pi}^{\pi} \cos nx \, dx = 2 \int_{0}^{\pi} \cos nx \, dx = (2/n) \left[\sin nx \right]_{0}^{\pi} = 0 \quad (4.1)$$

$$\int_{-\pi}^{\pi} \sin nx \, dx = 0, \ (\sin x \text{ being an odd function of } x) \quad (4.2)$$

we evaluate the Fourier coefficients:

$$\int_{-\pi}^{\pi} f(x) \, dx = a_0 \int_{-\pi}^{\pi} 1 \, dx = 2a_0 \int_{0}^{\pi} 1 \, dx = 2\pi a_0$$

$$\Rightarrow \qquad a_0 = (1/2\pi) \int_{-\pi}^{\pi} f(x) \, dx. \qquad (4.3)$$

4.2. Multiplying the series in Eq. (3.2) by $\cos mx$ and integrating the result so obtained over the interval $[-\pi, \pi]$, we get

$$\int_{-\pi}^{\pi} f(x). \cos mx. \, dx = a_0. \int_{-\pi}^{\pi} \cos mx. \, dx$$

$$+ \sum_{n=1}^{\infty} a_n. \int_{-\pi}^{\pi} \cos mx. \cos nx. \, dx + \sum_{n=1}^{\infty} b_n \int_{-\pi}^{\pi} \cos mx. \sin nx \, dx. \quad (4.4)$$

The integrands $\cos mx. \cos nx$ (respectively $\cos mx. \sin nx$) being even (resp. odd) functions, for Eqs. (4.1), (4.2) and the properties of definite integrals, Eq. (4.4) simplifies to

$$\int_{-\pi}^{\pi} f(x).\cos mx\ dx = \sum_{n=1}^{\infty} a_n \int_{0}^{\pi} \{\cos (m-n)\,x + \cos (m+n)\,x\ dx.(4.5)$$

$$= \sum_{n=1}^{\infty} a_n \left[\{\sin (m-n)x\}/(m-n) + \{\sin (m+n)x\}/(m+n)\right]_{0}^{\pi} = 0,\ (4.6)$$

when $n \neq m$. On the other hand, when $n = m$, Eq. (4.5) determines

$$\int_{-\pi}^{\pi} f(x) \cos mx\ dx = a_m \int_{0}^{\pi} \{1 + \cos 2mx)\ dx$$

$$= a_m \left[x + (\sin 2mx)/2m\right]_{0}^{\pi} = \pi a_m$$

$$\Rightarrow \qquad a_m = (1/\pi) \int_{-\pi}^{\pi} f(x).\cos mx\ dx. \qquad (4.7)$$

4.3. Multiplying the series in Eq. (3.2) by $\sin mx$ and integrating the result so obtained over the interval $(-\pi, \pi)$, we also find

$$\int_{-\pi}^{\pi} f(x) \sin mx\ dx = a_0 \int_{-\pi}^{\pi} \sin mx + \sum_{n=1}^{\infty} a_n \int_{-\pi}^{\pi} \sin mx.\cos nx\ dx$$

$$+ b_n \int_{-\pi}^{\pi} \sin mx.\sin nx\ dx.$$

Again, for Eq. (4.2) and the properties of definite integrals, above equation simplifies to

$$\int_{-\pi}^{\pi} f(x).\sin mx\ dx = \sum_{n=1}^{\infty} b_n \int_{0}^{\pi} \{\cos (m-n)\,x - \cos (m+n)\,x\ dx\ (4.8)$$

$$= \sum_{n=1}^{\infty} b_n \left[\{\sin (m-n)x\}/(m-n) - \{\sin (m+n)x\}/(m+n)\right]_{0}^{\pi} = 0,$$

when $n \neq m$. On the other hand, when $n = m$, Eq. (4.8) determines

$$\int_{-\pi}^{\pi} f(x) \sin mx\ dx = b_m \int_{0}^{\pi} (1 - \cos 2mx)\ dx$$

$$= b_m \left[x - (\sin 2mx)/2m\right]_{0}^{\pi} = \pi b_m$$

$$\Rightarrow \qquad b_m = (1/\pi) \int_{-\pi}^{\pi} f(x). \sin mx \, dx, \qquad (4.9)$$

Putting from Eqs. (4.3), (4.7) and (4.9), the Fourier series in Eq. (3.2) for the function $f(x)$ assumes the form:

$$f(x) = (1/2\pi) \int_{-\pi}^{\pi} f(x) \, dx + (1/\pi) \sum_{n=1}^{\infty} \left[\left\{ \int_{-\pi}^{\pi} f(x) \cos nx. \, dx \right\} \cos nx \right.$$

$$\left. + \left\{ \int_{-\pi}^{\pi} f(x). \sin nx. \, dx \right\}. \sin nx \right] \qquad (4.10)$$

$$= (1/2\pi) \int_{-\pi}^{\pi} f(t) \, dt + (1/\pi) \sum_{n=1}^{\infty} \int_{-\pi}^{\pi} f(t) \left\{ \cos nt. \cos nx \right.$$

$$\left. + \sin nt. \sin nx \right\} dt$$

$$= (1/2\pi) \int_{-\pi}^{\pi} f(t) \, dt + (1/\pi) \sum_{n=1}^{\infty} \int_{-\pi}^{\pi} f(t). \cos \left\{ n \left(t - x \right) \right\} dt. \quad (4.11)$$

4.4. Particular cases

(i) When $f(x)$ is an even function of x, i.e. $f(-x) = f(x)$, the Fourier coefficients reduce to

$$a_0 = (1/\pi) \int_{0}^{\pi} f(x) \, dx, \quad a_n = (2/\pi) \int_{0}^{\pi} f(x). \cos nx. \, dx, \quad b_n = 0, \quad (4.12)$$

for the odd character of the integrand in Eq. (4.9) causing vanishing of the integral. Consequently, the Fourier series given by Eq. (4.10), for such a function, reduces to

$$f(x) = a_0 + \sum_{n=1}^{\infty} a_n \cos nx = (1/\pi) \left[\int_{0}^{\pi} f(x) \, dx \right.$$

$$\left. + 2 \sum_{n=1}^{\infty} \left\{ \int_{0}^{\pi} f(x). \cos nx \, dx \right\} \cos nx \right] \qquad (4.10a)$$

(ii) On the other hand, if $f(x)$ is an odd function of x, i.e. $f(-x) = -f(x)$, the Fourier coefficients reduce to

$$a_0 = 0, \quad a_n = 0, \quad b_n = (2/\pi) \int_{0}^{\pi} f(x). \sin nx \, dx \quad (4.13)$$

again for the odd character of the integrand in Eq. (4.7) causing vanishing of the integral. Consequently, the Fourier series given by Eq. (4.10), for an odd function, reduces to

$$f(x) = \sum_{n=1}^{\infty} b_n \sin nx = (2/\pi) \sum_{n=1}^{\infty} \{ \int_0^{\pi} f(x). \sin nx \, dx \}. \sin nx]. \quad (4.10b)$$

Thus, we conclude that:

(i) The Fourier series for an *even* function consists of a constant term a_0 and the *cosine* series;

(ii) The Fourier series for an *odd* function consists of only a *sine* series.

Example 4.1. Find the Fourier series for the function $f(x) = x$ defined in the interval $[-\pi, \pi]$.

Solution. The function being odd, the Fourier coefficients can be evaluated by (4.13): $a_0 = 0$, $a_n = 0$, and

$$b_n = (2/\pi) \int_0^{\pi} x. \sin nx. \, dx = (2/n\pi)\{ [-x.\cos nx]_0^{\pi} + \int_0^{\pi} \cos nx. \, dx \}$$

$$= (2/n\pi) \{ -\pi. \cos n\pi + (1/n). [\sin nx]_0^{\pi} \}$$

$$= (2/n\pi)\{ -\pi. \cos n\pi + (1/n) \sin n\pi \} = -(2/n). \cos n\pi = (2/n)(-1)^{n+1}.$$

Hence, the Fourier series in Eq. (3.2) reduces to

$$f(x) \equiv x = \sum_{n=1}^{\infty} (2/n)(-1)^{n+1}. \sin nx$$

$$= 2\{(\sin x)/1 - (\sin 2x)/2 + (\sin 3x)/3 - (\sin 4x)/4 + \dots \infty \}. // \quad (4.14)$$

§ 5. Fourier series in any interval

5.1. Interval $[-l, l]$: The equation (4.11) gives a Fourier series expansion for the function $f(x)$ in the interval $(-\pi, \pi)$. Analogously, for some function $F(y)$ also defined in the same interval $[-\pi, \pi]$, Eq. (4.11) reads as

$$F(y) = (1/2\pi) \int_{-\pi}^{\pi} F(t)\, dt + (1/\pi) \sum_{n=1}^{\infty} \int_{-\pi}^{\pi} F(t)\cos\{n(t-y)\}dt. \quad (5.1)$$

Applying the change of parameters:

$$t = \pi u / l \quad \Rightarrow \quad dt = (\pi.\,du)/l, \quad \text{and} \quad y = \pi x / l. \quad (5.2)$$

Eq. (5.1) assumes the form

$$F(\pi x / l) = (1/2l) \int_{-l}^{l} F(\pi u / l)\, du$$

$$+ (1/l) \sum_{n=1}^{\infty} \int_{-l}^{l} F(\pi u / l)\cos\{n\pi(u-x)/l\}.\,du$$

Writing $F(\pi x / l)$ as $f(x)$, and similarly $F(\pi u / l)$ as $f(u)$ above relation further alters as

$$f(x) = (1/2l) \int_{-l}^{l} f(u)\, du + (1/l) \sum_{n=1}^{\infty} \int_{-l}^{l} f(u)\cos\{n\pi(u-x)/l\}.du$$

$$= (1/2l) \int_{-l}^{l} f(u)\, du + (1/l) \sum_{n=1}^{\infty} \int_{-l}^{l} f(u)\{\cos(n\pi u/l)\cos(n\pi x/l)$$

$$+ \sin(n\pi u / l)\sin(n\pi x / l)\}\, du, \quad (5.3)$$

which is the desired Fourier series expansion of $f(x)$ in interval $[-l, l]$ together with the Fourier coefficients:

$$a_0 = (1/2l) \int_{-l}^{l} f(u)\, du, \; a_n = (1/l) \int_{-l}^{l} f(u)\cos(n\pi u/l)\, du \quad \Big\}$$

$$b_n = (1/l) \int_{-l}^{l} f(u)\sin(n\pi u / l)\, du. \quad \Big\} \quad (5.4)$$

5.2. Interval $[0, \pi]$: Let a function $f(x)$ be defined on the interval $[0, \pi]$. In order to obtain a *cosine* series expansion for it on this interval, we construct a new function, say $g(x)$ on the extended interval $[-\pi, \pi]$ as follows:

$$g(x) = f(-x) \text{ (when } -\pi \le x < 0), \text{ and } f(x) \text{ (when } 0 \le x \le \pi). \quad (5.5)$$

As such, the new function is even on the extended interval and agrees with the given function $f(x)$ on the interval $[0, \pi]$. Thus, $g(x)$ can be expanded in a Fourier series as discussed in Sub-section 4.4, Case (i). The Fourier coefficients for the same are

$$a_0 = (1/2\pi) \int_{-\pi}^{\pi} g(x)\, dx = (1/\pi) \int_{0}^{\pi} g(x)\, dx = (1/\pi) \int_{0}^{\pi} f(x)\, dx$$

$$a_n = (1/\pi) \int_{-\pi}^{\pi} g(x) \cos nx.\, dx = (2/\pi) \int_{0}^{\pi} g(x).\cos nx.\, dx$$

$$= (2/\pi) \int_{0}^{\pi} f(x).\cos nx\, dx,$$

and

$$b_n = (1/\pi) \int_{-\pi}^{\pi} g(x).\sin nx\, dx = 0,$$

as the integrand $g(x).\sin nx$ then becomes an odd function of x causing vanishing of the last integral. We note that above values of the Fourier coefficients are in accordance with Eq. (4.12). Consequently, the Fourier series for $g(x)$ on the interval $[-\pi, \pi]$ turns out to be a purely *cosine* series given by Eq. (4.10a) on $[-\pi, \pi]$. For $g(x)$ being same as $f(x)$ on the interval $[0, \pi]$ the same *cosine* series is valid for $f(x)$ too on the interval $[0, \pi]$.

Note 5.1. Defining $g(x)$ as even function on the interval $[-\pi, \pi]$ in analogy with Eq. (5.5), we note that all the Fourier coefficients vanish for $\cos x$ except a_1:

$$a_0 = (1/\pi) \int_{0}^{\pi} \cos x.\, dx = (1/\pi)\left[\sin x\right]_{0}^{\pi} = 0,$$

$$a_n = (2/\pi) \int_{0}^{\pi} \cos x.\cos nx.\, dx$$

$$= (1/\pi) \int_{0}^{\pi} \{\cos(n+1)x + \cos(n-1)x\}\, dx$$

$$= (1/\pi)\left[\{\sin(n+1)x\}/(n+1) + \{\sin(n-1)x\}/(n-1)\right]_{0}^{\pi} = 0, \text{ when } n \neq 1;$$

$$a_1 = (1/\pi) \int_{0}^{\pi} (\cos 2x + 1)\, dx = (1/\pi)\left[(\sin 2x)/2 + x\right]_{0}^{\pi} = 1$$

and $b_n = 0$. Thus, a *cosine* series for $\cos x$ reduces to just $\cos x$.

5.3. A sine series in the interval $[0, \pi]$: To obtain a *sine* series expansion for a function $f(x)$ defined on the interval $[0, \pi]$, we must construct a new function, say $g(x)$ on the extended interval $[-\pi, \pi]$ which should be odd:

$$g(x) = f(x) \text{ (when } 0 \leq x \leq \pi) \text{ and } -f(x) \text{ (when } -\pi \leq x < 0), \quad (5.6)$$

and agree with the given function $f(x)$ on the interval $[0, \pi]$. The Fourier coefficients for the same are given by Eq. (4.13). Accordingly, its Fourier series expansion turns out to be purely a *sine* series as in Eq. (4.10b). Since $g(x)$, as per definition, agrees with $f(x)$ on the interval $[0, \pi]$ its Fourier series in the interval $[0, \pi]$ becomes the same as for $f(x)$.

5.4. The interval $[0, l]$: As seen in the Sub-section 5.2, a function can be expanded in the interval $[0, \pi]$ in a *cosine* series of the form in Eq. (4.10a). Re-writing it for a function $F(y)$:

$$F(y) = (1/\pi) \left[\int_0^\pi F(t)\, dt + 2 \sum_{n=1}^\infty \left\{ \int_0^\pi F(t) \cos nt.\, dt \right\} \cos ny \right]. \quad (5.7)$$

Applying the change of parameters as in Eq. (5.2), above result assumes the form:

$$F(\pi x / l) = (1/l) \left[\int_0^l F(\pi u / l)\, du \right.$$

$$\left. + 2 \sum_{n=1}^\infty \left\{ \int_0^l F(\pi u / l).\cos (n\pi u / l)\, du \right\}. \cos (n\pi x / l) \right]$$

Writing $F(\pi x/l) \equiv f(x)$, and $F(\pi u/l) \equiv f(u)$, above relation further alters as

$$f(x) = (1/l) \left[\int_0^l f(u)\, du \right.$$

$$\left. + 2 \sum_{n=1}^\infty \left\{ \int_0^l f(u) \cos (n\pi u / l)\, du \right\} \cos (n\pi x / l) \right], \quad (5.8)$$

which is a *cosine* series.

The *sine* series expansion of a function $f(x)$ in the interval $[0, \pi]$ is discussed in the Sub-section 5.3 and it is given by Eq. (4.10b). Re-writing it for a function $F(y)$:

$$F(y) = (2/\pi) \sum_{n=1}^{\infty} \{\int_0^{\pi} F(t) \sin nt. \, dt\}. \sin ny, \qquad (5.9)$$

and applying the change of parameters given by Eq. (5.2), above equation reduces to

$$F(\pi x / l) = (2 / l) \sum_{n=1}^{\infty} \{\int_0^l F(\pi u/l) \sin (n\pi u/l) \, du\} \sin (n\pi x/l)$$

or, $\quad f(x) = (2/l) \sum_{n=1}^{\infty} \{\int_0^l f(u) \sin (n\pi u / l) \, du\} \sin (n\pi x / l). \; // \quad (5.10)$

5.5. The interval $[0, 2\pi]$: Putting $t = u - \pi$ so that $dt = du$ and $y = x - \pi$ in Eq. (5.1) we deduce

$$F(x - \pi) = (1/2\pi) \int_{u=0}^{2\pi} F(u - \pi) \, du$$

$$+ (1/\pi) \sum_{n=1}^{\infty} \int_{u=0}^{2\pi} F(u - \pi) \cos \{n(u - x)\} du.$$

Writing $F(x - \pi)$ as $f(x)$, and $F(u - \pi)$ as $f(u)$ above relation further alters as

$$f(x) = (1/2\pi) \int_{u=0}^{2\pi} f(u) \, du + (1/\pi) \sum_{n=1}^{\infty} \int_{u=0}^{2\pi} f(u) \cos \{n(u - x)\} \, du$$

$$(5.11a)$$

$$= (1 / 2\pi) \int_{u=0}^{2\pi} f(u) \, du$$

$$+ (1/\pi) \sum_{n=1}^{\infty} \int_{u=0}^{2\pi} f(u) \{\cos nu. \cos nx + \sin nu. \sin nx)\} \, du$$

$$= (1/2\pi) \int_{u=0}^{2\pi} f(u) \, du + (1/\pi) \sum_{n=1}^{\infty} [\{\int_{u=0}^{2\pi} f(u).\cos nu. \, du\} \cos nx$$

$$+ \{\int_{u=0}^{2\pi} f(u). \sin nu \, .du\} \sin nx]. \qquad (5.11b)$$

Note 5.2. Comparing the results given by Eqs. (5.11a) and (5.11b) with Eqs. (4.11) and (4.10) respectively, we note that the difference lies in the limits only.

5.6. The interval $[0, 1]$: Rewriting the Fourier series expansion given by Eq. (5.11a) for a function $F(y)$ defined in the interval $[0, 2\pi]$:

$$F(y) = (1/2\pi) \int_{t=0}^{2\pi} F(t)\, dt$$

$$+ (1/\pi) \sum_{n=1}^{\infty} \int_{t=0}^{2\pi} F(t) \cos \{n(t-y)\} dt, \tag{5.12}$$

and applying the change of parameters $t = 2\pi u$ so that $dt = 2\pi\, du$ and $y = 2\pi x$, we derive

$$F(2\pi x) = \int_{u=0}^{1} F(2\pi u)\, du + 2 \sum_{n=1}^{\infty} \int_{u=0}^{1} F(2\pi u) \cos \{2n\pi(u-x)\} du.$$

Setting $F(2\pi x) = f(x)$ and $F(2\pi u) = f(u)$, above expansion reads as

$$f(x) = \int_{u=0}^{1} f(u)\, du + 2 \sum_{n=1}^{\infty} \int_{u=0}^{1} f(u) \cos \{2n\pi(u-x)\}\, du$$

$$= \int_{u=0}^{1} f(u)\, du + 2 \sum_{n=1}^{\infty} [\{\int_{u=0}^{1} f(u) \cos (2n\pi u).\, du\} \cos (2n\pi x)$$

$$+ \{\int_{u=0}^{1} f(u) \sin (2n\pi u).du\}.\sin (2n\pi x)]. \tag{5.13}$$

Particularly, the *cosine* (respectively *sine*) series expansions for a function defined in the interval [0, 1] can be found in analogy with Eq. (5.8) respectively Eq. (5.10):

$$f(x) = \int_{u=0}^{1} f(u)\, du$$

$$+ 2 \sum_{n=1}^{\infty} \{\int_{u=0}^{1} f(u) \cos (n\pi u)\, du\}.\cos (n\pi x) \tag{5.14}$$

(respectively)

$$f(x) = 2 \sum_{n=1}^{\infty} \{\int_{u=0}^{1} f(u) \sin (n\pi u)\, du\}.\sin (n\pi x). \,// \tag{5.15}$$

5.7. The interval [a, b]: Applying the change of parameters

$t = 2\pi(u-a)/(b-a) \Rightarrow dt = 2\pi\, du/(b-a)$ and $y = 2\pi(x-a)/(b-a)$,

the equation (5.20) reads as

$$F\{2\pi(x-a)/(b-a)\} = [\int_{u=a}^{b} F\{2\pi(u-a)/(b-a)\}\, du$$

$$+ 2 \sum_{n=1}^{\infty} \int_{u=a}^{b} F \{2\pi (u-a)/(b-a)\} \cos \{2n\pi (u-x)/(b-a)\} du]/(b-a)$$

or, $$f(x) = [\int_{u=a}^{b} f(u)\, du$$

$$+ 2 \sum_{n=1}^{\infty} \int_{u=a}^{b} f(u) \cos \{2n\pi (u-x)/(b-a)\} du] / (b-a). \quad (5.16)$$

§ 6. Fourier series for piecewise defined functions

Let a function $f(x)$ be defined on some closed interval $[a, b]$ at points of the interval except for a finite number of points, say $x_1, x_2,...,$ x_n. It is called *piecewise continuous* on the interval $[a, b]$ if:

(i) it is continuous on each sub–intervals $(a, x_1), (x_1, x_2),..., (x_{n-1},$ $x_n), (x_n, b)$;

(ii) it possesses a finite limit from the right at the (left) end $x = a$, a finite limit from the left at the (right) end $x = b$, and both left and right limits at each point $x_i, i = 1, 2, ..., n$.

We shall obtain Fourier series expansions for such functions in the following. The method is demonstrated by means of some examples.

Example 6.1. Find the Fourier series for the function defined on the interval $[-2, 2]$:

$$f(x) = 0 \text{ (when } -2 \le x < 0), \ 1 \ (\text{ when } 0 \le x < 1), \ 2 \text{ (when } 1 \le x \le 2).$$

Solution. A Fourier series expansion of a function defined over the interval $[-l, l]$ is obtained by Eq. (5.3) and the corresponding Fourier coefficients by Eq. (5.4). Thus, for $l = 2$, we have

$$a_0 = (1/4) \int_{-2}^{2} f(x)\, dx = (1/4) \{\int_{-2}^{0} + \int_{0}^{1} + \int_{1}^{2} \} f(x)\, dx$$

$$= (1/4) \{ 0 + \int_{0}^{1} dx + 2 \int_{1}^{2} dx \} = (1/4) \{ [x]_0^1 + 2 [x]_1^2 \}$$

$$= (1/4) (1 + 2) = 3/4,$$

$$a_n = (1/2) \int_{-2}^{2} f(x) \cos (n\pi x/2) \, dx$$

$$= (1/2) \left\{ \int_{-2}^{0} + \int_{0}^{1} + \int_{1}^{2} \right\} . f(x) \cos (n\pi x/2) \, dx$$

$$= (1/2) \left\{ \int_{0}^{1} \cos (n\pi x/2) \, dx + 2 \int_{1}^{2} \cos (n\pi x/2) \, dx \right\}$$

$$= (1/n\pi) \left\{ \left[\sin (n\pi x/2) \right]_{0}^{1} + 2 \left[\sin (n\pi x/2) \right]_{1}^{2} \right\}$$

$$= (1/n\pi) [\sin (n\pi/2) + 2\{\sin n\pi - \sin (n\pi/2)\}] = -\{\sin (n\pi/2)\}/n\pi,$$

and

$$b_n = (1/2) \int_{-2}^{2} f(x) \sin (n\pi x/2) \, dx$$

$$= (1/2) \left\{ \int_{-2}^{0} + \int_{0}^{1} + \int_{1}^{2} \right\} . f(x) \sin (n\pi x/2) \, dx$$

$$= (1/2) \left\{ \int_{0}^{1} \sin (n\pi x/2) \, dx + 2 \int_{1}^{2} \sin (n\pi x/2) \, dx \right\}$$

$$= - (1/n\pi) \left\{ \left[\cos (n\pi x/2) \right]_{0}^{1} + 2 \left[\cos (n\pi x/2) \right]_{1}^{2} \right\}$$

$$= - (1/n\pi) [\cos (n\pi/2) - 1 + 2 \{ \cos (n\pi) - \cos (n\pi/2) \}]$$

$$= (1/n\pi) \{ 1 + \cos (n\pi/2) - 2. \cos n\pi \}.$$

Thus, the desired Fourier series for above function is

$$f(x) = 3/4 + (1/n\pi) . \sum_{n=1}^{\infty} [- \sin (n\pi/2) . \cos (n\pi x/2)$$

$$+ \{ 1 + \cos (n\pi/2) - 2.\cos n\pi \} . \sin (n\pi x/2)]. \; //$$

CHAPTER 14

GEOMETRY (COORDINATE: 2-dimensional)

§ 1. Rectangular Cartesian coordinates

Let O*x* and O*y* be two mutually perpendicular straight lines intersecting each other in the point O. They determine a unique plane called *xOy*- (or *xy*-) plane. The lines O*x* and O*y* are called the coordinate axes of *x* and *y* respectively and the point O as origin. The combination (O, O*x*, O*y*) may be called a *coordinate frame*.

Let P be any point in this plane at distances *x* (from *y*-axis) and *y* (from *x*-axis) then the pair (*x*, *y*) determines the rectangular Cartesian coordinates of P with respect to above coordinate frame. It is customary to write *x* (called the *abscissa*) as the first member of the pair while the second member *y* is the *ordinate* of P.

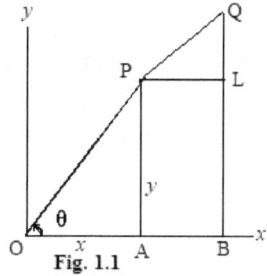

Fig. 1.1

Since *y* always remains zero along *x*-axis the latter is represented by equation

$$y = 0. \tag{1.1}$$

Similarly, *y*-axis has the equation

$$x = 0. \tag{1.2}$$

Lines parallel to the coordinate axes O*x* and O*y* are

$$y = a \quad (1.3); \quad \text{and} \quad x = b \tag{1.4}$$

respectively.

§ 2. Distance between two points

Let P(x_1, y_1) and Q (x_2, y_2) be two points the distance between them, by Pythagoras theorem, is given by (cf. Fig. 1.1)

$$(PQ)^2 = (PL)^2 + (LQ)^2 = (OB - OA)^2 + (BQ - BL)^2$$

$$= (x_2 - x_1)^2 + (y_2 - y_1)^2. \tag{2.1}$$

Particularly, the distance of P from origin O (0,0) is the positive square root:

$$OP = \sqrt{(x_1^2 + y_1^2)}. \qquad (2.2)$$

§ 3. Point dividing a line in a given ratio

Let the point P divide the join of two points P_1 (x_1, y_1) and P_2 (x_2, y_2) in the ratio $m: n$. The coordinates of P are

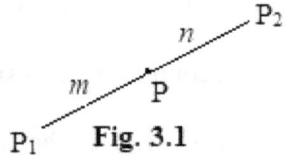
Fig. 3.1

$$x = (mx_2 + nx_1) / (m + n), \quad y = (my_2 + ny_1) / (m + n). \qquad (3.1)$$

The mid-point M of P_1P_2 bisects the line P_1P_2 and, thus, has the coordinates $\{(x_1 + x_2) / 2, (y_1 + y_2) / 2\}$.

§ 4. Equation of a straight line

4.1. Slope-intercept form: Let a straight line AB make an angle α to the x-axis and an intercept OC of length c on y-axis. If P (x, y) is any point on it with ordinate LP and CM perpendicular to LP, there holds

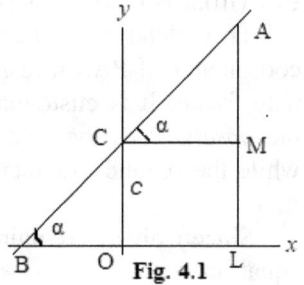
Fig. 4.1

$$\tan \alpha = MP / CM = (LP - LM) / OL = (y - OC) / x = (y - c) / x.$$

Thus, the Cartesian equation to AB is

$$y = mx + c \qquad (4.1), \quad \text{where } m \equiv \tan \alpha \qquad (4.2)$$

is called the *slope* (or *gradient*) of AB.

Note 4.1. Parallel lines have the same slope. Slopes of the axes Ox and Oy are zero and ∞ respectively.

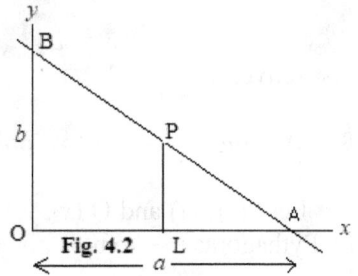
Fig. 4.2

4.2. Intercepts on axes: Let a straight line AB make intercepts OA and OB of lengths a and b on the respective coordinate axes. Taking a point P (x, y) on AB with ordinate LP there follows

$$PL / OB = LA / OA = (OA - OL) / OA = 1 - OL / OA,$$

$\Rightarrow$

$$OL \,/\, OA + PL \,/\, OB \,=\, 1, \qquad \text{i.e. } x\,/\,a + y\,/\,b \,=\, 1, \qquad (4.3)$$

from the similarity of triangles PLA and BOA.

4.3. Perpendicular form: Let AB be a line to which the perpendicular ON from O make angle α to the x-axis and be of length p. As such, the line AB makes angle $90° + \alpha$ (measured anti-clockwise) to the x-axis and intercept OB $= p$ cosec α on the y-axis. Hence, the equation to AB, by (4.1), is

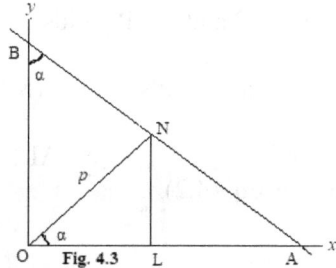

Fig. 4.3

$$y \,=\, x \tan (90 + \alpha) + p \text{ cosec } \alpha = \, -x \cot \alpha + p \text{ cosec } \alpha,$$

or,

$$x \cos \alpha + y \sin \alpha \,=\, p. \qquad (4.4)$$

Theorem 4.1. A linear equation in x, y

$$ax + by + c \,=\, 0, \qquad (4.5)$$

always represents a straight line provided the coefficients a and b do not vanish simultaneously, i.e. $a^2 + b^2 > 0$.

There arise three different cases for various choices of a and b:

(i) Both a, b being non-zero, Eq. (4.5) may be rewritten as

$$x\,/\,(-c/a) + y\,/\,(-c/b) \,=\, 1,$$

which, by Eq. (4.3), represents a straight line in the intercept form.

(ii) If $a = 0$, $b \neq 0$, Eq. (4.5) takes the form $y = -c/b$ representing a line parallel to x-axis.

(iii) If $a \neq 0$, $b = 0$, Eq. (4.5) reduces to the form $x = -c/a$ giving a straight line parallel to y-axis.

Thus, in either case Eq. (4.5) represents a straight line. //

Note 4.2. The line with Eq. (4.1) passes through origin if $c = 0$.

4.4. Line through a given point: Let A $(x_1,$ $y_1)$ be a given point and a line AB through A make angle α to the x-axis. Taking P (x, y) any current point on AB with ordinate PM and AN as the perpendicular from A to PM, similarity of right triangles ANP and CMP yields

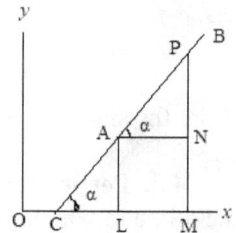
Fig. 4.4

$$\tan \alpha = PN \,/\, AN = (MP - MN) \,/\, LM$$

$$= (y - AL) \,/\, (OM - OL) \;=\; (y - y_1) \,/\, (x - x_1),$$

or, for Eq. (4.2),

$$y - y_1 \;=\; m\,(x - x_1). \tag{4.6}$$

Alternately, allowing the line in Eq. (4.5) to pass through A there results the relation

$$ax_1 + by_1 + c \;=\; 0; \tag{4.7}$$

which on subtraction from Eq. (4.5) yields

$$a\,(x - x_1) + b\,(y - y_1) \;=\; 0.$$

Putting $m = -\,a/b$, above equation assumes the form of Eq. (4.6).

4.5. Line through two points: If A_1 (x_1, y_1) and A_2 (x_2, y_2) be two points on a line (4.5) there follow the relations (4.7) and

$$ax_2 + by_2 + c \;=\; 0. \tag{4.8}$$

Eliminating the unknown coefficients a, b, c from Eqs. (4.5), (4.7) and (4.8), we get the equation of the line in a determinant form:

$$\begin{vmatrix} x & y & 1 \\ x_1 & y_1 & 1 \\ x_2 & y_2 & 1 \end{vmatrix} = 0. \tag{4.9}$$

Subtracting the second row from the remaining rows:

$$\begin{vmatrix} x - x_1 & y - y_1 & 0 \\ x_1 & y_1 & 1 \\ x_2 - x_1 & y_2 - y_1 & 0 \end{vmatrix} = 0,$$

and expanding the resulting determinant along its last column, we derive

$$(x_2 - x_1)(y - y_1) - (y_2 - y_1)(x - x_1) = 0,$$

or,

$$y - y_1 = \{(y_2 - y_1) / (x_2 - x_1)\}(x - x_1). \tag{4.10}$$

Alternately, as the line with Eq. (4.6) also passes through A_2 (x_2, y_2), there results

$$y_2 - y_1 = m(x_2 - x_1). \tag{4.11}$$

Elimination of m from Eqs. (4.6) and (4.11) yields the desired equation.

§ 5. Angle between two lines

5.1. Let two lines AB and AC having slopes $m_1 = \tan\alpha_1$ and $m_2 = \tan\alpha_2$ intersect each other in the point A. An angle between them is given by

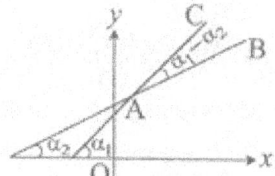

Fig. 5.1

$$\tan(\alpha_1 \sim \alpha_2) = (\tan\alpha_1 \sim \tan\alpha_2) / (1 + \tan\alpha_1 . \tan\alpha_2)$$

$$= (m_1 \sim m_2) / (1 + m_1 . m_2). \tag{5.1}$$

Accordingly,

$$\cos(\alpha_1 \sim \alpha_2) = (1 + m_1 m_2) / \sqrt{\{(1 + m_1{}^2)(1 + m_2)^2\}}. \tag{5.2}$$

Thus, lines are parallel (respectively perpendicular) to each other iff

$$m_1 = m_2 \quad (5.3) \quad (\text{resp.}) \quad m_1 m_2 = -1. \tag{5.4}$$

5.2. If the lines are represented by their general equations

$$a_1 x + b_1 y + c_1 = 0, \qquad a_2 x + b_2 y + c_2 = 0, \tag{5.5}$$

with their slopes $m_i = \tan\alpha_i = -a_i / b_i$, $i = 1, 2$, the formula (5.2) reduces to

$$\cos(\alpha_1 \sim \alpha_2) = (1 + a_1 a_2 / b_1 b_2) / \sqrt{[\{1 + (a_1 / b_1)^2\}\{1 + (a_2 / b_2)^2\}]}$$

$$= (a_1 a_2 + b_1 b_2) / \sqrt{\{(a_1{}^2 + b_1{}^2)(a_2{}^2 + b_2{}^2)\}}. \tag{5.6}$$

Accordingly, the conditions for their *parallelism* (resp. *orthogonality*) reduce to

$$a_1 / b_1 = a_2 / b_2 \qquad (5.7) \qquad (\text{resp.}) \qquad a_1 a_2 + b_1 b_2 = 0. \qquad (5.8)$$

§ 6. Perpendicular from a point to a line

6.1. First, we consider the line with Eq. (4.4) to which the perpendicular ON from origin is of length say p. Let P (x_1, y_1) be a point outside the line and the perpendicular PQ $= p_1$ (say) be drawn from it to the line. Draw a line PI through P parallel to line (4.4). So, it has the same gradient as line (4.4) and the perpendicular from the origin to it is of length $p + p_1$. Following Eq. (4.4), its equation is

$$x \cos \alpha + y \sin \alpha = p + p_1. \qquad (6.1)$$

Since P (x_1, y_1) lies on it there follows the desired length

$$p_1 = x_1 \cos \alpha + y_1 \sin \alpha - p. \qquad (6.2)$$

6.2. Now, we consider the general equation (4.5) of a line. Rewriting the same in the perpendicular form:

$$ax / \sqrt{(a^2 + b^2)} + by / \sqrt{(a^2 + b^2)} = -c / \sqrt{(a^2 + b^2)},$$

and comparing it with Eq. (4.4), we note that

$$\cos \alpha = a / \sqrt{(a^2 + b^2)}, \quad \sin \alpha = b / \sqrt{(a^2 + b^2)}, \quad p = -c / \sqrt{(a^2 + b^2)}.$$

Hence, Eq. (6.2) assumes the form

$$p_1 = (ax_1 + by_1 + c) / \sqrt{(a^2 + b^2)}, \qquad (6.3)$$

§ 7. Intersection of two lines

Any two lines in a common plane are either parallel to each other or they intersect in a unique point.

Theorem 7.1. Three points $A_i (x_i, y_i)$, $i = 1, 2, 3$, are *collinear* if and only if their coordinates satisfy

$$\begin{vmatrix} x_1 & y_1 & 1 \\ x_2 & y_2 & 1 \\ x_3 & y_3 & 1 \end{vmatrix} = 0 \qquad (7.1)$$

Proof. Let Eq. (4.5) represent the line through the given points so there also holds

$$ax_3 + by_3 + c = 0, \qquad (7.2)$$

together with Eqs. (4.7) and (4.8). Eliminating a, b, c from Eqs. (4.7), (4.8) and (7.2), we immediately get the condition (7.1).

Conversely, let there hold the condition (7.1). The same, on comparison with Eq. (4.9), locates the third point A_3 too on the line (4.9). This makes all three points collinear. //

§ 8. Circle

Definition 8.1. Locus of a point that moves in E_3 keeping a constant distance from a fixed point is called a *circle*. The fixed point is called the *centre* and the constant distance the *radius* of the circle.

Thus, the circle centered at the point A (x_i, y_i) and of radius r is represented by the equation

$$(x - x_1)^2 + (y - y_1)^2 = r^2, \qquad (8.1)$$

or,

$$x^2 + y^2 - 2(xx_1 + yy_1) + (x_1^2 + y_1^2 - r^2) = 0,$$

or,

$$x^2 + y^2 + 2gx + 2fy + c = 0, \qquad (8.2)$$

where $g = -x_1$, $f = -y_1$ and $c = x_1^2 + y_1^2 - r^2$.

Note 8.1. Equation (8.2) is the general form of the equation of a circle centered at $(-g, -f)$ and radius

$$r = \sqrt{(x_1^2 + y_1^2 - c)}. \qquad (8.3)$$

In particular, a circle with centre at origin and radius a has the equation

$$x^2 + y^2 = a^2, \qquad (8.4)$$

resulting from Eq. (8.1).

§ 9. Parabola

Definition 9.1. Locus of points moving in E_3 keeping equal distances from a fixed point (called the *focus*) and a fixed line (called the *directrix*) is called a *parabola*.

Taking the focus S at the point $(a, 0)$ and the line PK with equation

$$x + a = 0, \qquad (9.1)$$

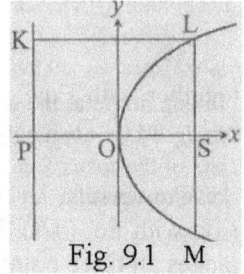

as the directrix the parabola is represented by the equation

$$y^2 = 4ax. \qquad (9.2)$$

It has its vertex at origin, axis along the x-axis and the tangent at the vertex along y-axis. The chord

Fig. 9.1   M

LM through the focus and perpendicular to the axis of the parabola is called the *latus rectum* of the curve and is of length $4a$.

Definition 9.2. The constant ratio of the distances of a moving point from the focus and from the directrix is called the *eccentricity* of the curve and is denoted by the letter e [2].

Thus, for a parabola, $e = 1$; whereas for a circle $e = 0$, as the focus of a circle coincides with its centre and the line of directrix is at infinite distance from the focus.

§ 10. Ellipse

Definition 10. 1. The path of a point P moving in E_2 with eccentricity less than 1 is called an *ellipse*.

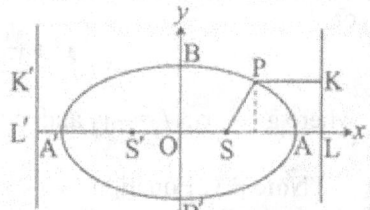

Fig. 10.1

Taking origin as the centre, $OA = a$ as the semi-major axis, $OB = b \,(< a)$ as the semi-minor axis, $e \,(< 1)$ the eccentricity, one of the focus (say S) at the point $(ae, 0)$ and the corresponding directrix with the equation $x = a/e$, the equation of the ellipse can be deduced as

$$x^2/a^2 + y^2/b^2 = 1, \qquad (10.1); \quad \text{where} \qquad b^2 = a^2(1 - e^2). \quad (10.2)$$

Note 10.1. If $e = 0$, Eq. (10.2) determines $b = a$, that reduces Eq. (10.1) to Eq. (8.4) representing a circle.

---

[2] Neither to be confused with the exponential value nor the identity element of an algebraic group.

§ 11. Hyperbola

When $e > 1$, the path traced by a point in E_2 becomes a hyperbola. Hence, for a hyperbola

$$1 - e^2 < 0. \qquad (11.1)$$

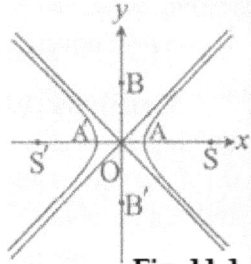

Fig. 11.1

Taking origin at the centre, transverse axis $AA' = 2a$ along x-axis, conjugate axis $BB' = 2b$ along y-axis and one of the focus S at the point $(ae, 0)$ the hyperbola has the equation

$$x^2/a^2 - y^2/b^2 = 1. \qquad (11.2)$$

Also, the Eq. (10.2), for Eq. (11.1), assumes the form

$$-b^2 = a^2(1 - e^2). \qquad (11.3)$$

Two asymptotes of the hyperbola are represented by the equation

$$x^2/a^2 - y^2/b^2 = 0; \quad \text{or,} \quad y = \pm(b/a)x. \qquad (11.4)$$

In particular, when $a = b$, the Eq. (11.2) assumes the form

$$x^2 - y^2 = a^2, \qquad (11.5)$$

and represents a *rectangular hyperbola*. Accordingly, its asymptotes with their equations $y = \pm x$ become mutually perpendicular. Rotating the coordinate axes through an angle $\pi/4$ anticlockwise (so that they coincide with the asymptotes) the equation of the rectangular hyperbola transforms to

$$xy = a^2/2 \equiv c^2. \qquad (11.6)$$

§ 12. Polar coordinates

Let O be a fixed point and a line OA through it be given. Let the line OA be rotated (anti-clockwise) about O through an angle θ in a given plane. Choosing a point P on the (rotated) line at a distance $OP = r$ from O the pair (r, θ) defines the polar coordinates of P with respect to O as pole and OA as the initial line. The linear distance r is called the *radius vector* and θ the *vectorial angle* of P with respect to above frame. Making this frame coincident with the rectangular Cartesian frame in Fig.

1.1, we find a relationship between the Cartesian coordinates (x, y) and the polar coordinates (r, θ):

$$OA = (OP) \cos \theta \quad \text{and} \quad AP = (OP) \sin \theta,$$

i.e.

$$x = r \cos \theta, \quad y = r \sin \theta \tag{12.1}$$

$\Rightarrow$

$$r = \sqrt{(x^2 + y^2)} \quad \text{and} \quad \theta = \tan^{-1}(y/x). \tag{12.2}$$

Hence, the Eq. (8.4) of a circle changes to polar coordinates as:

$$r = a. \tag{12.3}$$

The respective coordinate axes Ox and Oy have polar equations

$$\theta = 0 \quad \text{and} \quad \theta = \pi/2. \tag{12.4}$$

§ 13. General equation of second degree in x and y

The equation
$$a x^2 + b y^2 + 2 h x y + 2 (u x + v y) + d = 0, \tag{13.1}$$

where a, b, h do not vanish simultaneously, is a general equation of second degree in the variables x and y. If there holds the condition

$$D \equiv \begin{vmatrix} a & h & u \\ h & b & v \\ u & v & d \end{vmatrix} = 0, \tag{13.2}$$

it represents a pair of straight lines else a conic section in E_2. Further, if there also hold

$$u = v = d = 0, \tag{13.3}$$

which evidently imply Eq. (13.2), the Eq. (13.1) becomes homogeneous in (x, y) :

$$a x^2 + b y^2 + 2 h x y = 0. \tag{13.4}$$

It then represents a pair of straight lines each passing through origin.

In the following, we consider some standard conic sections:

13.1. When $a = b \neq 0$, but $h = 0$, the Eq. (13.1) assumes the form of Eq. (8.2) and represents a *circle*.

13.2. When $ab = h^2$, i.e. the second degree terms form a perfect square the conic section is a *parabola*.

Theorem 13.1. Eq. (13.1) represents an *ellipse* (respectively *hyperbola*) if there holds

$$ab > h^2 \qquad (13.5a); \quad (resp.) \qquad ab < h^2. \qquad (13.5b)$$

Proof. Rotating the coordinate axes Ox and Oy anticlockwise through an angle θ the coordinates (x, y) of a point P in the plane xOy transform to (X, Y) according to

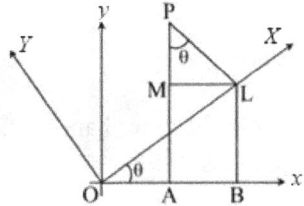
Fig. 13.1

$$x = OA = OB - AB = OB - ML$$

$$= (OL) \cos \theta - (LP) \sin \theta = X \cos \theta - Y \sin \theta, \qquad (13.6a)$$

and

$$y = AP = AM + MP = BL + MP$$

$$= (OL) \sin \theta + (LP) \cos \theta = X \sin \theta + Y \cos \theta. \qquad (13.6b)$$

Accordingly, the rotation of axes transforms the Eq. (13.1) to

$$(a \cos^2 \theta + b \sin^2 \theta + h \sin 2\theta) X^2 + (a \sin^2 \theta + b \cos^2 \theta - h \sin 2\theta) Y^2$$

$$+ \{(b - a) \sin 2\theta + 2h \cos 2\theta\} X Y$$

$$+ 2(u \cos \theta + v \sin \theta) X + 2(- u \sin \theta + v \cos \theta) Y + d = 0. \quad (13.7)$$

Choosing θ so that the coefficient of $X Y$ in above equation vanishes:

$$\tan 2\theta = 2h / (a - b); \qquad (13.8)$$

$\Rightarrow$

$$\sin 2\theta = 2h / k, \quad \text{and} \quad \cos 2\theta = (a - b) / k,$$

where

$$k^2 = (a - b)^2 + 4h^2; \qquad \left.\right\} \quad (13.9)$$

the coefficients of X^2 and Y^2 in Eq. (13.7) simplify to $(a + b + l) / 2$

and $(a + b - 1)/2$ respectively, where

$$l = (a - b) \cos 2\theta + 2h \sin 2\theta = \{(a - b)^2 + 4h^2\}/k = k,$$

by Eq. (13.9). Thus, the coefficients of X^2 and Y^2 are

$$(a + b + k)/2 \text{ and } (a + b - k)/2.$$

These are of same signs if $a + b > a + b$; or, by Eq. (13.9),

$$(a + b)^2 - (a - b)^2 > 4h^2,$$

reducing to Eq. (13.5a). Accordingly, the conic becomes an ellipse.

On the other hand, Eq. (13.7) represents a hyperbola if above coefficients are of different signs, i.e. when $a + b < k$ yielding Eq. (13.5b). //

Note 13.1. Solvng the simultaneous Eqs. (13.6), the inverse transformation of above is also obtained by

$$X = x \cos \theta + y \sin \theta, \text{ and } Y = -x \sin \theta + y \cos \theta. \tag{13.10}$$

Example 13.1. Show that the equation

$$2x^2 + 2y^2 - 4xy + x + y = 0 \tag{13.11}$$

represents a parabola. Find its vertex, focus, axis and the tangent at the vertex.

Solution. Since the determinant

$$D \equiv \begin{vmatrix} 2 & -2 & \frac{1}{2} \\ -2 & 2 & \frac{1}{2} \\ \frac{1}{2} & \frac{1}{2} & 0 \end{vmatrix} = -2,$$

contradicts Eq. (13.2) and the second degree terms form a perfect square $2(x - y)^2$ the conic section represented by Eq. (13.11) is a parabola. Rewriting it as

$$(x - y)^2 = -x/2 - y/2,$$

or,

$$(x - y + \lambda)^2 = (2\lambda - 1/2)x - (2\lambda + 1/2)y + \lambda^2.$$

Further, reducing it in the normal form:

$$\left[\frac{x-y+\lambda}{\sqrt{(1)^2+(-1)^2}}\right]^2 = \frac{1}{4}\left[\frac{(4\lambda-1)x - (4\lambda+1)y + 2\lambda^2}{\sqrt{(4\lambda-1)^2 + (4\lambda+1)^2}}\right]\cdot\sqrt{(4\lambda-1)^2 + (4\lambda+1)^2}$$

(13.12)

where λ is so chosen that the axis of the parabola represented by the equation

$$x - y + \lambda = 0,$$
(13.13)

and the tangent at the vertex represented by Eq.

$$(4\lambda - 1)x - (4\lambda + 1)y + 2\lambda^2 = 0,$$
(13.14)

are mutually perpendicular. This requires

$$(4\lambda - 1)/(4\lambda + 1) = -1 \quad \Rightarrow \quad \lambda = 0.$$

Hence, Eq. (13.12) reduces to

$$Y^2 = -4AX,$$
(13.15)

where

$$Y = (x - y)/\sqrt{2}, \qquad X = (x + y)/\sqrt{2},$$
(13.16)

together with $A = \sqrt{2}/16$. Also, the Eqs. (13.13) and (13.14) reduce to

$$x - y = 0, \qquad x + y = 0, \qquad (13.17)$$

and determine the axis and the tangent at the vertex respectively. The vertex being the point of intersection of these is obtained by solving them:

$$x = y = 0.$$

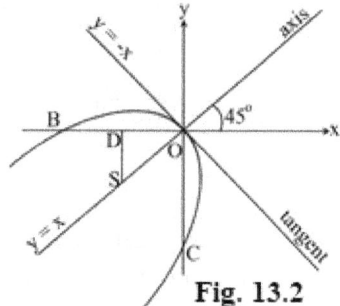

Fig. 13.2

The negative sign in the right side of Eq. (13.15) shows that the curve lies on the left side of the tangent at the vertex. The focus S being one–fourth of the

$$\text{latus rectum} = (1/4).\,4A = A = \sqrt{2}/16$$

from the vertex lies on the left side of the axis. The parabola meets the x-axis at the point $B\left(-\frac{1}{2},0\right)$ also; and it intersects the y-axis in the point

$C\ (0, -\frac{1}{2})$ in addition to the origin. The axis of the parabola makes $45°$ angle with x-axis. Therefore, in the right triangle $\triangle$ ODS, we find

$$OD = (OS) \cos 45° = \tfrac{\sqrt{2}}{16} \cdot \tfrac{1}{\sqrt{2}} = \tfrac{1}{16}, \text{ and } \ DS = (OS) \sin 45° = 1/16.$$

Thus, the focus has coordinates $(-\ 1/16, -\ 1/16)$.The directrix of the parabola, being parallel to the tangent at the vertex and lying in the right side of the tangent at a distance $OS' = OS = \sqrt{2}/16$ is represented by the equation

$$x + y + k = 0, \tag{13.18}$$

where k is determined by Eq. (15.6.3):

$$\frac{0.1 + 0.1 + k}{\sqrt{(1^2 + 1^2)}} = \frac{\sqrt{2}}{16} \qquad \Rightarrow \ k = 1/8.$$

Thus, Eq. (13.18) reduces to $x + y + 1/8 = 0$. //

Example 13.2. Reduce the equation

$$x^2 + x\,y + y^2 = 3 \tag{13.19}$$

to a standard form and show that it represents an ellipse. Find its centre and foci.

Solution. Comparing Eq. (13.19) with Eq. (13.1), we note that

$$a = b = 1, \qquad h = 1/2, \qquad u = v = 0, \qquad d = -3.$$

Thus, in order to get rid of the non-square term $x\,y$ from Eq. (13.19), we rotate the coordinate axes anticlockwise through an angle θ given by Eq. (13.8):

$$\tan 2\theta = 1/0 = \infty \qquad\qquad \Rightarrow \qquad \theta = \pi/4.$$

Accordingly, for Eq. (13.9), the Eq. (13.19) transforms to

$$(3/2)\,X^2 + Y^2/2 = 3, \qquad \text{or} \qquad X^2/2 + Y^2/6 = 1,$$

which represents an ellipse. The lengths of semi-major and semi-minor axes are $\sqrt{6}$ and $\sqrt{2}$ respectively. Also, putting for θ in Eq. (13.10), their equations are derived:

$X \equiv (x + y)/\sqrt{2} = 0$, and $Y \equiv (-x + y)/\sqrt{2} = 0$.

The centre of the ellipse being their point of intersection lies at the origin. The eccentricity of the conic, being given by Eq. (15.10.2), is $e = \sqrt{(2/3)}$. The foci S and S' are at distances OS = OS' = $\sqrt{6}.e = 2$ from the centre O and they lie on the major axis.

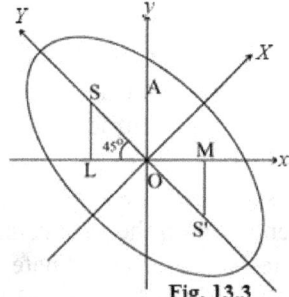
Fig. 13.3

Since OL = LS = (OS)/$\sqrt{2}$ = $\sqrt{2}$, the coordinates of the foci are S $(-\sqrt{2}, \sqrt{2})$ and S' $(\sqrt{2}, -\sqrt{2})$. //

Example 13.3. Reduce the equation

$$x^2 + y^2 + 4xy + 2x + 2y + 1 = 0 \qquad (13.20)$$

to a standard form and show that it represents a hyperbola. Find its centre and the foci.

Solution. Comparing Eq. (13.20) with Eq. (13.1), we have

$$a = b = u = v = d = 1, \quad \text{and} \quad h = 2.$$

Rotating the coordinate axes anticlockwise through an angle θ given by Eq. (13.8):

$$\tan 2\theta = 4/0 = \infty \qquad \Rightarrow \qquad \theta = \pi/4,$$

the Eq. (13.20), for Eqs. (13.7) and (13.9), transforms to

$$3X^2 - Y^2 + 2\sqrt{2}\,X + 1 = 0; \quad \text{or} \quad (X + \sqrt{2}/3)^2 - Y^2/3 = 1/9,$$

i.e.

$$\frac{(X + \sqrt{2}/3)^2}{(1/3)^2} - \frac{Y^2}{(1/\sqrt{3})^2} = 1, \qquad (13.21)$$

which represents a hyperbola. The lengths of semi-transverse and semi conjugate axes are $1/3$ and $1/\sqrt{3}$ respectively. Putting for θ in Eq. (13.10) their equations are obtained:

$$X + \sqrt{2}/3 \equiv (x + y)/\sqrt{2} + \sqrt{2}/3 = 0; \text{ or, } x + y + 2/3 = 0;$$

and

$$\left.\begin{array}{c} \\ \\ \end{array}\right\} \quad (13.22)$$

$$Y \equiv (-x + y)/\sqrt{2} = 0; \text{ or, } -x + y = 0.$$

The centre C of the conic being their point of intersection has the coordinates $(-1/3, -1/3)$. The eccentricity of the conic given by Eq. (15.11.3) is $e = 2/\sqrt{3}$. The foci S and S′ being at distances $CS = CS' = e/\sqrt{3} = 2/3$ from the centre lie on the transverse axis. Also, this axis intersects the coordinate axes Ox and Oy in points K $(-2/3, 0)$ and K′ $(0, -2/3)$, so that $CK = CK' = \sqrt{2}/3$. Since $CA = CA' = 1/\sqrt{3}$, we note that

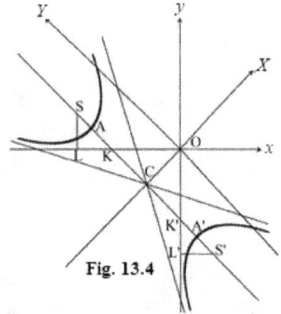

Fig. 13.4

$$CK < CA < CS, \text{ and } CK' < CA' < CS'.$$

Thus, A lies between K and S; and A′ lies between K′ and S′. Therefore,

$$KS = CS - CK = (2 - \sqrt{2})/3 = K'S'.$$

As the transverse axis makes angle $\angle SKL = 45°$ with Ox, we get

$$KL = LS = (KS)/\sqrt{2} = (\sqrt{2} - 1)/3 = K'L' = L'S'.$$

Hence,

$$OL = OK + KL = (\sqrt{2} + 1)/3 = OL'.$$

Thus, the foci S and S′ have the coordinates

$$S(-OL, LS) = \{-(\sqrt{2} + 1)/3, (\sqrt{2} - 1)/3\},$$

and

$$S'(L'S', -OL') = \{(\sqrt{2} - 1)/3, -(\sqrt{2} + 1)/3\}.$$

Asymptotes of the conic are obtained by replacing the right member of Eq. (13.21) by zero:

$$Y = \pm\{\sqrt{3}X + \sqrt{(2/3)}\}.$$

Putting for X and Y from Eqs. (13.22), above equation simplifies to

$$(2 \pm \sqrt{3})x + y + (1 \pm 1/\sqrt{3}) = 0. \; //$$

CHAPTER 15

GEOMETRY (COORDINATE: 3-dimensional)

§ 1. Rectangular Cartesian coordinates

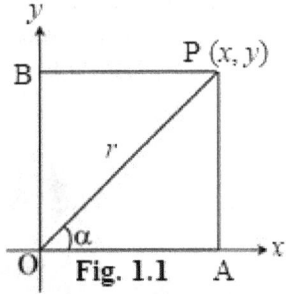

1.1. Let Ox and Oy be two mutually per-pendicular straight lines, called the *coordinate axes*, intersecting in the point O (called the origin). If P (x, y) is any point in the plane determined by these lines at distances OA = x (called the abscissa) and OB = y (called the ordinate) from the axes then the real numbers x and y form the rectangular Cartesian coordinates of P with respect to the system (O, Ox, Oy). If OP = r makes angle α with Ox and, therefore, $90° - α$ with Oy the coordinates x and y are seen to be the projections of OP along Ox and Oy respectively. So, there hold the relations

$$x = r \cos α, \quad y = r \sin α \quad (1.1)$$

$$\Rightarrow$$

$$x^2 + y^2 = r^2, \quad (1.2)$$

for

$$\cos^2 α + \sin^2 α = 1 \quad (1.3)$$

1.2. On the other hand, three mutually perpendicular straight lines Ox, Oy, Oz intersecting each other in the point O in an Euclidean space E$_3$ of dimension three determine the rectangular Cartesian coordinate axes in E$_3$. These axes together with the origin O form a rectangular Cartesian coordinate system (O, Ox, Oy, Oz) in E$_3$. Let P be any point in the space at a distance OP = r from origin with L as the perpendicular drawn from P to the plane determined by Ox and Oy. The points O and L being in the xOy-plane determine the line OL in this plane and, therefore, $\angle$ OLP = 90°. The projections OA and OB of OL on the axes Ox and Oy are actually the projections of OP along the respective coordinate axes:

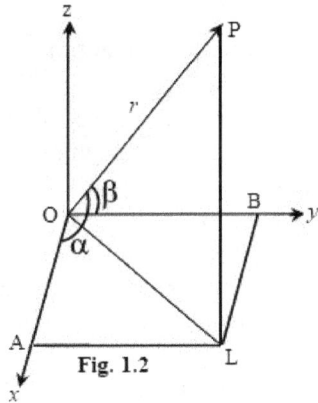

Fig. 1.2

$$OA = (OP) \cos α, \qquad OB = (OP) \cos β, \qquad (1.4)$$

where α, β are the inclinations of OP to the respective coordinate axes

Ox and Oy. Marking a point C on Oz so that OC = LP determines the projection OC of OP on Oz:

$$OC = (OP) \cos \gamma, \qquad (1.5)$$

γ being the angle of inclination of OP to Oz. The relations (1.4) and (1.5) then determine the *rectangular Cartesian coordinates* (x, y, z) of the point P with respect to the system (O, Ox, Oy, Oz) in E;

$$x = r \cos \alpha, \qquad y = r \cos \beta, \qquad z = r \cos \gamma. \qquad (1.6)$$

Analogous to Eq. (1.2) there holds the relation

$$x^2 + y^2 + z^2 = r^2, \qquad (1.7)$$

together with

$$\cos^2 \alpha + \cos^2 \beta + \cos^2 \gamma = 1. \qquad (1.8)$$

Indeed, in an E_2, $\beta = 90° - \alpha$ and $\gamma = 90°$ as OP lies in the xOy-plane; and Oz is perpendicular to this plane. Consequently, Eq. (1.8) assumes the form of Eq. (1.3).

§ 2. Slope and direction cosines of a line

Referring to Fig. 1.1, the numbers $\cos \alpha$ and $\sin \alpha$ determine the slope (or gradient) $\tan \alpha = \sin \alpha / \cos \alpha$ of the line OP in E_2. Contrary to this, cosines of angles of inclination of OP to the respective coordinate axes in E_3 (cf. Fig. 1.2); namely, $\cos \alpha$, $\cos \beta$, $\cos \gamma$ are called the *direction cosines* of the line OP satisfying Eq. (1.8). Multiples of the direction cosines by a common non-zero factor are called the *direction ratios* or *direction numbers* of the line.

§ 3. Line through a given point

Let A (x_1, y_1) be a given point in E_2 and a line through A make angle α to Ox. If P (x, y) is any current point on the line the projections of AP = r on the coordinate axes are:

Fig. 3.1

$$AN = (AP) \cos \alpha, \qquad NP = (AP) \sin \alpha;$$

or,

$$x - x_1 = r \cos \alpha, \qquad y - y_1 = r \sin \alpha.$$

These relations determine the equation of the line AP:

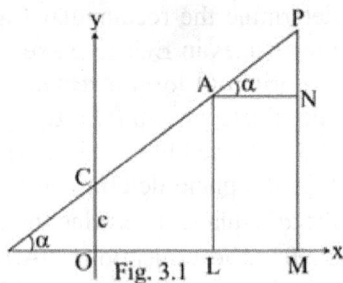

$$(x - x_1) / \cos \alpha = (y - y_1) / \sin \alpha, \tag{3.1}$$

or,

$$y - y_1 = (\tan \alpha) . (x - x_1), \tag{3.2}$$

which is in agreement with Eq. (14.4.6) for $m = \tan \alpha$.

Note 3.1. If the line makes an intercept OC of length c on the y-axis so that the coordinates of the point C are $(0, c)$, the Eq. (3.2) assumes the form

$$y = mx + c. \tag{3.3}$$

On the other hand, if A (x_1, y_1, z_1) and P (x, y, z) are points in E_3 and a line through them making angles α, β, γ to the respective coordinate axes is considered, then the projections of AP $= r$ on the coordinate axes are:

$$x - x_1 = (AP) \cos \alpha, \quad y - y_1 = (AP) \cos \beta, \quad z - z_1 = (AP) \cos \gamma. \tag{3.4}$$

These relations determine the equation

$$(x - x_1) / \cos \alpha = (y - y_1) / \cos \beta = (z - z_1) / \cos \gamma, \tag{3.5}$$

of the line AP, which is a direct generalization of Eq. (3.1).

§ 4. Line through two points

Let A (x_1, y_1) and B (x_2, y_2) be two points in E_2 determining a unique line AB through them. If P(x, y) is any current point on AB the coordinates of these points determine two right triangles $\triangle$ ACB and $\triangle$ ADP which are similar. Hence, there follows

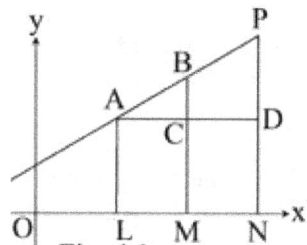
Fig. 4.1

$$\frac{AD}{AC} = \frac{PD}{BC}, \quad \text{or,} \quad \frac{x - x_1}{x_2 - x_1} = \frac{y - y_1}{y_2 - y_1} \tag{4.1}$$

giving equation of the line AB as alternateively Eq. (15.4.10).

Note 4.1. Above equation can also be deduced by eliminating α from Eq. (3.1) and

$$\frac{x_2 - x_1}{\cos \alpha} = \frac{y_2 - y_1}{\sin \alpha} = r, \quad \text{(for line AB passing through B).}$$

Extending the discussion to E_3 equation of the line through two points A (x_1, y_1, z_1) and B (x_2, y_2, z_2) will read as

$$\frac{x - x_1}{x_2 - x_1} = \frac{y - y_1}{y_2 - y_1} = \frac{z - z_1}{z_2 - z_1}. \tag{4.2}$$

Note 4.2. This equation can also be deduced by eliminating the direction cosines between Eq. (3.4) and

$$\frac{x_2 - x_1}{\cos \alpha} = \frac{y_2 - y_1}{\cos \beta} = \frac{z_2 - z_1}{\cos \gamma}.$$

§ 5. The coordinate axes and the coordinate planes

5.1. It is evident from the forgoing discussion that in E_2, y coordinate remains constant (indeed zero) throughout the x-axis and vice–versa. Hence, the relation

$$y = 0 \quad \text{(respectively } x = 0) \tag{5.1}$$

represents eqn. of the coordinate axis Ox (resp. Oy) and their combined equation is

$$x \cdot y = 0. \tag{5.2}$$

5.2. On the other hand in E_3, along any coordinate axis either of two remaining coordinates remain constant (zero). Thus, the equations

$$y = 0 = z, \qquad z = 0 = x, \qquad x = 0 = y \tag{5.3}$$

determine the respective coordinate axes Ox, Oy, Oz. Either of the coordinate axes pass through the origin O $(0,0,0)$ and are inclined at right angles to the remaining axes so that their direction cosines are

$$1, 0, 0; \qquad\qquad 0, 1, 0; \qquad\qquad 0, 0, 1.$$

respectively, Hence, their equations could also be deduced from Eq. (3.4):

Ox:
$$\frac{x - 0}{1} = \frac{y - 0}{0} = \frac{z - 0}{0}, \tag{5.4}$$

Oy:
$$\frac{x - 0}{0} = \frac{y - 0}{1} = \frac{z - 0}{0}, \tag{5.5}$$

and

Oz: $$\frac{x-0}{0} = \frac{y-0}{0} = \frac{z-0}{1},\qquad (5.6)$$

which are same as Eqs. (5.3).

5.3. Unlike E_2 having only one coordinate plane xOy determined by the coordinate axes Ox and Oy there exist three coordinate planes xOy, yOz, zOy determined by the pairs Ox, Oy; Oy; Oz; and Oz, Ox of coordinate axes in E_3. We note that only one coordinate remains constant (zero) on these planes. Thus, contrary to the discussion in Subsection 5.1, in E_3, the equations

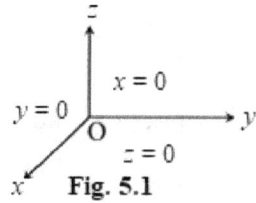

$$x = 0, \qquad y = 0, \qquad z = 0. \qquad (5.7)$$

determine the planes yOz, zOx and xOy respectively. Accordingly, their combined equation is

$$x . y . z = 0. \qquad (5.8)$$

§ 6. A linear equation

6.1. We consider a linear equation

$$a x + b y + d = 0, \qquad (6.1)$$

together with

$$a^2 + b^2 > 0, \qquad (6.2)$$

in two variables x and y. The condition (6.2) states that the coefficients a and b do not vanish simultaneously. However, at most one of them can be zero. There arise the following cases:

(i) when $a = 0$, $b \neq 0$, the Eq. (6.1) assumes the form $y = -d / b$ and represents a straight line (in E_2) parallel to the x-axis.

(ii) when $a \neq 0$, $b = 0$ the Eq. (6.1) reduces to $x = -d / a$

and so it represents a line (in E_2) parallel to y-axis.

(iii) when $a \neq 0$, $b \neq 0$ but $d = 0$, the Eq. (6.1) reduces to the form $y = (-a / b) x$ and represents a straight line (in E_2) through origin.

(iv) when neither of the coefficients a, b, d is zero, Eq. (6.1) takes the form

$$\frac{x}{-d/a} + \frac{y}{-d/b} = 1,$$

and represents a straight line (in E₂) making intercepts of lengths $-d/a$ and $-d/b$ on the respective coordinate axes.

Thus, in all situations the Eq. (6.1) represents a straight line in E₂.

6.2. Now we consider a linear equation

$$a x + b y + c z + d = 0 \tag{6.3}$$

together with

$$a^2 + b^2 + c^2 > 0. \tag{6.4}$$

The condition (6.4) now states that at most two of the coefficients a, b, c can vanish. Analogous to Subsection 6.1, there may arise various choices for the constants a, b, c, d but in all situations the Eq. (6.3) can be seen to represent a *plane* in E₃.

§ 7. Angle between two lines / planes

7.1. Let two lines AB and AC having the slopes $m_1 = \tan \alpha_1$ and $m_2 = \tan \alpha_2$ intersect each other in the point A. An angle between them is given by

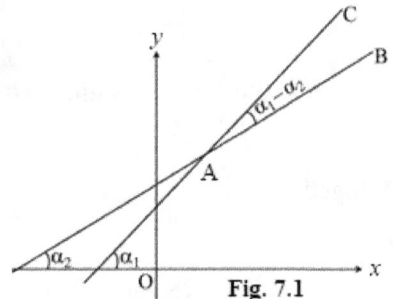

Fig. 7.1

$$\tan (\alpha_1 \sim \alpha_2) = \frac{\tan \alpha_1 - \tan \alpha_2}{1 + \tan \alpha_1 \tan \alpha_2} = \frac{m_1 \sim m_2}{1 + m_1 \, m_2}. \tag{7.1}$$

Accordingly,

$$\cos (\alpha_1 \sim \alpha_2) = \frac{1 - m_1 \cdot m_2}{\sqrt{\{(1 + m_1^2)(1 + m_2^2)\}}}. \tag{7.2}$$

Thus, lines are parallel iff

$$m_1 = m_2, \tag{7.3}$$

and perpendicular to each other iff

$$m_1 \, m_2 = -1. \tag{7.4}$$

7.2. If the lines are represented by their general equations

$$a_i x + b_i y + d_i = 0, \tag{7.5}$$

$i = 1, 2$, with their slopes $m_i = \tan \alpha_i = -a_i / b_i$, the formula (7.2) reduces to

$$\cos (\alpha_1 \sim \alpha_2) = \frac{1 + a_1 a_2 / b_1 b_2}{\sqrt{[\{1 + (a_1 / b_1)^2\}.\{1 + (a_2 / b_2)^2\}]}}$$

$$= \frac{a_1 a_2 + b_1 b_2}{\sqrt{\{(a_1^2 + b_1^2).(a_2^2 + b_2^2)\}}}. \tag{7.6}$$

Accordingly, the conditions for their parallelism and orthogonality reduce to

$$a_1 / b_1 = a_2 / b_2, \tag{7.7}$$

and

$$a_1 a_2 + b_1 b_2 = 0. \tag{7.8}$$

7.3. Similarly, by Eq. (3.5), two lines in E_3 are represented by

$$\frac{x - x_i}{l_i} = \frac{y - y_i}{m_i} = \frac{z - z_i}{n_i}, \tag{7.9}$$

where l_i, m_i, n_i are their direction numbers and $i = 1, 2$, the formula (7.6) extends to

$$\cos (\alpha_1 \sim \alpha_2) = \frac{l_1 l_2 + m_1 m_2 + n_1 n_2}{\sqrt{\{(l_1^2 + m_1^2 + n_1^2).(l_2^2 + m_2^2 + n_2^2)\}}}. \tag{7.10}$$

Consequently, the lines become parallel iff

$$l_1 / l_2 = m_1 / m_2 = n_1 / n_2, \tag{7.11}$$

and orthogonal iff

$$l_1 l_2 + m_1 m_2 + n_1 n_2 = 0. \tag{7.12}$$

7.4. Normal to a plane

As remarked in the Subsection 6.2, a general linear Eq. (6.3) represents a plane. Therefore, a plane through origin and parallel to the one given by Eq. (6.3) is represented by

$$a x + b y + c z = 0. \tag{7.13}$$

If P (x_1, y_1, z_1) is any point in the plane the line OP with its direction ratios x_1, y_1, z_1 lies in the plane and Eq. (7.13) gets satisfied by the coordinates of P:

$$a x_1 + b y_1 + c z_1 = 0. \qquad (7.14)$$

Interpreting the left side of above equation as the scalar product of two vectors (a, b, c) and (x_1, y_1, z_1), Eq. (7.14) justifies their orthogonal character. Thus, the coefficients a, b, c in Eq. (7.13) determine the direction ratios of a line which is perpendicular to the vector $\overrightarrow{OP} = (x_1, y_1, z_1)$. Since the choice of P in the plane represented by Eq. (7.13) is arbitrary the vector

$$\overrightarrow{ON} = (a, b, c) \qquad (7.15)$$

is orthogonal to the plane with Eq. (7.13) – hence called the *normal vector to the plane*. The planes represented by Eqs. (6.3) and (7.13) are parallel, the vector $\overrightarrow{ON}$ is also normal to the plane with Eq. (6.3).

Example 7.1. Normals to the coordinate planes with Eqs. (5.7) at O are represented by Eqs. (5.4), (5.5) and (5.6) respectively.

7.5. Angle between two planes

Following discussion of Subsection 7.4, the lines with Eq. (7.9) represent normals to the planes

$$l_i x + m_i y + n_i z - p_i = 0, \qquad i = 1, 2. \qquad (7.16)$$

Thus, the angle between the planes at any of their point of intersection is the angle between their normals at that point. The same is given by the formula (7.10).

§ 8. Perpendicular from a point

8.1. First we consider a line AB, in E_2, to which the perpendicular ON of length, say p, is dropped from the origin O. If ON makes angle α to the x-axis its gradient is $\tan \alpha$. Accordingly, the gradient of the line AB (being perpendicular to ON), by Eq. (7.4), is $-1 / \tan \alpha = -\cot \alpha$. Also, the co-ordinates of N are $(p \cos \alpha, p \sin \alpha)$. Therefore, the line AB (passing thr-

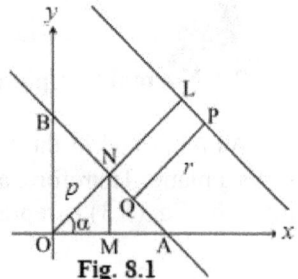

Fig. 8.1

ough N and having gradient $-1/\tan \alpha$ has equation

$$y - p \sin \alpha = (-\cot \alpha). (x - p \cos \alpha),$$

or,

$$x \cos \alpha + y \sin \alpha = p. \tag{8.1}$$

This represents the *perpendicular form* of equation to line AB.

8.2. Let P (x_1, y_1) be any point (not on the line AB) and PQ $= r$ be the perpendicular drawn from P to the line. We consider a line through P parallel to AB. So, by Eq. (7.3), such a line through P also has the same gradient $-\cot \alpha$ as that of AB; and the length of perpendicular from O to it is OL $=$ ON $+$ NL $= p + r$. In view of Eq. (8.1), it is represented by

$$x \cos \alpha + y \sin \alpha = p + r. \tag{8.2}$$

The point P lying on it, its Eq. (8.2) gets satisfied by the coordinates of P:

$$x_1 \cos \alpha + y_1 \sin \alpha = p + r$$

$\Rightarrow$

$$PQ = r = x_1 \cos \alpha + y_1 \sin \alpha - p. \tag{8.3}$$

8.3. Rewriting the general equation (6.1) of a line in the perpendicular form:

$$\{a / \sqrt{(a^2 + b^2)}\}x + \{b / \sqrt{(a^2 + b^2)}\}y = -d / \sqrt{(a^2 + b^2)},$$

the length of perpendicular from P to the line is obtained by Eq. (8.3):

$$PQ = (a x_1 + b y_1 + d) / \sqrt{(a^2 + b^2)}. \tag{8.4}$$

8.4. Extending above discussion to E$_3$, lengths of the perpendiculars drawn from a point P (x_1, y_1, z_1) to the planes

$$x \cos \alpha + y \cos \beta + z \cos \gamma = p \tag{8.5}$$

(respectively the one with Eq. (6.3)) are given by

$$r = x_1 \cos \alpha + y_1 \cos \beta + z_1 \cos \gamma - p \tag{8.6}$$

(respectively)

$$r = (a x_1 + b y_1 + c z_1 + d) / \sqrt{(a^2 + b^2 + c^2)}. \tag{8.7}$$

§ 9. Intersection of lines / planes

9.1. Intersection of two straight lines is a unique point. Also, as per the Postulate of Euclidean Geometry [cf. 25], two distinct points determine a unique straight line. The condition for collinearity of three points in E_2 has been discussed vide Theorem 14.7.1.

9.2. On the other hand, two planes (in E_3), if they intersect each other, they do so in a unique straight line. Also, there exists a unique plane through two distinct straight lines. Consequently, three non-collinear points determines a unique plane.

§ 10. Shortest distance between two lines

Definition 10.1. Length of the common perpendicular between two non-intersecting lines is called their *shortest distance*.

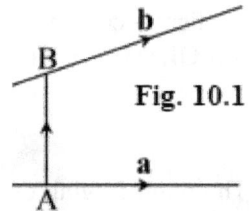

Fig. 10.1

If **a** and **b** are the vectors along two lines passing through points A and B respectively then the shortest distance between the lines is the projection of $\overrightarrow{AB}$ on the unit vector $(\mathbf{a} \times \mathbf{b}) / |\mathbf{a} \times \mathbf{b}|$.

Theorem 10.1. Shortest distance between two lines represented by Eqs. (7.9) is

$$\begin{vmatrix} x_1 - x_2 & y_1 - y_2 & z_1 - z_2 \\ l_1 & m_1 & n_1 \\ l_2 & m_2 & n_2 \end{vmatrix}$$

$$\div \sqrt{(m_1 n_2 - m_2 n_1)^2 + (n_1 l_2 - n_2 l_1)^2 + (l_1 m_2 - l_2 m_1)^2} \qquad (10.1)$$

where (l_i, m_i, n_i), $i = 1, 2, 3$ are the 'direction numbers' of two lines forming vectors along the respective lines.

§ 11. Parametric equations of curves

Definition 11.1. Locus of a point, whose coordinates are functions of a single parameter, on a plane is a *curve* in E_2.

Thus, equations

$$x = x(t), \qquad y = y(t), \qquad (11.1)$$

where t is a parameter, represent a *curve* in E_2.

Example 11.1. The parametric equations

$$x = a \cos t, \qquad y = a \sin t \qquad\qquad (11.2)$$

determine a *circle* with centre at origin and radius a having Catesian equation (14.8.4).

On the other hand, if a point moves in the space E_3 with its coordinates as functions of a single parameter:

$$x = x(t), \quad y = y(t), \quad z = z(t). \qquad\qquad (11.3)$$

its locus becomes a curve in E_3. Such a curve is called a *space curve*. If all its points lie on a common plane it is a *plane curve* else a *twisted curve*.

Note 11.1. In particular, if one of the coordinates, say z, remains constant throughout the curve:
$$z = k, \qquad\qquad (11.4)$$

Eqs. (11.3) together with Eq. (11.4) determine a plane curve lying on the plane represented by Eq. (11.4). For example, Eqs. (11.1) and (11.2) represent curves on the plane $z = 0$.

Definition 11.2. The limiting position of a chord PQ of a curve, when Q $\rightarrow$ P, defines the *tangent* to the curve at P. A line through P and perpendicular to the tangent is called the *normal* to the curve at P.

A curve, irrespective of being plane or twisted, has a unique tangent at a given point. Also, there exist a unique normal to a curve in E_2 but a curve in E_3 has infinitely a large number of normals at a point. However, only two of these (normals) forming a right handed system of mutually perpendicular vectors with the tangent vector are called *principal normal* and *binormal* to the curve at P.

Note 11.2. Some standard curves in E_2 such as circle, parabola, ellipse, and hyperbola have been discussed in the previous chapter.

§ 12. Sphere

If a point moves in the space E_3 maintaining a constant distance from a fixed point, its path becomes a *sphere*. As in case of a circle, the fixed point is called the *centre* and the constant distance the *radius* of the sphere.

For example, a sphere with centre (x_1, y_1, z_1) and radius a is represented by

$$(x - x_1)^2 + (y - y_1)^2 + (z - z_1)^2 = a^2. \tag{12.1}$$

or,

$$x^2 + y^2 + z^2 + 2(ux + vy + wz) + d = 0, \tag{12.2}$$

where $u = -x_1$, $v = -y_1$, $w = -z_1$ and

$$a = \sqrt{(x_1^2 + y_1^2 + z_1^2 - d)} = \sqrt{(u^2 + v^2 + w^2 - d)}. \tag{12.3}$$

Particularly, a sphere with centre at origin and radius a has equation

$$x^2 + y^2 + z^2 = a^2. \tag{12.4}$$

§ 13. More general surfaces in E_3

Definition 13.1. Locus of a point moving in E_3 so that its coordinates are expressible in terms of two independent parameters is called a *surface*.

Thus, the equations

$$x = x(u, v), \quad y = y(u, v), \quad z = z(u, v), \tag{13.1}$$

where u, v are two independent parameters, represent a surface in E_3. It can be a flat (i.e. plane) or curved and is of dimension two.

Example 13.1. The relations

$$x = a \cos u \sin v, \quad y = a \sin u \sin v, \quad z = a \cos v. \tag{13.2}$$

determine *sphere* with Cartesian equation (12.4).

Note. 13.1. A comparison of Eq. (12.4) with Eq. (14.8.4) shows that a sphere is a true generalization of a circle from E_2 to E_3.

13.1. Ellipsoid and hyperboloid: The equation

$$x^2/a^2 + y^2/b^2 + z^2/c^2 = 1 \qquad (13.3)$$

represents an *ellipsoid* with centre at origin and three axes of length $2a$, $2b$, $2c$ along the coordinate axes x, y, z respectively.

Note 13.2. When $a = b = c \neq 0$, the Eq. (13.3) reduces to Eq. (12.4). Thus, a sphere is a special case of an ellipsoid.

If at least one of the terms in the left side of Eq. (13.3) is negative the surface become a hyperboloid. Thus,

$$x^2/a^2 + y^2/b^2 - z^2/c^2 = 1 \qquad (13.4)$$

represents a *hyperbolpoid of one sheet*; whereas the equation

$$x^2/a^2 - y^2/b^2 - z^2/c^2 = 1 \qquad (13.5)$$

represents a *hyperboloid of two sheets*.

13.2. Equations (13.3) – (13.5) are the special cases of equation

$$a x^2 + b y^2 + c z^2 = 1, \qquad (13.6)$$

where a, b, c are non-zero real numbers. It represents an ellipsoid (resp. imaginary ellipsoid) when a, b, c are all positive (resp. negative) else a hyperboloid. For any point P (x, y, z) on the surface there also exist a point P $'(-x, -y, -z)$ on the surface. Both P and P $'$ are equidistant from the origin. Hence, the origin becomes the centre of the surface and the latter is called a *central conicoid*.

13.3. Finally, we consider a second degree equation (in the variables x, y, z) of the form

$$a x^2 + b y^2 = 2c z, \qquad (13.7)$$

where a, b, c are again non-zero real numbers. If a and b have the same signs the surface is called an *elliptic paraboloid* else a *hyperbolic paraboloid*.

Note 13.3. If the parameters u and v are not independent to each other Eqs. (12.1) reducing to the form of Eqs. (10.3) represent a curve on the surface.

In particular, $u = $ const. implying

$$x = x\,(v), \quad y = y\,(v), \quad z = z\,(v), \qquad (13.8)$$

and $v = $ const. implying

$$x = x\,(u), \quad y = y\,(u), \quad z = z\,(u), \qquad (13.9)$$

determine the *parametric curves* on the surface.

13.4. In view of Definition 11.2, there exists infinitely large number of tangent (lines) to a surface at a given point. These tangents lie in a plane called the *tangent plane* to the surface. Also, there exists a unique normal to this plane at the point of contact. The same is called the *normal to the surface* at the point.

§ 14. General equation of second degree in three variables

The equation

$$F\,(x, y, z) \equiv$$

$$ax^2 + by^2 + cz^2 + 2\,(fyz + gzx + hxy) + 2(ux + vy + wz) + d = 0 \quad (14.1)$$

is a general equation of second degree in the variables x, y, z provided the coefficients a, b, c, f, g and h do not vanish simultaneously. In particular, when $u = v = w = d = 0$, the Eq. (14.1) reducing to the form

$$H\,(x, y, z) \equiv ax^2 + by^2 + cz^2 + 2\,(fyz + gzx + hxy) = 0, \qquad (14.2)$$

is a homogeneous equation of second degree in x, y, z. Accordingly, for any real t, Eq. (14.2) implies

$$H\,(t\,x, t\,y, t\,z) = 0. \qquad (14.3)$$

Theorem 14.1. The equation (14.2) represents a pair of planes through origin iff there holds the condition

$$D \equiv \begin{vmatrix} a & h & g \\ h & b & f \\ g & f & c \end{vmatrix} = 0. \qquad (14.4)$$

Proof. Let Eq. (14.2) represent two planes (through origin):

$$l\,x + m\,y + n\,z \,=\, 0, \qquad l'\,x + m'\,y + n'\,z \,=\, 0,$$

having their combined equation

$$(l\,x + m\,y + n\,z)\,.\,(l'\,x + m'\,y + n'\,z) = l\,l'\,x^2 + m\,m'\,y^2 + n\,n'\,z^2$$

$$+ (m\,n' + m'\,n)\,y\,z + (n\,l' + n'\,l)\,z\,x + (l\,m' + l'\,m)\,x\,y \,=\, 0.$$

Comparing the coefficients of like terms in Eqs. (14.1) and above, we obtain

$$l\,l'/a \,=\, m\,m'/b \,=\, n\,n'/c \,=\, (mn' + m'n)\,/\,2f \,=\, (nl' + n'l\,)\,/\,2g$$

$$= (lm' + l'm)\,/\,2h \,=\, k\,\text{(say)}. \tag{14.5}$$

Multiplying two vanishing determinants and putting from Eqs. (14.5), we derive

$$0 \,=\, \begin{vmatrix} l & l' & 0 \\ m & m' & 0 \\ n & n' & 0 \end{vmatrix} \cdot \begin{vmatrix} l' & m' & n' \\ l & m & n \\ 0 & 0 & 0 \end{vmatrix}$$

$$= \begin{vmatrix} ll' + l'l & lm' + l'm & ln' + l'n \\ ml' + m'l & mm' + m'm & mn' + m'n \\ nl' + n'l & nm' + n'm & nn' + n'n \end{vmatrix} = k^3 . \begin{vmatrix} 2a & 2h & 2g \\ 2h & 2b & 2f \\ 2g & 2f & 2c \end{vmatrix};$$

yielding the desired result.

Conversely, if Eq. (14.4) holds, its a th multiple yields

$$a^2 b\,c - a\,c\,h^2 - a\,b\,g^2 - a^2 f^2 + 2a f g h \,=\, 0,$$

or,

$$g^2 h^2 - a\,c\,h^2 - a\,b\,g^2 + a^2 b\,c \,=\, g^2 h^2 + a^2 f^2 - 2a f g h,$$

or,

$$h^2 (g^2 - a\,c) - a\,b\,(g^2 - a\,c) \,=\, (g\,h - a f)^2,$$

or,

$$(g^2 - a\,c)\,(h^2 - a\,b) \,=\, (g\,h - a f)^2. \tag{14.6}$$

On the other hand, a th multiple of Eq. (14.2) gives

$$a^2 x^2 + 2a\,h\,x\,y + 2a\,g\,z\,x \,=\, -a\,b\,y^2 - a\,c\,z^2 - 2a f y\,z,$$

or,

$$(a\,x + h\,y + g\,z)^2 = (h^2 - a\,b)\,y^2 + (g^2 - a\,c)\,z^2 + 2\,(g\,h - af)\,y\,z$$

$$= (h^2 - a\,b)\,y^2 + (g^2 - a\,c)\,z^2 + 2\sqrt{(g^2 - a\,c)}\,\sqrt{(h^2 - a\,b)}\,y\,z$$

$$= \{\sqrt{(h^2 - a\,b)}\,y + \sqrt{(g^2 - a\,c)}\,z\}^2, \quad \text{by Eq. (14.6)};$$

or,

$$[a\,x + h\,y + g\,z + \{\sqrt{(h^2 - a\,b)}\,y + \sqrt{(g^2 - a\,c)}\,z\,\}].$$

$$.\,[a\,x + h\,y + g\,z - \{\sqrt{(h^2 - a\,b)}\,y + \sqrt{(g^2 - a\,c)}\,z\,\}] = 0,$$

representing a pair of planes

$$a\,x + \{h + \sqrt{(h^2 - a\,b)}\,\}\,y + \{g + \sqrt{(g^2 - a\,c)}\,\}\,z = 0,$$

and

$$a\,x + \{h - \sqrt{(h^2 - a\,b)}\,\}\,y + \{g - \sqrt{(g^2 - a\,c)}\,\}\,z = 0.$$

This establishes sufficiency of the condition. //

In the following we consider the cases when $D \neq 0$.

Theorem 14.2. Every homogeneous equation of second degree in variables x, y, z represents a cone with vertex at origin.

Proof. Evidently the origin lies on the surface represented by Eq. (14.2). If $P\,(x_1, y_1, z_1)$ is any point on the surface there holds

$$H\,(x_1, y_1, z_1) = 0; \tag{14.7}$$

while the line OP is represented by the equation

$$x/x_1 = y/y_1 = z/z_1 = t \text{ (say)}. \tag{14.8}$$

Any point Q on this line having coordinates ($t\,x_1$, $t\,y_1$, $t\,z_1$) also lies on the surface as its coordinates satisfy $H\,(tx_1, ty_1, tz_1) = 0$, that results from Eqs. (14.3) and (14.7) as Eq. (14.2) is homogeneous. Thus, the whole line OP lies on the surface. As such, it becomes a generator of the cone with vertex at O. //

Theorem 14.3. A general equation of second degree in variables x, y, z represents a cone if

$$\begin{vmatrix} a & h & g & u \\ h & b & f & v \\ g & f & c & w \\ u & v & w & d \end{vmatrix} = 0. \qquad (14.9)$$

Proof. Let A (x_1, y_1, z_1) be the vertex of the cone represented by the Eq. (14.1). Transferring the origin to A, so that the variables x, y, z transform as:

$$x \to x + x_1, \quad y \to y + y_1, \quad z \to z + z_1.$$

Hence, the Eq. (14.1) transforms to

$$\Sigma a (x + x_1)^2 + 2 \Sigma f (y + y_1)(z + z_1) + 2 \Sigma u (x + x_1) + d = 0,$$

or,

$$H (x, y, z) + 2 (ax_1 + hy_1 + gz_1 + u) x + 2 (hx_1 + by_1 + fz_1 + v) y$$

$$+ 2 (gx_1 + fy_1 + cz_1 + w) z + F (x_1, y_1, z_1) = 0,$$

which, in view of Theorem 14.2, should be homogeneous. So, the terms other than $H (x, y, z)$ should vanish:

$$\left. \begin{aligned} ax_1 + hy_1 + gz_1 + u &= 0, \\ hx_1 + by_1 + fz_1 + v &= 0, \\ gx_1 + fy_1 + cz_1 + w &= 0, \end{aligned} \right\} \qquad (14.10)$$

and,

$$F (x_1, y_1, z_1) = (ax_1 + hy_1 + gz_1 + u) x_1 + (hx_1 + by_1 + fz_1 + v) y_1$$

$$+ (gx_1 + fy_1 + cz_1 + w) z_1 + (ux_1 + vy_1 + wz_1 + d)$$

$$= ux_1 + vy_1 + wz_1 + d = 0, \qquad (14.11)$$

by Eqs. (14.1) and (14.10). Eliminating x_1, y_1, z_1 from Eqs. (14.10) and (14.11), we immediately get the result. //

§ 15. Some standard cases

15.1 If the second degree terms in Eq. (14.1) form a perfect square the surface represented by the equation is a *paraboloid*.

15.2. If $a = b = c \neq 0$ but $f = g = h = 0$, the Eq. (14.1) assumes the form as in Eq. (11.2) and it represents a *sphere*.

Note 15.1. Other conicoids represented by the Eq. (14.1) will be discussed later in Section 21.

§ 16. Intersection of a line and a conicoid

Let a line

$$\frac{x - x_1}{l} = \frac{y - y_1}{m} = \frac{z - z_1}{n} = r \tag{16.1}$$

through a given point A (x_1, y_1, z_1) with direction numbers l, m, n intersect the conicoid represented by Eq. (14.1) in some point(s). Thus, for a common point between the line and the conicoid the coordinates of any point on the line should satisfy Eq. (14.1):

$$F(x_1 + lr, y_1 + mr, z_1 + nr)$$

$$\equiv \Sigma a (x_1 + lr)^2 + 2 \Sigma f(y_1 + mr)(z_1 + nr) + 2 \Sigma u (x_1 + lr) + d = 0,$$

or,

$$\{a l^2 + b m^2 + c n^2 + 2 (f m n + g n l + h l m)\} r^2$$

$$+ 2\{alx_1 + bmy_1 + cnz_1 + f(y_1 n + m z_1) + g(z_1 l + nx_1)$$

$$+ h(x_1 m + ly_1) + ul + vm + wn\} r + \{\Sigma ax_1^2 + 2\Sigma f y_1 z_1 + 2\Sigma ux_1 + d\} = 0,$$

or,

$$H(l, m, n) r^2 + 2 \{l (ax_1 + hy_1 + gz_1 + u) + m (hx_1 + by_1 + fz_1 + v)$$

$$+ n (gx_1 + f y_1 + cz_1 + w)\} r + F(x_1, y_1, z_1) = 0, \tag{16.2}$$

in view of Eqs. (14.1) and (14.2). Considering partial derivatives of the function $F(x, y, z)$, we note that

$$\begin{aligned} (\partial F / \partial x)_A &= 2(ax_1 + hy_1 + gz_1 + u), \\ (\partial F / \partial y)_A &= 2(hx_1 + by_1 + fz_1 + v), \\ (\partial F / \partial z)_A &= 2(gx_1 + f y_1 + cz_1 + w). \end{aligned} \tag{16.3}$$

Accordingly, Eq. (16.2) reduces to

$$H(l, m, n) r^2 + \{l (\partial F / \partial x)_A + m (\partial F / \partial y)_A + n (\partial F / \partial z)_A\} r$$

$$+ F(x_1, y_1, z_1) = 0. \tag{16.4}$$

Being a quadratic this equation determines two values of r. Hence, the line with Eq. (16.1) intersects the conicoid represented by Eq. (14.1) in two real and distinct points if

$$\{ \Sigma l (\partial F / \partial x)_A \}^2 > 4 H(l, m, n) . F(x_1, y_1, z_1). \tag{16.5}$$

In case, above inequality turns to equality both roots of Eq. (16.4) are equal causing the points of intersection to coincide and the line touches the conicoid. Thus, the *condition for tangency* between the line and the conicoid is

$$\{ \Sigma l (\partial F / \partial x)_A \}^2 = 4 H(l, m, n) . F(x_1, y_1, z_1). \tag{16.6}$$

§ 17. Tangent plane

If the point A (x_1, y_1, z_1) itself becomes the point of contact between the line and the conicoid both roots of Eq. (16.4) become zero and there holds

$$F(x_1, y_1, z_1) = 0. \tag{17.1}$$

Accordingly, Eq. (16.6) reduces to

$$l (\partial F / \partial x)_A + m (\partial F / \partial y)_A + n (\partial F / \partial z)_A = 0, \tag{17.2}$$

for arbitrary choices of l, m, n. So, there exist infinitely large number of tangents to the conicoid at the point A. However, all these (infinite) tangent lines lie in a plane – called the *tangent plane* to the conicoid at A. Its equation can be obtained by eliminating l, m, n from Eqs. (17.2) and (16.1):

$$(x - x_1) (\partial F / \partial x)_A + (y - y_1) (\partial F / \partial y)_A + (z - z_1) (\partial F / \partial z)_A = 0, \tag{17.3}$$

or,

$$x (\partial F / \partial x)_A + y (\partial F / \partial y)_A + z (\partial F / \partial z)_A$$

$$= x_1 (\partial F / \partial x)_A + y_1 (\partial F / \partial y)_A + z_1 (\partial F / \partial z)_A. \tag{17.4}$$

Making the function $F(x, y, z)$ homogeneous by introducing fourth variable, say t, which is to be equated to 1 after differentiation of F with respect to t, above equation further reduces to

$$x\,(\partial F\,/\,\partial x)_{\text{A}} + y\,(\partial F\,/\,\partial y)_{\text{A}} + z\,(\partial F\,/\,\partial z)_{\text{A}} + t\,(\partial F\,/\,\partial t) \;=\; 0, \quad (17.5)$$

by Euler's theorem on homogeneous functions.

Putting from Eqs. (16.3) and (14.1), the right member of Eq. (17.4) also simplifies to $-2\,(ux_1 + vy_1 + wz_1 + d\,)$. Hence, the tangent plane is also represented by the equation

$$x\,(\partial F/\partial x)_{\text{A}} + y\,(\partial F/\partial y)_{\text{A}} + z\,(\partial F/\partial z)_{\text{A}} + 2(ux_1 + vy_1 + wz_1 + d) = 0. \quad (17.6)$$

Theorem 17.1. Tangent plane to the conicoid represented by Eq. (13.6) at the point A (x_1, y_1, z_1) is represented by

$$T\,(x, y, z) \equiv a\,x\,x_1 + b\,y\,y_1 + c\,z\,z_1 - 1 = 0. \quad (17.7)$$

Proof. Putting

$$F\,(x, y, z) \equiv a\,x^2 + b\,y^2 + c\,z^2 - 1, \quad (17.8)$$

and evaluating its derivatives at the point of contact:

$$(\partial F\,/\,\partial x)_{\text{A}} = 2a\,x_1, \quad (\partial F\,/\,\partial y)_{\text{A}} = 2b\,y_1, \quad (\partial F\,/\,\partial z)_{\text{A}} = 2\,c\,z_1, \quad (17.9)$$

the Eq. (17.3) reduces to

$$2\,a\,(x - x_1)\,x_1 + 2b\,(y - y_1)\,y_1 + 2\,c\,(z - z_1)\,z_1 \;=\; 0,$$

or, to the form of Eq. (17.7) by Eqs. (17.1) and (17.8). //

Corollary 17.1. Equation to the tangent plane to the ellipsoid represented by Eq. (13.3) at the point A (x_1, y_1, z_1) is

$$T\,(x, y, z) \equiv x\,x_1 /\,a^2 + y\,y_1 /\,b^2 + z\,z_1 /\,c^2 - 1 \;=\; 0; \quad (17.10)$$

and the normal to the ellipsoid there is represented by equation

$$\frac{x - x_1}{x_1 /a^2} = \frac{y - y_1}{y_1 /b^2} = \frac{z - z_1}{z_1 /c^2}. \quad (17.11)$$

Example 17.1. Find the tangent plane to the surface

$$x\,y = c\,z \quad (17.12)$$

at a point P (x', y', z').

Solution. Putting $F(x, y, z) \equiv 2xy - 2cz$, we derive

$$(\partial F / \partial x)_P = 2y', \quad (\partial F / \partial y)_P = 2x', \quad (\partial F / \partial z)_P = -2c.$$

So, the Eq. (17.3) representing the tangent plane reduces to

$$2(x - x')y' + 2(y - y')x' - 2c(z - z') = 0,$$

or,

$$xy' + yx' - cz = 2x'y' - cz' = cz' = x'y',$$

by Eq. (17.12). The same can also be written as

$$x/x' + y/y' - z/z' = 1. \; // \tag{17.13}$$

Exercise 17.2. Show that the plane

$$lx + my + nz = p \tag{17.14}$$

touches the conicoid represented by Eq. (17.12) iff

$$clm + np = 0. \tag{17.15}$$

Proof. If the plane with Eq. (17.14) touches the conicoid at some point say P (x', y', z') then the Eqs. (17.13) and (17.14) represent the same plane. Hence, a comparison of coefficients of like terms in these equations yields

$$\frac{l}{1/x'} = \frac{m}{1/y'} = \frac{n}{-1/z'} = \frac{p}{1},$$

determining the point of contact P:

$$x' = p/l, \qquad y' = p/m, \qquad z' = -p/n. \tag{17.16}$$

Point P lying on the conicoid its coordinates satisfy the equation of the conicoid:

$$p^2/lm = -cp/n \qquad \Rightarrow \qquad pn = -clm,$$

as $p \neq 0$. Thus, Eq. (17.15) becomes a necessary condition. Conversely, when there holds Eq. (17.15), the Eq. (17.13) of the tangent plane to the conicoid at the point given by Eqs. (17.14) reduces to

$$\frac{x}{p/l} + \frac{y}{p/m} - \frac{z}{-p/n} = 1,$$

which is the same as represented by Eq. (17.14). This establishes the sufficiency of the condition. //

§ 18. Polar plane of a point with respect to a conicoid

Let a secant drawn from any point A (x_1, y_1, z_1) intersect the conicoid with Eq. (14.1) in two points P and Q. If R is a point on the secant such that AR is the Harmonic Mean of AP and AQ:

$$AR = \frac{2\,AP.\,AQ}{AP + AQ}, \qquad (18.1)$$

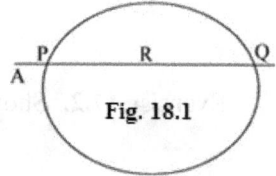

Fig. 18.1

the locus of R is then called the *polar plane* of A with respect to the conicoid. Assuming l, m, n as the direction cosines of the line APQ the distances AP $= r_1$ and AQ $= r_2$ are given by the quadratic equation (16.4). If AR $= \rho$, Eq. (18.1) determines

$$\rho\,(r_1 + r_2) = 2\,r_1 r_2. \qquad (18.2)$$

But, from Eq. (16.4)

$$r_1 + r_2 = -\{\Sigma\,l\,(\partial F/\partial x)_A\,/H\,(l, m, n)\}, \quad r_1 r_2 = F\,(x_1, y_1, z_1)/H\,(l, m, n).$$

Hence, Eq. (18.2) reduces to

$$\rho\,\{\,\Sigma\,l\,(\partial F\,/\,\partial x)_A\,\} + 2\,F\,(x_1, y_1, z_1) = 0. \qquad (18.3)$$

On the other hand, the coordinates (x', y', z') of R are determined by the Eq. (16.1):

$$x' = x_1 + \rho\,l, \qquad y' = y_1 + \rho\,m, \qquad z' = z_1 + \rho\,n.$$

Multiplying these relations by $(\partial F\,/\,\partial x)_A$, $(\partial F\,/\,\partial y)_A$ and $(\partial F\,/\,\partial z)_A$ respect–tively, and adding the results so obtained, we deduce

$$(x' - x_1)\,(\partial F\,/\,\partial x)_A + (y' - y_1)\,(\partial F\,/\,\partial y)_A + (z' - z_1)\,(\partial F\,/\,\partial z)_A$$

$$= \rho\,\Sigma\,l\,(\partial F\,/\,\partial x)_A = -2\,F\,(x_1, y_1, z_1),$$

by Eq. (18.3). Thus, the locus of R is obtained by changing its coordinates into current coordinates in above equation:

$$(x - x_1) (\partial F / \partial x)_A + (y - y_1) (\partial F / \partial y)_A + (z - z_1) (\partial F / \partial z)_A$$

$$+ 2 F (x_1, y_1, z_1) = 0. \tag{18.4}$$

Note 18.1. If the point A lies on the conicoid, so that there holds Eq. (17.1), the Eq. (18.4) reduces to Eq. (17.3).

Thus, we have the

Theorem 18.1. The tangent plane acts as the polar plane of the point of contact.

Theorem 18.2. Equation (17.7) also represents the polar plane of the point A (x_1, y_1, z_1) with respect to the conicoid with Eq. (13.6).

Proof. As in Theorem 17.1, for the derivatives of the function F, the Eq. (18.4) yields

$$2a (x - x_1) x_1 + 2b(y - y_1) y_1 + 2c(z - z_1) z_1 + 2(ax_1^2 + by_1^2 + cz_1^2 - 1) = 0,$$

which simplifies to Eq. (17.7). //

Example 18.1. The polar plane of the point P (x', y', z') with respect to the conicoid

$$x^2 + 2yz = a^2 \quad (18.5); \qquad \text{is} \qquad x x' + y z' + z y' = a^2. \quad (18.6)$$

Proof. Put $F (x, y, z) \equiv x^2 + 2yz - a^2$, and evaluate its partial derivatives at the point P:

$$(\partial F / \partial x)_P = 2 x', \quad (\partial F / \partial y)_P = 2 z', \quad (\partial F / \partial z)_P = 2y'. \quad (18.7)$$

Also,

$$F (x', y', z') = (x')^2 + 2 y' z' - a^2. \tag{18.8}$$

Putting from Eqs. (18.7) and (18.8) in Eq. (18.4), the equation of polar plane of the point P reduces to

$$2 x' (x - x') + 2 z' (y - y') + 2 y' (z - z') + 2 \{(x')^2 + 2 y' z' - a^2\} = 0,$$

that simplifies to Eq. (18.6). //

§ 19. Enveloping cone of a conicoid

We have seen in the Section 16 that a line with Eq. (16.1) becomes a tangent to the conicoid represented by Eq. (14.1) if its direction ratios satisfy the condition vide Eq. (16.6). Totality of all such (infinite) tangents drawn from an external point A to the conicoid is called the *enveloping cone* of the conicoid with A as the vertex. Eliminating l, m, n from Eq. (16.6) by means of Eq. (16.1) the equation of such a cone may be derived:

$$\{ (x - x_1)(\partial F / \partial x)_A + (y - y_1)(\partial F / \partial y)_A + (z - z_1)(\partial F / \partial z)_A \}^2$$

$$= 4\, H\,(x - x_1, y - y_1, z - z_1) \,.\, F\,(x_1, y_1, z_1). \tag{19.1}$$

Since

$$F\,(x, y, z) \equiv F\{ (x - x_1) + x_1, (y - y_1) + y_1, (z - z_1) + z_1 \}$$

$$= \Sigma\, a\, \{ (x - x_1) + x_1 \}^2 + 2\,\Sigma f\{ (y - y_1) + y_1 \}\{ (z - z_1) + z_1 \}$$

$$+ 2\,\Sigma\, u\, \{ (x - x_1) + x_1 \} + d$$

$$= \; \{ \Sigma\, a\,(x - x_1)^2 + 2\,\Sigma f(y - y_1)(z - z_1) \}$$

$$+ 2\,[\, \Sigma\, ax_1\,(x - x_1) + \Sigma f\{ y_1\,(z - z_1) + (y - y_1)\, z_1 \} + 2\,\Sigma\, u\,(x - x_1)\,]$$

$$+ \{ \Sigma\, ax_1^2 + 2\,\Sigma f y_1 z_1 + 2\,\Sigma\, u\, x_1 + d \}$$

$$= H\,(x - x_1, y - y_1, z - z_1) + \Sigma\,(x - x_1)\,(2ax_1 + 2hy_1 + 2gz_1 + 2u)$$

$$+ F\,(x_1, y_1, z_1)$$

$$= H\,(x - x_1, y - y_1, z - z_1) + \Sigma\,(x - x_1)\,(\partial F / \partial x)_A + F\,(x_1, y_1, z_1),$$

by Eq. (16.3). Therefore,

$$H\,(x - x_1, y - y_1, z - z_1) = F\,(x, y, z) - \Sigma\,(x - x_1)\,(\partial F / \partial x)_A - F\,(x_1, y_1, z_1). \tag{19.2}$$

Putting from Eq. (19.2) in Eq. (19.1) we, thus, obtain

$$\{ \Sigma\,(x - x_1)\,(\partial F / \partial x)_A \}^2$$

$$= 4\{ F\,(x, y, z) - \Sigma\,(x - x_1)\,(\partial F / \partial x)_A - F\,(x_1, y_1, z_1) \} \,.\, F\,(x_1, y_1, z_1),$$

or,

$$\{ \Sigma\,(x - x_1)\,(\partial F / \partial x)_A \}^2 + 4 F^2\,(x_1, y_1, z_1) + 4 F(x_1, y_1, z_1)\Sigma\,(x - x_1).(\partial F / \partial x)_A$$

$$= 4\, F\,(x, y, z)\,.\,F\,(x_1, y_1, z_1),$$

or,

$$\Sigma\,(x - x_1)\,(\partial F/\partial x)_A + 2F\,(x_1, y_1, z_1)\}^2 = 4F\,(x, y, z).F\,(x_1, y_1, z_1).\ (19.3)$$

As seen in Note 18.1, above equation reduces to Eq. (18.3), when

when the point A lies on the conicoid. Accordingly, the enveloping cone with the point of contact as its vertex becomes the tangent plane.

Example 19.1. Enveloping cone of the conicoid given by Eq. (13.6) with vertex at A (x_1, y_1, z_1) is represented by

$$T^2 = F.\,F_1, \tag{19.4}$$

where the symbols T and F are as per Eqs. (17.7), (17.8) and

$$F_1 \equiv F\,(x_1, y_1, z_1) = a\,x_1{}^2 + b\,y_1{}^2 + c\,z_1{}^2 - 1. \tag{19.5}$$

Solution. As seen in Theorem 18.2, the Eqs. (17.9) and (19.5) simplify Eq. (19.3) to the form

$$(a\,x\,x_1 + b\,y\,y_1 + c\,z\,z_1 - 1)^2 = F.\,F_1,$$

which is same as Eq. (19.4). //

§ 20. Centre of a conicoid

Definition 20.1. A point A is called the *centre* of the conicoid if every chord (of the conicoid) through A gets bisected at A.

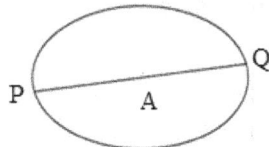

Fig. 20.1

Let PAQ be any chord of the conicoid with Eq. (14.1) and A the centre of the conicoid. As per definition, the distances PA and AQ are then equal in magnitude but opposite in sign. Accordingly, the Eq. (16.2) has equal roots with opposite signs. So, there must hold

$$r_1 + r_2 \equiv -2\,\{\,\Sigma\,l\,(ax_1 + hy_1 + gz_1 + u)\,\}\,/\,H\,(l, m, n) = 0,$$

for arbitrary choices of l, m, n. This implies Eqs. (14.10) determining the centre of the conicoid.

Example 20.1. Centre of the conicoid represented by Eq. (13.6) is at origin.

Solution. The statement follows immediately from Eqs. (14.10) and (17.9). //

Example 20.2. The point $(-1/6, -5/3, -13/6)$ is the centre of the conicoid

$$3x^2 + 5y^2 + 3z^2 - 2yz + 2zx - 2xy + 2(x + 6y + 5z + 10) = 0. \qquad (20.1)$$

Solution. The Eqs. (14.10), for the function given by Eq. (20.1), reduce to

$$3x_1 - y_1 + z_1 + 1 = 0, \quad -x_1 + 5y_1 - z_1 + 6 = 0, \quad x_1 - y_1 + 3z_1 + 5 = 0.$$

Solving these simultaneous equations for x_1, y_1, z_1, we get the centre. //

§ 21. Reduction of equation (14.1) into standard forms

In continuation with Section 15, other forms of conicoids represented by Eq. (14.1) are discussed here. The reduction of Eq. (14.1) into a standard form is carried in the following steps.

Step 1: First, we check if the function $H(x, y, z)$ representing the second degree terms forms a perfect square of a linear form not proportional to the one already present in Eq. (14.1); i.e when

$$f^2 = bc, \qquad g^2 = ca, \qquad h^2 = ab. \qquad (21.1)$$

In such a case, the conicoid becomes a *paraboloid*.

Lemma 21.1. A paraboloid admits Eq. (14.4).

Proof. Relations (21.1) imply $fgh = abc$. Hence,

$$D \equiv abc + 2fgh - af^2 - bg^2 - ch^2 = 0. \text{ //}$$

Step 2: If $H(x, y, z)$ does not form a perfect square we write the discriminating cubic

$$\begin{vmatrix} a-\lambda & h & g \\ h & b-\lambda & f \\ g & f & c-\lambda \end{vmatrix} = 0, \tag{21.2}$$

and solve it for λ.

Step 3: If all the three roots, say λ_1, λ_2, λ_3, are non-zero the general Eq. (14.1) reduces to the form

$$\lambda_1 x^2 + \lambda_2 y^2 + \lambda_3 z^2 + (u x_1 + v y_1 + w z_1 + d) = 0, \tag{21.3}$$

where (x_1, y_1, z_1) is the centre of the conicoid obtainable from Eqs. (14.10). As seen in the Subsection 13.3, the Eq. (21.3) represents an *ellipsoid* or a *hyperboloid*.

Step 4: If Eq. (21.2) has one zero root, say $\lambda_3 = 0$, we evaluate the direction cosines l_3, m_3, n_3 of the principal direction corresponding to this zero root by solving any two of the equations

$$\partial H(l_3, m_3, n_3)/\partial l_3 = 0, \; \partial H(l_3, m_3, n_3)/\partial m_3 = 0, \; \partial H(l_3, m_3, n_3)/\partial n_3 = 0. \tag{21.4}$$

If

$$k \equiv u l_3 + v m_3 + w n_3 \tag{21.5}$$

is not zero, the Eq. (14.1) reduces to the form

$$\lambda_1 x^2 + \lambda_2 y^2 + 2k z = 0, \tag{21.6}$$

and represents a *paraboloid.*

Step 5: In case $k = 0$ there exists a line of centres given by Eqs. (14.10). Choosing any point (x_1, y_1, z_1) on this line as centre the Eq. (14.1) reduces to

$$\lambda_1 x^2 + \lambda_2 y^2 + (u x_1 + v y_1 + w z_1 + d) = 0; \tag{21.7}$$

and it represents a *cylinder.* If λ_1, λ_2 are of same sign the cylinder is *elliptic* otherwise *hyperbolic*.

Theorem 21.1. Discriminating cubic of a paraboloid has two zero roots.

Proof. Expanding the determinant in Eq. (21.2):

$$-\lambda^3 + (a + b + c)\, \lambda^2 + \{(f^2 - bc) + (g^2 - ca) + (h^2 - ab)\}\lambda + D = 0,$$

putting from Eqs. (21.1) and applying Lemma 21.1, above Eq. reduces to

$$-\lambda^3 + (a + b + c)\, \lambda^2 = 0.$$

Clearly, this equation has two vanishing roots. //

Note 21.1. We thus note that the case of two vanishing roots of the discriminating cubic is already covered in the Step 1; and, as such, the conicoid becomes a paraboloid.

Step 6: In addition to Step 3, if two roots, say λ_1, λ_2, of Eq. (21.2) are equal, the Eq. (14.1) reduces to

$$\lambda_1\, (x^2 + y^2) + \lambda_3\, z^2 + (u\, x_1 + v\, y_1 + w\, z_1 + d) = 0, \tag{21.8}$$

which represents a *surface of revolution*: ellipsoid or hyperboloid or cone.

Step 7: As a special case of the Step 4, if $\lambda_1 = \lambda_2$, so that Eq. (21.6) assumes the form

$$\lambda_1\, (x^2 + y^2) + 2k\, z = 0, \tag{21.9}$$

and it represents a *paraboloid of revolution*.

Step 8: On the other hand, when $\lambda_1 = \lambda_2$, the Eq. (21.7) reduces to

$$\lambda_1\, (x^2 + y^2) + (u\, x_1 + v\, y_1 + w\, z_1 + d) = 0, \tag{21.10}$$

and it represents a *circular cylinder*.

Note 21.2. As a special case, when the coefficients $u = v = w = 0$ and $d\,/\,\lambda_1$ is negative, say $-a^2$, above equation reducing to Eq. (14.8.4) represents a *right circular cylinder*.

Example 21.1. The equation

$$4x^2 + 9y^2 + 36z^2 - 36yz + 24zx - 12xy - 10x + 15y - 30z + 6 = 0 \tag{21.11}$$

represents a pair of planes

$$2x - 3y + 6z = 2, \qquad \text{and} \qquad 2x - 3y + 6z = 3. \qquad (21.12)$$

Solution. Although the second degree terms in Eq. (21.11) form a perfect square of a linear factor $2x - 3y + 6z$ yet it cannot represent a paraboloid as the first degree terms also contain the same factor. Indeed, it is a quadratic equation in this factor:

$$(2x - 3y + 6z)^2 - 5(2x - 3y + 6z) + 6 = 0, \qquad (21.13)$$

having roots

$$2x - 3y + 6z = (5 \pm \sqrt{25 - 24})/2 = 3, 2.$$

Thus, factorizing the Eq. (21.13), the combined equation of two planes is:

$$(2x - 3y + 6z - 2) \cdot (2x - 3y + 6z - 3) = 0;$$

and the individual planes are represented by Eqs. (21.12). //

Example 21.2. The equation

$$9x^2 + 4y^2 + 4z^2 + 8yz + 12zx + 12xy + 4x + y + 10z + 1 = 0, \qquad (21.14)$$

represents a paraboloid.

Solution. Since the second degree terms in the equation form a perfect square:

$$H(x, y, z) \equiv (3x + 2y + 2z)^2,$$

while the first degree terms do not contain the same factor, viz. $3x + 2y + 2z$, it represents a paraboloid. Rewriting the Eq. (21.14) as

$$(3x + 2y + 2z)^2 = -(4x + y + 10z) - 1,$$

or, in normal form

$$\left\{ \frac{3x + 2y + 2z + \lambda}{\sqrt{3^2 + 2^2 + 2^2}} \right\}^2 = \frac{1}{17} \left[\frac{(6\lambda - 4)x + (4\lambda - 1)y + (4\lambda - 10)z + \lambda^2 - 1}{\sqrt{(6\lambda - 4)^2 + (4\lambda - 1)^2 + (4\lambda - 10)^2}} \right].$$

$$\cdot \sqrt{(6\lambda - 4)^2 + (4\lambda - 1)^2 + (4\lambda - 10)^2}, \qquad (21.15)$$

where λ is such that the planes

$$3x + 2y + 2z + \lambda = 0, \tag{21.16}$$

and

$$(6\lambda - 4)\, x + (4\lambda - 1)\, y + (4\lambda - 10)\, z + \lambda^2 - 1 = 0 \tag{21.17}$$

are perpendicular to each other. Therefore, by Eq. (7.12) and the Subsection 7.5, we have

$$3(6\lambda - 4) + 2\,(4\lambda - 1) + 2\,(4\lambda - 10) = 0 \quad\Rightarrow\quad \lambda = 1.$$

Hence, setting $\xi \equiv (3x + 2y + 2z + 1)/\sqrt{17}$ and $\eta \equiv (2x + 3y - 6z)\,/\,7$, the Eq. (21.15) transforms as

$$\xi^2 = (7/17)\,\eta. \tag{21.18}$$

The Eqs. (21.16) and (21.17) representing the plane through axis of the paraboloid and the tangent plane at the vertex reduce to

$$3x + 2y + 2z + 1 = 0, \quad \text{and} \quad 2x + 3y - 6z = 0$$

respectively. //

Example 21.3. The equation

$$3x^2 + 7y^2 + 3z^2 + 10\,y\,z - 2z\,x + 10\,x\,y + 4\,x - 12\,y - 4\,z + 1 = 0 \tag{21.19}$$

represents a hyperboloid of two sheets.

Solution. We note that the second degree terms in Eq. (21.19) do not form a perfect square. The discriminating cubic of the conicoid is

$$\begin{vmatrix} 3-\lambda & 5 & -1 \\ 5 & 7-\lambda & 5 \\ -1 & 5 & 3-\lambda \end{vmatrix} \;\overset{\substack{R_1-R_3,\\ R_2+5R_3}}{=\!=\!=}\; \begin{vmatrix} 4-\lambda & 0 & -4+\lambda \\ 0 & 32-\lambda & 20-5\lambda \\ -1 & 5 & 3-\lambda \end{vmatrix}$$

$$\overset{C_1+C_3}{=\!=\!=} \begin{vmatrix} 0 & 0 & -4+\lambda \\ 20-5\lambda & 32-\lambda & 20-5\lambda \\ 2-\lambda & 5 & 3-\lambda \end{vmatrix} = (\lambda-4)\,(100-25\lambda-64+34\lambda-\lambda^2)$$

$$= (\lambda-4)\,(36+9\lambda-\lambda^2) = (\lambda-4)\,(\lambda+3)\,(12-\lambda) = 0,$$

having roots $-3, 4, 12$. Also, the Eqs. (13.10) are

$3x_1 + 5y_1 - z_1 + 2 = 0$, $5x_1 + 7y_1 + 5z_1 - 6 = 0$, $-x_1 + 5y_1 + 3z_1 - 2 = 0$,

determining the centre of the conicoid: $x_1 = 1/3$, $y_1 = -1/3$, $z_1 = 4/3$. Hence,

$$u x_1 + v y_1 + w z_1 + d = 2/3 + 2 - 8/3 + 1 = 1.$$

Thus, following Step 3, the Eq. (20.19) reduces to

$$-3 x^2 + 4 y^2 + 12 z^2 + 1 = 0; \quad \text{or,} \quad 3 x^2 - 4 y^2 - 12 z^2 = 1,$$

representing a hyperboloid of two sheets. //

Example 21.4. The equation

$$2x^2 + 20y^2 + 18z^2 - 12 y z + 12 x y + 22 x + 6 y - 2 z - 2 = 0 \quad (21.20)$$

represents an elliptic paraboloid.

Solution. The second degree terms do not form a perfect square. Its discriminating cubic:

$$\begin{vmatrix} 2-\lambda & 6 & 0 \\ 6 & 20-\lambda & -6 \\ 0 & -6 & 18-\lambda \end{vmatrix} \equiv (2-\lambda).\,(360 - 38\,\lambda + \lambda^2 - 36) - 36\,(18-\lambda)$$

$$\equiv -\lambda^3 + 40\,\lambda^2 - 364\,\lambda = 0,$$

has roots $\lambda = 14, 26, 0$. Next, following Step 4, we evaluate the direction cosines l_3, m_3, n_3 for the vanishing root. Putting

$$H(l_3, m_3, n_3) = 2\,l_3^2 + 20\,m_3^2 + 18\,n_3^2 - 12\,m_3\,n_3 + 12\,l_3\,m_3,$$

Eqs. (21.4) reduce to

$$l_3 + 3\,m_3 = 0, \quad 3\,l_3 + 10\,m_3 - 3\,n_3 = 0, \quad \text{and} \; -m_3 + 3\,n_3 = 0.$$

Their solutions are $l_3/(-3) = m_3 = 3n_3$, or,

$$\frac{l_3}{-9} = \frac{m_3}{3} = \frac{n_3}{1} = \sqrt{\{l_3^2 + m_3^2 + n_3^2\}} / \sqrt{\{(-9)^2 + 3^2 + 1^2\}} = \frac{1}{\sqrt{91}}.$$

$$\Rightarrow \qquad l_3 = -\tfrac{9}{\sqrt{91}}, \; m_3 = \tfrac{3}{\sqrt{91}} \; \text{and} \; n_3 = \tfrac{1}{\sqrt{91}}.$$

Hence, by Eq. (21.5),

$$k = (-99 + 9 - 1)/\sqrt{(91)} = -\sqrt{(91)}$$

and the Eq. (21.20) reduces to the form

$$14\,x^2 + 26\,y^2 - 2\,\sqrt{(91)}\,z = 0, \quad \text{i.e.} \quad 7\,x^2 + 13\,y^2 = \sqrt{(91)}\,z,$$

which is an elliptic paraboloid. //

BIBLIOGRAPHY

1. Anton, Howard and Herr, Albert: Calculus with Analytic Geometry, *John Wiley & Sons, New York (USA)*, 5th ed., 2009.

2. Ayres, Frank: Theory and Problems of Differential and Integral Calculus, Schaum Outline Series, *Schaum Publishing Co., New York*, 1992.

3. Bose, S.K. and Mishra, R.S.: Dynamics of a Particle, *Prakashan Kendra, Lucknow (India)*, 3rd ed., 1962.

4. Copson, E.T.: Introduction to the Theory of Functions of a Complex Variable, Paperback ed., 1970.

5. Grewal, B. S.: Higher Engineering Mathematics, *Khanna Publishers, New Delhi (India)*, 34th ed., 1998.

6. Kapur, N.M.: A text-book of Differential Equations, *Pitambar Book Depot, Delhi (India)*, 2nd ed., 1979.

7. Kreyszig, E.: Advanced Engineering Mathematics, *John Wiley and Sons, USA*, 7th ed. 1993.

8. Mishra, R.S.: Differential Calculus, *Prakashan Kendra, Lucknow (India)*, 1969.

9. Misra, R.B.: Concepts of Plane and Solid Geometry (Math 211), *University of Asmara*, 1999, pp. 1 - 84.

10. Misra, R.B. : Fundamental Concepts of Geometry (Math 311), *University of Asmara*, 2000, pp. 1-72.

11. Misra, R.B.: Analytical Geometry of Planes and Solids, *Hardwari Publications, Allahabad (India)*, pp. xiv + 500, 2004, ISBN 81-88574-01-5.

12. Misra, R.B.: Theory of Sets, Groups, Rings, Fields, Integral Domains, Vector Spaces, Metric Spaces and Topological Spaces, *Lam-

bert Academic Publishers, Saarbrücken (Germany), 2010, ISBN 978-3-8383-9943-0.

13. Misra, R.B.: A First Course on Calculus with Applications to Diffrential Equations, *ibid*, 2010, ISBN 978-3-8433-7871-0.

14. Misra, R.B.: A Text-book of Classical Mechanics, *ibid*, 2010, ISBN 978-3-8433-8306-6.

15. Misra, R.B.: Laplace Transform, Differential Equations and Fourier Series, *ibid*, 2010, ISBN 978-3-8433-8328-8.

16. Misra, R.B.: Basic Mathematics at a Glance, *ibid*, 2010, ISBN 978-3-8433-8696-8.

17. Misra, R.B.: Complex Analysis, *ibid*, 2010, ISBN 978-3-8433-8859-7.

18. Misra, R.B.: Engineering Mathematics, 2011, *ibid*, ISBN 978-3-8433-8931-0.

19. Misra, R.B.: Glossary of Mathematics, *ibid*, 2011, ISBN 978-3-8443-0203-5.

20. Misra, R.B.: Advanced Integral Calculus, *ibid*, 2011, ISBN 978-3-8443-1916-3.

21. Misra, R.B.: Advanced Applied Mathematics, *Central West Publishing, Orange, NSW (Australia)*, 2018, xi + 260, ISBN 978-1-925823-11-0.

22. Misra, R.B.: Mathematics for Engineers and Physicists, Part 1, *ibid*, 2019, pp. xiv + 306. ISBN (print): 978-1-925823-51-6, ISBN (e-book): 978-1-925823-50-9.

23. Misra, R.B.: Mathematics for Engineers and Physicists, Part 2, *ibid*, 2019, pp. xiv + 326. ISBN (print): 978-1-925823-53-0,

ISBN (e-book): 978-1-925823-52-3.

24. Misra, R.B.: Glossary of Mathmatical Terms and Concepts, Part 2, *ibid*, 2019.

25. Misra, R.B.: Glossary of Mathmatical Terms and Concepts, Part 3, *ibid*, 2019.

26. Misra, R.B.: Glossary of Mathmatical Terms and Concepts, Part 4, *ibid*, 2019.

27. Pati, T.: Functions of a Complex Variable, *Pothishala Pvt. Ltd., Allahabad*, 1971; reprinted by *Prentice Hall of India Pvt. Ltd., New Delhi (India)*, 2000.

28. Prasad, Chandrika: A text-book of Algebra and Theory of Equations, *Pothishala Pvt. Ltd., Allahabad (India)*, 10th ed., 1985.

29. Prasad, Gorakh: Text-book on Integral Calculus and Elementary Differential Equations (revised by Chandrika Prasad), *ibid*, 8th ed. 1964.

30. Prasad, Gorakh: Text-book on Differential Calculus (revised by Chandrika Prasad), *ibid*, 11th ed. 1968.

31. Raisinghania, M.D.: Advanced Differential Equations, *S. Chand Pvt. Ltd, New Delhi (India)*, 2009.

32. Ray, M.: Text-book on Dynamics, *ibid*, 2003.

33. Sinha, R.S. and Srivastava, B.K.: A text-book of Differential Equations, *Chandra Prakashan, Gorakhpur (India)*, 2nd ed., 2005.

34. Spivak, Michael: Calculus, *Publish or Perish, Inc. Houston (USA)*, 3rd ed., 1994 (updated, 2008).

35. Thomas, George B.: Calculus (revised by Ross B. Finney and Mauri-

ce D. Weir), *Addison Wesley Publishing Co.*, *New York*, 10th ed., 2002.

36. Titschmarsch, E.C.: The Theory of Functions, *Oxford University Press, London*, 1932.

INDEX

A

abelian group 13
absolute maxima of a fn. 55
- minima of a fn. 55
absolutely convergent 79, 128
acceleration 41, 222
- normal 228, 231
- radial 227, 231
- tangential 228, 231
- transverse 227, 231
- uniform 221
algebraic
- group 13
- laws 13
- structures 13
algebraic functions 44
- derivatives of 44
- properties of derivatives 44
amplitude of motion 236
analytic function 120
- singular point of 120
angle, vectorial 76
- between vectors 2
angle of projection 239
angular velocity 230
anti-derivation 87
Anton, Howard 317
applications of derivation 66
areas 87, 97, 103,
- surface 103
arithmetic mean 6
- progression 5, 6
associative law 13
asymptotes 147
- horizontal 30
- oblique 31
- vertical 30
Average 6
Ayres, Frank 317

B

bearing of ship 253
bilinear transformation 144
- fixed / invariant pts. of 145
binary law / operation 13
binomial coefficients 8
- theorem 8
Bose, S.K. 317

C

Cardioid 76
Cartesian coordinates 73
Cauchy, A.L.
- integral formula 126
- integral theorem 126
- residue theorem 131
Cauchy-Euler /Euler diff.
  eqn.176
Cauchy-Riemann eqns. 120
- polar form of 121
centre of group 15
chain rule of differentiation 47
change of order of integration
  104
characteristic curve 198
- system of eqns. 189
circle 1, 275
- unit 133
circular motion 230
- trigonometric fns. 46
clopen interval 40, 60
closed interval 35
coefficient of restitution 248
cofactors 157
collinear points 273
combination 8
commutative
- group 13

- law 13
- semigroup 13
complex function 119
- conjugate 172
- derivative of 120
- integration of 124
- potential 151
- real & imaginary parts of 119
- roots 172
complex number(s) 1, 13
- conjugate of 172
- log of 1
- modulus of 143
components of vector 1
compound interest 28
conformal mapping/transf. 139
- critical pt. of 140
- special 146
- standard 141
- transformation 139
conjugate function 120
conservation of linear momentum 247
conservative force 245
continuous function 34
- but not differentiable 43
- properties of 34
contour integration 133
convergence, interval of 79
- radius of 79
convergent, absolutely 129
- series 129
continuity 32
- one-sided 37
continuous fn. of 2 variables 69
coordinates, polar 76
- Cartesian 73
coplanar lines 245
Copson, E.T. 317
cost price 27
critical number of function 65,
- point 140

cross product of 2 vectors 2
- three vectors 3
cross-ratio of 4 points 145
curl (or *rot*) of vector 151
cyclic group 16
- finite 16
- generator of 16
cylinder 311

D

decreasing monotonic fn. 61,
- strictly 61
definite integral(s) 88, 92
- properties of 93
- theorems on 94
del operator 3
derivation / differentiation
- algebraic properties of 44
- application of 66
- chain rule for 47
- first principle of 41
- of algebraic fn. 44
- of complex function 120
- of exponential fn. 45
- of hyperbolic fns. 47
- of implicit fns. 49
- of inverse hyperbolic fns. 49
- of inverse trigonometric fns.48
- of logarithmic fn. 45
- of standard functions 44
- of trigonometric fns. 46
- substitution method for 35, 47
- successive 49
determinant 157
- properties of 158
differentiable function 42
differential eqn. 161
- Bernoulli's 165
- Cauchy-Euler 175
- complementary fn. of 170, 176
- degree of 161

- Euler 176
- homogeneous 163
- integrating factor of 163, 165
- linear 164, 171
- linear partial 188, 209
- normal form of 184
- of any degree 161, 167
- of any order 162
- of Clairaut's form 169
- of exact form 166
- of first degree 161 - 163
- of first order 161, 162, 167
- of general form 169
- of nth *order* 162, 171
- of second order 181
- order of 161
- ordinary 161
- partial 185
- particular integral of 173, 177
- reducible to homogeneous
    form 164, 179
- reducible to linear form 165
- separable variable form 163
- solvable for p 167
- solvable for x 168
- solvable for y 168
- simultaneous 178
- singular solution of 169
diffusion equation 218,
direct impact of 2 spheres 251
direction cosines 286
Dirichlet's integral 111
displacement 41, 100
distance b/n 2 points 269
distributive laws 19
divergence of a vector 152
dot product of 2 vectors 2
- three vectors 3
- four vectors 3

E

eigen functions / values 212

element, identity 14
- inverse 14
- unit 14
ellipse 1, 276
ellipsoid 296
energy 244
- mechanical 245
- potential 244, 245
envelope of family 73
 - 1-parameter family 73
 - 2-parameters family 74
 - (general method) 75
 - in polar coordinates 76
equal fractions 5
equation(s)
- Cauchy-Riemann 120, 122
- heat 205, 217
- indicial 170
- intercepts form 270
- Laplace 122, 205
- of circle 1
- of motion 223
- of plane 290
- of straight line 270
- perpendicular form of 271
- Poisson 205
- Schrödinger 205
- slope-intercept form of 270
- wave 205, 210
equi-potential lines 151, 152
Euler's theorem 70
evolute 77
exact differential eqn. 166
expansion of logarithmic fn. 11
exponential function 9
- derivatives of 45
- series 9
- transformation 148
extension of Cauchy's thm. 126
extremum value theorem 53

F

factorial 7
1st derivative test for local
    extrema 62
1st fundamental thm. 95
flux function 154
formula, Cauchy's integral 126
force 226
- work done by 100
formulae, reduction 92
Fourier coefficients 257
- particular cases 259
Fourier series 255, 256
- as *cosine* series 260,261,263
- as *sine* series 213, 263
- for even function 259
- for odd function 259
- for piecewise fns. 266
- in interval [*a*, *b*] 265
- in interval [– *l*, *l*] 260
- in interval [0, *l*] 263
- in interval [0, 1] 264
- in interval [0, π] 261, 263
- in interval [0, 2π] 264
Fourier transform 255
- properties 255, 256
frequency of S.H.M. 236
function, algebraic 44
- analytic 120
- complex 119
- conjugate 172
- continuous 34
- derivative of 40
- differentiable 42
- expansion of 11
- exponential 9
- eigen 212
- flux 154
- Gamma 110
- gradient of 3
- harmonic 123, 151
- homogeneous 71
- hyperbolic 47
- integration of 87
- limit of 29
- local extrema of 54
- local maxima of 54
- local minima of 54
- logarithmic 11
- monotonic 61
- partial derivation of 66
- regular 120
- residue of 131
- stream 152
- total derivative of 69
- trigonometric 46
fundamental thm.
- 1st (integral calculus) 95
- for local extrema 64
- of calculus 18
- on homomorphism 18

G

Gamma function 110
Gauss divergence thm. 218
geometric mean 6
___ progression 6
grad *f* 3, 264
gradient of scalar fn. 3
Grewal, B.S. 317
group, 13
- abelian / commutative 13
- algebraic 13
- centre of 15
- cyclic 16
- homomorphic 17
- kernel of 17
- order of 16
- permutation 16
- properties of 14
- semi- 13
- sub 14

- symmetric 17
- transformation 16
- types of 16
groupoid 13

H

harmonic function 123
- mean 7
- progression 6
heat equation 205, 217
heat flow lines 153
homogeneous linear ODE 175
horizontal asymptote 30
hyperbola 1, 277
- rectangular 1, 146
hyperbolic functions 47
- derivatives of 47
- inverse 49
hyperboloid 296

I

identity element 14
- law 14
impact 248
-, direct 251
-, oblique 249
-, perpendicular 248
impulse 247
indefinite integrals 87
indenting a contour 137
indicial equation 170
infinite series 9
inner product of 2 vectors 2
integral(s)
- definite 88, 92
- Dirichlet's 111
- indefinite 87
- mean value thm. of 58
- theorems on definite 94
integrating factor of linear ODE

163, 165
integration
- around rectangle 135
- around semi-circle 134
- around unit circle 133
- by parts method 89
- change of order of 104
- contour 133
- methods of 90
- of complex function 124
- of power series 128
- substitution method for 90
interest, compound 28
- simple 28
intermediate value thm. 35
interval, clopen 40, 60
- closed 35
- of convergence 79, 128
- open 37
- open-close 60
inverse element 14
- law 13
- of hyperbolic fns. 49
- of trigonometric fns. 48
inversion 142
involute 77
irrotational motion 151
isolated singular point 131
isothermal lines 155

J

Jacobian 114
Joukowski's transformation 150

K

Kapur, N.M. 317
kernel of homomorphism 18
kinematics in 2 dimensions 223
kinetic energy 244
knot 252

Kreyszig, E. 317

L

Laplace equation 122, 151, 205
Laplacian operator 3
Laurent's series 129
- integration of 128
law, algebraic 13
- associative 13
- binary 13
- commutative 13
- distributive 19
- identity 14
- inverse 14
- of indices 5
-(s) of motion 221
Leibnitz theorem 51
lemniscate 143
limit(s) of a function 29
- from left 29
- from right 29
- properties of 30
- some important 31
line integral 124
line through a point 272
linear differential eqn. 164, 171
- homogeneous 175
- reducible to homogeneous
   form 166
- reducible to linear form 164,
   179
- simultaneous 178
- with const. coefficients 170
linear partial diff. eqn. 209
lines, equi-potential 151, 152
- heat flow 153
- isothermal 155
- of force 153
- orthogonal 273
- parallel 273
- stream 152

local extrema of a fn. 54
- fundamental thm. of 64
local maxima of a fn.54
- minima of a fn. 54
logarithm of complex no. 1
- change of base 11
- common 10
- natural 10
- rules for 10
logarithmic differentiation 47
logarithmic function 11
- derivation of 45
- expansion of 11

M

Maclaurin's series / thm. 80
matrix, Vandermonde 159
maxima of a function 53
- absolute 55
- fundamental thm. of 64
- local 54
maximum range on
- horizontal plane 239
- inclined plane 241
mean, arithmetic 6
- geometric 6
- harmonic 7
mean value theorem 58
- for integrals 95
mechanical energy 245
- principle of conservation 245
method(s) of integration 90
- by parts 89
- substitution 90
minima of a function 53
- absolute 55
- fundamental thm. of 64
- local 54
Mishra, R.S. 317
Misra, R.B. 317, 318
Möbius, August Ferdinand 144

- transformations 144
momentum 221
- conservation of linear 247
monoid 13
monotonic function 61
- decreasing 61
- increasing 61
- strictly 61
- strictly decreasing 61
- strictly increasing 61
motion, laws of 221
- along horizontal circle 229
- downward 224
- in circle with const. vel. ω 230
- in straight line 221
- in vertical circle
- relative 252
- under gravity 224
- upward 225

N

nabla operator 3
natural logarithm 10
- numbers 7
Newton 221
Newton's laws of motion 221
normal
- acceleration 228, 231
- binormal 295
- form of ODE 184
- principal 295
- to a curve 295
- to plane 291
- to surface 298
- velocity 228, 231
number, complex 1

O

oblique asymptote 31
- impact 249

one-sided continuity 37
open interval 37
- close interval 60
operation, binary 13
operator, *del* 3
order of group 16
orthogonal lines 140
orthonormal vectors 2
oscillatory series 9

P

parabola 74, 275
parabolic transformation 145
paraboloid 297
- of revolution 312
parallel lines 270
- vectors 2
parametric curves 298
partial derivation 66
particular integral of ODE 173, 177
Pati, T. 319
PDE, 185
- 1st order 185, 188
- homogeneous 207
- linear 188
- non-linear 188, 194
- quasi-linear 188
- second order 197, 204
- semi-linear 188
- subsidiary equations 191, 195
percentage 27
periodic time of S.H.M. 235
permutation 8
- group 16
perpendicular impact 248
plane, equation of 290
Poisson equation 205
point, critical 140
- dividing a line 270
- singular 120

polar coordinates 76
polar form of C-R eqs. 121
pole of order n 131
- simple 131
polynomial, rational 35
potential, complex 151
- energy 244, 245
- theory 123
- velocity 140, 151
power series 78, 128
- absolutely convergent 79,128
- convergent 129
- integration of 128
- interval of convergence 79,
  128
- radius of convergence 78
Prasad, Chandrika 319
- Gorakh 319
product of two vectors 2
- cross / vector 2
- dot / inner / scalar 2
product of three vectors
- dot / scalar 3
- cross / vector 3
product of four vectors
- dot / scalar 3
product method for soln. of
  PDE 207
profit 27
projectile on horizontal plane
  237
- angle of projection 239
- horizontal range 238
- max. range 239
projectile on inclined plane 240
- maximum range 241
properties of definite integrals
  93
Pythagoras theorem 269

R

radial acceleration 227, 231
- velocity 227, 231
radius of convergence 78
radius vector 105
Raisinghania, M.D. 319
rate of change 66
rational polynomial 35
Ray, M. 319
real line 36
rectangular hyperbola 1, 146
reduction formulae 92
reflection 142
regular function 120
relative motion 252
residue class 15
residue of a function 131, 132
resolved part of vector 1
resultant of 2 foces 1
reversal rule 15
Rolle's theorem 56
rotation 141
rule, abelian 13
- reversal 15

S

scalar product of
- four vectors 3
- three vectors 3
- two vectors 2
Schrödinger equation 205
second derivative test for local
  extrema 63
second fundamental thm. 95
semigroup 13
- commutative 13
- with identity 13
series, exponential 9
- Fourier 194
- infinite 9

- Laurent's 129
- Maclaurin's 80
- of complex terms 128
- oscillatory 9
- power 78, 128
- Taylor's 82
set of
- complex numbers 19
- even numbers 19
- Gaussian integers 19
- integers 19
- rational numbers 19
- real numbers 19
- of residue classes 15
$\sum n$, $\sum n^2$, $\sum n^3$ 7
simple harmonic motion 234
- amplitude 236
- frequency 236
- periodic time 235
simple interest 28
simultaneous linear ODEs 178
- in 3 variables 179
sine series in interval $[0, \pi]$ 262
singular point 120
- essential 131
- isolated 131
- simple 131
Sinha, R.S. 319
solids of revolution 102
- volume of 102
solution of ODE with
- const. coefficients 170, 181
- variable coeff. 183
- variation of parameters 180
solution of PDE
- Charpit's method 194
- Lagrange method 191
- separation of variables 207
Spivak, Michael 319
Srivastava, B.K. 319
straight line, d.c. of 286
- equation 270

stream function 152
- lines 152
strictly decreasing monotonic
  fn. 61
- increasing monotonic fn. 61
subgroup 14
substitution
- method for integration 90
- rule for derivation 35, 47
successive derivation 113
superposition principle 209
surface, area of 103
- normal to 298
- of revolution 312
symmetric group 17

T

tangential acceleration 228, 231
- velocity 228, 231
Taylor's series 82, 129
theorem, binomial 8
- Cauchy's integral 126
- Cauchy's residue 131
- on definite integrals 94
- Euler's theorem 70
- extremum value 53
- first fundamental 95
- Gauss divergence 218
- intermediate value 35
- Leibnitz 51
- Maclaurin's 80
- mean value (derivatives) 58
- mean value (integrals) 95
- Pythagoras 269
- Rolle's 56
Thomas, George B. 319
Titschmarsch, E.C. 319
total derivative of fn. 69
transform, Fourier 255
transformation
- bilinear / Möbius 144

- conformal 139
- exponential ($w = e^z$) 148
- group 16
- Joukowski's 150
- parabolic 145
- special conformal 146
- $w = \cosh z$ 148
- $w = z^n$ 147
- $w = z^2$ 146
translation 141
transverse
- acceleration 227, 231
- direction 229
- velocity 227, 231
trigonometric functions 46
- derivatives of 46
- inverse 48

U

unit circle 133
unit element (of semi-group) 14
unit vector(s) 2
- along coordinate axes 1, 2

V

Vandermonde matrix 159
variation of parameters 180
vectors
- angle between 2
- cross product of two 2
- cross product of three 3
- components of 1

- dot / inner product of two 2
- dot/inner product of three 3
- magnitude of 1
- orthogonal 2
- orthonormal 3
- parallel 3
- radius 105
- resolved part of 1
- scalar product of four 3
- scalar product of three 3
- scalar product of two 2
- unit 1
vector triple product 3
vectorial angle 76
velocity 41
- angular 229, 230
- in circle 231
- normal 228, 231
- potential 140, 151
- radial 226, 231
- tangential 228, 231
- transverse 226, 231
vertical asymptote 30
volume/single integration 102
- of solids of revolution 102
- of sphere 102

W

wave equation 205, 210
work done by a force 100
work problem 28
w-plane 119

www.ingramcontent.com/pod-product-compliance
Lightning Source LLC
Chambersburg PA
CBHW061203220326
41597CB00015BA/1306